Experiments, Models, Paper Tools

WRITING SCIENCE

EDITORS Timothy Lenoir and Hans Ulrich Gumbrecht

Experiments, Models, Paper Tools

CULTURES OF ORGANIC CHEMISTRY IN THE NINETEENTH CENTURY

Ursula Klein

STANFORD UNIVERSITY PRESS
STANFORD, CALIFORNIA
2003

Stanford University Press
Stanford, California

Printed in the United States of America,
on acid-free, archival-quality paper.

Library of Congress Cataloging-in-Publication Data

Klein, Ursula.

Experiments, models, paper tools : cultures of organic chemistry in the nineteenth century / Ursula Klein.
p. cm. — (Writing science)
Includes bibliographical references and index.
ISBN 978-0-8047-4359-4
1. Chemistry—Notation—History—19th century. 2. Chemical structure—History—19th century. 3. Chemistry, Organic—History—19th century. I. Title: Cultures of organic chemistry in the nineteenth century. II. Title. III. Series.
QD291 .K57 2002
547′.009034—dc21

2002003144

Original Printing 2003

Last figure below indicates year of this printing:
11 10 09 08 07 06 05 04 03
Typeset by G&S Typesetters, Inc., in 10/14 Sabon

For Wolfgang

CONTENTS

PREFACE

This book is an epistemological and semiotic inquiry into the various functions of sign systems and their manipulations on paper in the laboratory sciences, coupled with detailed historical analyses of scientists' experimental and classificatory performances. It tackles questions such as, Why is it that a particular sign system—chemical formulas—became just as emblematic of a laboratory science like nineteenth-century chemistry as did instruments, manipulative skills, and experimental interventions? Why did chemists prefer the algebraic mode of Berzelian formulas over their more pictorial Daltonian alternatives? Why was *organic* chemistry the first place in which chemists applied Berzelian chemical formulas as productive tools, an area considered an impenetrable "jungle" in the early nineteenth century? Was there any causal connection between the acceptance and propagation of Berzelian chemical formulas in organic chemistry from the late 1820s onward and the simultaneous transformation of that chemical domain at the deep structural and communal level of a scientific culture between the late 1820s and the early 1840s?

The organization of the book reflects its interdisciplinary approach. Chapters organized along a historical narrative, which offer detailed historical analyses and descriptions, alternate with others that summarize the historical details and link them to the ongoing discourse in the philosophy and semiotics of science. Readers interested mainly in the latter can concentrate on Chapters 1, 2, 8, and 9 and read only the summaries of the other chapters, whereas readers with a strong predilection for historical issues may benefit more from Chapters 3 through 7, which also provide the bulk of empirical evidence for my broader contentions.

There are only a few places in the world where research for a book at the intersection between the history, philosophy, and semiotics of science would have been encouraged. The Max Planck Institute for the History of Science in Berlin is such as place. I am particularly grateful to its three directors, Lorraine Daston, Jürgen Renn, and Hans-Jörg Rheinberger, for their constant support.

Research for several chapters was done while I was a visiting fellow in the Department for the History of Science at Harvard University from 1996 until 1998 and a resident senior fellow at the Dibner Institute for the History of Science and Technology from 1997 until 1998. I thank Everett Mendelsohn as well as the two directors of the Dibner Institute, Jed Buchwald and Evelyn Simhar, for their kind invitation and support of my work. I also thank Jürgen Mittelstrass for his constant encouragement and support of my "Habilitation" at the University of Constance, which was based on the much longer German version of this book. The entire book manuscript, or the earlier German version of it, has been read by David Bloor, William Brock, Wolfgang Lefèvre, Alan Rocke, Steven Weininger, and Gereon Wolters. I owe an immense debt to their scholarly help and suggestions for improvements. I am also grateful to Andrew Pickering for many important comments on a draft of Chapter 3. For the chapters on semiotics, I received important insights in discussions with participants of a colloquium on modes of representation at the Institute of Philosophy of the Free University in Berlin during 1999, organized by Andreas Arndt and Wolfgang Lefèvre. I also benefited immensely from stimulating conversations on subjects covered by this book with Sam Schweber, Jed Buchwald, and Peter Galison during my stay at Harvard University and the Dibner Institute. Furthermore, I thank my colleagues and guests at the Max Planck Institute, in particular John Dettloff, Eric Francoeur, Mary Jo Nye, Peter Ramberg, Andrea Woody, and Sara Vollmer, for many helpful conversations and suggestions. Special thanks to Gisela Marquardt, who prepared the electronic form of the manuscript and redrew and formatted all the figures contained in this book. I also thank Nathan MacBrien, Janna Palliser and Tony Hicks from Stanford University Press for their kind support and help.

In several chapters of this book, I have drawn on material that I have previously published in the following articles: "Paving a Way through the Jungle of Organic Chemistry—Experimenting within Changing Systems of Order" (in *Experimental Essays—Versuche zum Experiment*, edited by Michael Heidelberger and Friedrich Steinle. Baden-Baden: Nomos Verlagsgesellschaft, 1998, 251–271); "Techniques of Modelling and Paper Tools in Classical Chemistry" (in *Models as Mediators: Perspectives on Natural and Social Sciences*, edited by Mary Morgan and Margaret Morrison. Cambridge: Cambridge University Press, 1999, 146–167); "Paper Tools in Experimental Cultures" (*Studies in the History and Philosophy of Science* 32, 2001, 1–40); "Berzelian Formulas as Paper Tools in Early Nineteenth-Century Chemistry" (*Foundations of Chemistry* 3 (1), 2001, 7–32); "The Creative Power of Paper Tools in Early Nineteenth-Century Chemistry" (in *Tools and Modes of Representation in the*

Laboratory Sciences, edited by Ursula Klein, Dordrecht: Kluwer Academic Publishers, Series: Boston Studies in the Philosophy of Science, 2001, 11–30). The kind permission of Cambridge University Press, Nomos Verlagsgesellschaft, Kluwer Academic Publishers, and Elsevier to republish this material in revised form is gratefully acknowledged.

Experiments, Models, Paper Tools

Introduction

The performative image of science epitomized as "science in action," "science as cultural activity," or "science as practice" has been most successfully substantiated during the past two or three decades in studies of experimental practices.[1] In many of these studies, the focus has been on instruments, experimental skill, and the interventionist side of experimentation. More recently, the "science as practice" approach has been enriched by the increasing sensibility to the semiotic and cognitive dimensions of experimental practices and other scientific activities. "Practices of theory,"[2] "semiotic turn,"[3] "models as autonomous agents,"[4] "conceptual practice,"[5] "theoretical technologies,"[6] and others are new labels that are redirecting our attention to inscriptions and conceptual development in the laboratory sciences and other areas of science, including theoretical subcultures, such as theoretical physics. Bruno Latour was one of the first scholars in science studies to remind us that there is "no reason to give up following scientists simply because they are handling paper and pencil instead of working in laboratories or traveling through the world."[7] Andrew Pickering has taken a similar stance by criticizing that "machines are located in a field of agency but concepts are not"[8]; and Peter Galison and Andrew Warwick have suggested that the "imbalance" between detailed studies of experimental practices and processes of theory formation be redressed.[9] Works like these have contributed to the growing conviction that the formation of scientific objects, inscriptions, and concepts in the laboratory sciences, as well as work on paper (and on the computer monitor) in general can be studied with the same kind of analytical resources that have been applied to instruments, experimental manipulation, and skill. "Treating theory, experiment and instrumentation practices on a symmetrical footing"[10] has now become one of the most challenging endeavors in the history of science and science studies.

This book continues that line of investigation by concentrating on interpretive modeling in the laboratory sciences and on the question of how new experimental objects and inscriptions and concepts bound up with them come into being and are shaped. It weds the analytical tools of semiotics with concepts and methods developed on the basis of the performative image of science to analyze and historically reconstruct a deep structural transformation in the nineteenth-century scientific landscape, which had a strong impact on physical sciences, on the life sciences, as well as on the broader socioeconomic culture. This was the transformation of the pluricentered culture of plant and animal chemistry, which included physiology and the study and production of pharmaceutically relevant materials, and the emergence of synthetic carbon chemistry, which, in an unforeseen way, was to become the bedrock of the chemical industry in the second half of the nineteenth century.

The various topics discussed in this book revolve around a challenging historiographical paradox. Experimentation in organic chemistry prior to the mid-nineteenth century has often been described as artful handicraft or playful cooking rather than as a scientific enterprise. Yet, it was precisely in the practice of organic chemistry that, from the late 1820s onward, a new kind of sign system was accepted by the European chemical community: chemical formulas, which signified invisible theoretical entities that overlapped (but were not identical) with "atoms" in the natural philosophical tradition. Chemical formulas, such as H_2O for water or H_2SO_4 for sulfuric acid, were introduced in 1813 by the Swedish chemist Jacob Berzelius. A few decades later, chemists used them in organic chemistry to create a multifaceted world on paper, which they correlated with experiments and the traces and inscriptions produced in test tubes, vessels, and with the balance. It was exactly in this most chaotic and most practice-oriented domain of chemistry where Berzelian formulas began to spread and to conquer the whole field of chemistry. Today, the striking duality of a world of signs and a world of laboratory manipulations and instruments, and the more or less obvious interaction of both worlds, has become emblematic of the experimental science of chemistry. Roald Hoffmann and Pierre Laszlo, both well-known chemists, for instance, write: "Chemical structures are among the trademarks of our profession, as surely chemical as flasks, beakers and distillation columns. When someone sees one of us busily scribbling formulas or structures, he or she has no trouble identifying a chemist."[11]

A central argument of the book, which I seek to support by careful analyses, is that chemists began applying chemical formulas not primarily to represent and to illustrate preexisting knowledge, but rather as productive tools on paper or "paper tools" for creating order in the jungle of organic chemistry. Berzelian chemical formulas made it possible to extend the chemical order established and stabilized in inorganic chemistry to the comparatively much more complex and chaotic area of organic matter. They were tools for experimentally investigating organic chemical reactions and for constructing models of reactions and of the invisible constitution of organic substances. The performance of modeling first took place within the collectively shared chemical order carried over from inorganic to organic chemistry. Yet, the manipulations of formulas on paper and the visual display of possible recombinations of signs had the suggestive power of introducing new significances, which chemists attempted to match up with experimental traces. In doing so, they tacitly modified the existing intellectual framework and introduced new concepts and research objects that opened up new avenues of research, such as the experimental practice of synthesizing archipelagos of organic derivatives by means of "substitutions."

The notion of "paper tools" is an analytical category that serves to focus the historical analysis and reconstruction on the performative, cultural, and material aspects of the development of inscriptions, models, concepts, and theories, without disregarding the goals and other intellectual preconditions of the human actors. On many levels, paper tools are fully comparable to physical laboratory tools or instruments and share several features with them.[12] Although both kinds of tools embody intellectual assumptions, they are no longer epistemically relevant in themselves but serve to investigate and represent new scientific objects. Similarly, as the technical design of a laboratory instrument and the possibilities for and constraints on its manipulations and physical interactions with an experimental target shape the immediate outcome of experiments, so too does the syntax of paper tools—its visual form, rules of construction and combination, and maneuverability—count in scientists' production of representations. I also argue that paper tools, like laboratory tools, are resources whose possibilities are not exhausted by the knowledge and intentions of their inventors and by scientists' attempts to achieve preset goals but rather whose application generates new goals, objects, inscriptions, and concepts linked to them. Similar approaches studying the alignment of laboratory instruments and tools of representation and the similarities between both kinds of devices so far have been pursued mainly in the history of physics.[13]

In a recent article on the emergence of particle physics from the early 1900s onward, Jed Buchwald points to the "set of conceptual, mathematical, and instrumental tools for handling microphysical entities."[14] "Like laboratory work," Buchwald writes, "paper work deploys certain kinds of tools in flexible ways, but, again, like laboratory work, it is also subject to constraints imposed by those tools."[15] Nancy Cartwright and others have examined "tools" for the building of models of superconductivity in the 1930s, and in this context they refer to scientists' various resources for modeling as their "toolbox."[16] In his *Image and Logic*, Peter Galison observes that laboratory instruments and apparatus are "dense with meaning, not only laden with their direct functions but also embodying strategies of demonstration, work relationships in the laboratory, and material and symbolic connections to the outside cultures in which these machines have roots."[17] He has convincingly argued that there are two principal modes of representing invisible physical objects and processes in twentieth-century particle physics and that these are linked to different types of instruments—the "homomorphic" (as part of the "mimetic tradition") and the "homologous" (as part of the "logic tradition"). In a recent article, Galison studied the role of Feynman diagrams in wartime Los Alamos for "the development of widely applicable elements of theory that could be combined the way modularised instruments could be engaged one way and then rearranged in another."[18] Also with respect to twentieth-century physics, Andrew Pickering studied "representational chains ascending and descending through layers of conceptual multiplicity and terminating in captures and framings of material agency."[19] Conceptual structures, Pickering argues, are like "precisely engineered valves" that have to be understood "as positioned in fields of disciplinary agency, much as machines are positioned in fields of material agency."[20] Furthermore, Andrew Warwick described Cambridge mathematical physics around 1900 as "skilled work" and "theoretical technology,"[21] and Jeff Hughes has added that in theoretical physics "mathematical tools are open to the same kind of manipulative and interpretive procedures as the experimentalists' instruments and other material resources."[22] Even though terms like conceptual or mathematical *tools*, *toolbox of science*, *theoretical technology*, and so on are sometimes used in a strictly metaphoric sense and hence do not attribute a material dimension to these entities, they have nevertheless narrowed the former gulf between manual and cognitive practices and called into question the dichotomy between hand and mind, and seeing and thinking.

In carefully following the ways chemists applied chemical formulas in the experimental and classifying practice of organic chemistry from the late 1820s onward, this book also pursues a second goal. This is the historical reconstruction of the trajectories that led to the disintegration of the pluricentered culture of plant and animal chemistry, which investigated "natural" organic materials extracted from plant and animal tissues and phenomena of life, and the formation of the experimental culture of synthetic carbon chemistry, which created a plethora of "artificial" organic compounds and chemical reactions that do not occur "in nature" outside the laboratory. This historical process on the communal level of European chemistry took place in the comparatively short period roughly between 1827 and 1840. As a framework for its analysis and historical reconstruction, I compare structural elements of the pluricentered culture of plant and animal chemistry and the experimental culture of carbon chemistry. Together with further detailed analyses of French, German, and other European chemists' experimentation, classification, and model construction from the late 1820s until roughly 1840, this comparison demonstrates that the transformation of organic chemistry occurred on a deep structural level. It comprised the type of experiments, the representational tools and modes of representation, the style of argumentation and justification, and the classification of organic substances. Moreover, it restructured the entire area of scientific objects and altered the notion of organic matter. The meaning of "organic" and the classification of organic substances before and after the transformation period were incommensurable.

Chemical formulas contributed considerably to this transformation process. Without them, it would have been impossible to establish the chemical order accepted in inorganic chemistry in the comparatively chaotic area of organic matter. Further, the application of chemical formulas as paper tools for building models that fit the extended chemical order and matched with experimental inscriptions initiated a dynamics that transcended the original constraints of the chemical order. The dialectic of collectively shared goals and tools, both laboratory and paper tools, propelled a historical process that eventually yielded results that none of the historical actors had foreseen.

The historical issues of this book are intertwined with but not identical to the semiological and epistemological issues. Consequently, some chapters discuss both issues, whereas others concentrate on only one of them. Chapter 1 begins with a semiotic analysis of Berzelian formulas and a short history of their application. I argue that Berzelian formulas had several layers of reference and meaning of older as well as of more recent origin. They were "compact" or "dense" signs in the sense of Dagognet, that is, able to convey simultaneously a plurality of information.[23] A new and particular relevant semantic feature was the concept of scale-independent bits or portions of elements, which overlapped but was not identical with the concept of "atom" in the philosophical and physical tradition. In the subsequent analysis of the syntax of Berzelian formulas, I question the widely held sharp distinction of Berzelian formulas as linguistic or logical signs from structural and stereochemical formulas as graphical or iconic signs. I assert that all the different kinds of chemical formulas introduced during the nineteenth century had a mixed logical and iconographic form and that there is only a difference of degree between them. Furthermore, I compare the syntax of Berzelian formulas with available alternatives—ordinary language and Daltonian diagrams—and I argue that the "graphic suggestiveness" and "maneuverability"[24] of Berzelian formulas was an important material precondition for their application as paper tools beginning in the late 1820s.

Chapter 2 contains a comparison of structural elements of the "pluricentered culture" of plant and animal chemistry and the "experimental culture" of synthetic carbon chemistry. The comparison revolves around four items: the field of scientific objects in its entirety, the reference and meaning of "organic" matter or substance as a distinctive scientific object, the classification of organic substances, and the types of experiments. The result of this comparison is to show that both forms of organic chemistry differed so profoundly that it is no exaggeration to conceive of them as two different cultures. The former overlapped with natural history, pharmacy, and chemical arts and is hence termed a *pluricentered* culture of organic chemistry, whereas carbon chemistry is defined as a culture that was purely *experimental.*

Chapter 3 follows the history of experiments performed by French, German, and other European chemists with alcohol, so-called ethers, and other derivatives of alcohol from 1794 until 1820. These substances and experiments initially were located at the periphery of vegetable chemistry. In the 1820s, they moved to the center of research, where they played the dual role of model objects and paradigmatic achievements in the transformation of organic chemistry. I describe how these experiments developed out of the context of commercial pharmacy and thence gained entry into the laboratories of academic chemists, who investigated the chemical reactions underlying the manufacture of ethers and their by-products. My focus is on the historical reconstruction of a continuous change in the style of quantitative experimentation and on the obstacles which confronted chemists in their rare attempts to study and represent chemical reactions of organic substances before they began to apply chemical formulas.

Chapters 4 through 7 scrutinize chemists' application of Berzelian formulas as paper tools for investigating organic chemical reactions and constructing interpretive models of them, for modeling the invisible constitution of organic compounds, and for classifying them. These detailed analyses, covering the period from 1827 to the mid-1830s, provide most of the empirical evidence for my assertion that Berzelian formulas functioned as productive paper tools and played a pivotal role in the structural transformation of organic chemistry. Furthermore, these analyses show that a particular group of organic compounds—alcohol, the so-called ethers, and other alcohol derivatives—played an outstanding role in chemists' research. They thus underpin my claim, discussed in Chapter 8, that this particular group of substances became "model objects" in the transformation of organic chemistry.

Chapter 4 takes up the thread of Chapter 3, in which experiments on the reactions of alcohol and its derivatives were depicted, and it goes on to reconstruct how Jean Dumas and Polydore Boullay used Berzelian formulas in 1827 to build a model of a chemical reaction underlying the production of ordinary ether and its by-products. It shows how Berzelian formulas were applied to construct schemata of balancing the masses of the initial substances and reaction products and how Berzelian formulas both substituted the measurement of the reacting masses, which was unfeasible on the experimental level, and functioned as tools for modeling the recombinations of the components of the initial substances into the reaction products. By distinguishing between independent parallel and successive reactions of the initial substances, Dumas and Boullay's formula model of reaction also solved a vexing problem that had arisen in experiments performed by German and French chemists seven years previously.

Chapter 5 studies the new mode of classifying organic substances proposed by Jean Dumas and Polydore Boullay in 1828 that was to become the model for all later classifications. It was based on experimental investigation of the composition and constitution of organic compounds and on the subsequent construction of formula models of constitution rather than on the observable properties of organic substances and their natural origins, which were the most important criteria of classification in the pluricentered culture of organic chemistry. The new mode of classification assimilated the classification of organic substances to the classification of inorganic ones. My analyses will reveal that Berzelian formulas were essential paper tools for this enterprise.

Chapter 6 reconstructs in detail the construction of formula models of constitution by Jean Dumas, Justus Liebig, Jacob Berzelius, and other French and German chemists in the late 1820s and 1830s, when these models became the subject of a heated controversy among European chemists. It shows how chemists, starting with inscriptions from the quantitative analysis of an organic compound, constructed a Berzelian raw formula, which in turn was manipulated to come up with a binary formula model of constitution. The latter was often performed without additional experimental support, so that models of the constitution of organic compounds were basically the result of manipulations of formulas. My scrutiny of these manipulations is linked to the reexamination of the prominent controversy known as the "controversy about radicals."

Chapter 7 exemplifies the dialectics between tools (both paper tools and laboratory tools) and goals that engendered a previously unintended parcel of scientific object, formula model, and concept: "substitution." I investigate the formation of "substitution" in 1834 in the context of experimentation and modeling, and I show how this formation was aligned with Berzelian formulas' graphic suggestiveness, their manipulations, and the superimposition of additional significance onto the manipulated signs. With the introduction and development of the concept of substitution between 1834 and 1840 and the associated practice of experimentation and modeling, the process of the transformation of plant and animal chemistry into the experimental culture of synthetic carbon chemistry accelerated. "Substitution" was more than a minor deviation from the original intentions, concepts, and models formed in analogy with inorganic substances. It was a change that radically questioned the adequacy of that analogy and a major step in the formation of the new experimental culture of synthetic carbon chemistry. The last section of Chapter 7 discusses a characteristic aspect of this new culture—the creation of "robust fits"[25] between Berzelian raw formulas and various kinds of formula models and the reflection of these fits in the new mode of justification, which reminds one of what Ian Hacking called the "self-vindication" of laboratory sciences.[26]

In Chapters 8 and 9, the two main issues of this book again part ways. Chapter 8 reflects on topics related to the historical transformation process. The analyses of chemists' performances in the transition period described in Chapters 4 through 7 showed that a particular group of substances, namely alcohol and its derivatives, played an outstanding role. They became "model objects" in European chemists' collective enterprise of extending the chemical order established in inorganic chemistry into the field of organic chemistry. At the same time, the experiments, classifications, and model constructions that revolved around these model objects became unintended "paradigmatic achievements" of the newly emerging experimental culture of synthetic carbon chemistry. I discuss the question of why the particular group of alcohol and alcohol derivatives rather than any other one functioned as model objects, and I summarize the features of chemists' performances that became unintended paradigmatic or exemplary achievements. Further, I reconstruct the overall trajectories of the historical transformation process and discuss questions posed by its dynamics.

Chapter 9 recapitulates the practical applications of Berzelian formulas, concentrating on their perhaps most crucial application as paper tools for representing chemical reactions and constructing simple, interpretive models of them. I summarize the dual function of Berzelian formulas in the experimental investigation of organic chemical reactions as tools that filled gaps left open by experimental inscription devices and as tools for the construction of interpretive models of reaction. In the second section of this last chapter, I discuss the notion of paper tools by exploring the advantages and limitations of Latour's concept of "chains of inscriptions," and by discussing the aspects of "paper tools" revealed by their comparison with laboratory tools or instruments.

CHAPTER ONE

The Semiotics of Berzelian Chemical Formulas

> To master the symbolic language of chemistry, so as to understand fully what it expresses, is a great step forward in mastering the science.
> —COOKE 1874, 149

Chemical formulas, such as H_2O for water or C_2H_6O for alcohol, were introduced by the Swedish chemist Jacob Berzelius in 1813 and 1814 to represent the composition of chemical compounds in a new theoretical fashion (Fig. 1.1).[1] After having been largely ignored for more than a decade,[2] the new sign system began to spread quickly in European and North American chemistry, first in organic chemistry, then in increasingly different forms in other areas of the science. Its rapid dissemination in the 1830s is striking, as is the fact that it was in organic chemistry that the chemical community first accepted the system.

By 1835, journal articles by German and French chemists concerned with organic chemistry were abundant with Berzelian formulas. In that same year, the British Association for the Advancement of Science explicitly recommended the use of Berzelian formulas, and in 1836 the Oxford chemistry professor Charles Daubeny argued for adoption of Berzelian formulas with the observation that "one who *is engaged in the analysis of organic compounds* will be more sensible of the utility of such symbols, than another who is conversant chiefly with a less complicated class of combinations."[3]

British chemists had been particularly reluctant to accept Berzelian formulas in the years before, and several British chemists had fiercely attacked them.[4] By 1835, however, Justus Liebig, Jean Dumas, and other continental chemists who had been employing Berzelian formulas in organic chemistry since the late 1820s had convincingly demonstrated how productive the Berzelian sign system was in this thriving area of chemistry. As Maurice Crosland pointed out, "In the 1830s the chemical symbols of Berzelius offered a new aid to organic chemistry. Even some of the more conservative British chemists, who were slow to make any general use of Berzelius's symbols, saw their value in organic chemistry."[5] William Brock added the following explanation: "It was the difficulty of dealing verbally or in Daltonian symbolism, with the increasingly

The chemical signs ought to be letters, for the greater facility of writing, and not to disfigure a printed book. Though this last circumstance may not appear of any great importance, it ought to be avoided whenever it can be done. I shall take, therefore, for the chemical sign, the *initial letter of the Latin name of each elementary substance:* but as several have the same initial letter, I shall distinguish them in the following manner:—1. In the class which I call *metalloids,* I shall employ the initial letter only, even when this letter is common to the metalloid and to some metal. 2. In the class of metals, I shall distinguish those that have the same initials with another metal, or a metalloid, by writing the first two letters of the word. 3. If the first two letters be common to two metals, I shall, in that case, add to the initial letter the first consonant which they have not in common: for example, S = sulphur, Si = silicium, St = stibium (antimony), Sn = stannum (tin), C = carbonicum, Co = cobaltum (cobalt), Cu = cuprum (copper), O = oxygen, Os = osmium, &c.

The chemical sign expresses always one volume of the substance. When it is necessary to indicate several volumes, it is done by adding the number of volumes: for example, the *oxidum cuprosum* (protoxide of copper) is composed of a volume of oxygen and a volume of metal; therefore its sign is Cu + O. The *oxidum cupricum* (peroxide of copper) is composed of 1 volume of metal and 2 volumes of oxygen; therefore its sign is Cu + 2 O. In like manner, the sign for sulphuric acid is S + 3 O; for carbonic acid, C + 2 O; for water, 2 H + O, &c.

When we express a compound volume of the first order, we throw away the +, and place the number of volumes above the letter: for example, Cu O + S $\overset{3}{\dot{\mathrm{O}}}$ = sulphate of copper, Cu $\overset{2}{\dot{\mathrm{O}}}$ + 2 S $\overset{3}{\dot{\mathrm{O}}}$ = persulphate of copper. These formulas have this advantage, that if we take away the oxygen we see at once the ratio between the combustible radicles. As to the volumes of the second order, it is but rarely of any advantage to express them by formulas as one volume; but if we wish to express them in that way, we may do it by using the parenthesis, as is done in algebraic formulas: for example, alum is composed of 3 volumes of sulphate of alumina and 1 volume of sulphate of potash. Its symbol is 3 (Al $\overset{2}{\dot{\mathrm{O}}}$ + 2 S $\overset{3}{\dot{\mathrm{O}}}$) + (P$\overset{2}{\mathrm{o}}$ + 2 S $\overset{3}{\dot{\mathrm{O}}}$). As to the organic volumes, it is at present very uncertain how far figures can be successfully employed to express their composition. We shall have occasion only in the following pages to express the volume of ammonia. It is 6 H + N + O, or $\overset{6}{\mathrm{H}}$ N O.

Figure 1-1 Excerpt from Berzelius's introduction of chemical formulas (Berzelius 1814a, 51 f.)

bulky continental research on organic chemistry which forced a reappraisal of Berzelius's symbols, and their expansion in the 1830s."[6] This insight has not, however, been an incentive to further and more detailed study of the application and functions of Berzelian formulas in the practice of organic chemistry. The popular view has gone unchallenged, that is, that Berzelian formulas were merely a shorthand for names, a convenient notation that—compared with the graphical formulas of the second half of the nineteenth century—was of little use for research.[7]

I disagree with this view, and one of the main purposes of this book is to prove that it is incorrect. A careful analysis of European chemists' published research papers on organic chemistry in the late 1820s and 1830s demonstrates that Berzelian formulas were not merely convenient means of representing and illustrating preexisting knowledge but were primarily tools for structuring organic chemistry and thus mastering its confusing chaos. The unambiguous denotation and clear demarcation of organic species depended on the representation of their composition by Berzelian formulas. The new mode of classifying organic substances introduced after 1827 also required Berzelian formulas to model the invisible constitution of organic substances. In the 1830s, these formula models were the subject of a fierce controversy involving leading chemists, such as Jacob Berzelius, Justus Liebig, and Jean Dumas.[8] Moreover, in the mid-1830s, experimental investigation of organic chemical reactions required the application of Berzelian formulas, both on the level of data gathering and processing and in constructing interpretive models of reactions. In the course of the 1830s and 1840s, Berzelian formulas became indispensable paper tools at literally all sites of European and even North American chemical practice. John Webster, professor of chemistry at Harvard University, gives testimony to the productive function of Berzelian formulas in the third edition of his textbook.[9] After mentioning that "the symbols contrived by Berzelius are now extensively used by chemists and mineralogists,"[10] he continues:[11]

> All the elements contained in a compound are thus visibly represented, and the chemist is able readily to trace all possible modes of combination, and to select that which is most in harmony with the facts and principles of his science. He may, and often does, thereby detect relations which might otherwise have escaped notice. Another advantage attributable to such formulae is, that they facilitate the comprehension of chemical changes.

My analyses of the practical application of chemical formulas in Chapters 4, 5, and 6 will confirm this judgment of a contemporary.

This first chapter concerns the semiotics of Berzelian formulas. In the semantic analysis in the first section, I argue that Berzelian formulas contain several layers of reference and meaning of older as well as of more recent origin. They were "compact" or "dense signs" in François Dagognet's sense of the term; that is, they conveyed a plurality of information simultaneously.[12] A new and particularly relevant feature was the concept of scale-independent bits or portions of elements, which overlapped but was not identical with the concept of "atom" in the philosophical and physical tradition. In the second section, in which I analyze the syntax of Berzelian formulas, I argue that chemical formulas of every kind have a mixed logical and iconographic form and that the difference between them is only one of degree. The iconographic features became stronger during the continuous evolution of formulas from Berzelian formulas to type formulas, structural formulas, and stereochemical formulas.

Excursus: Stoichiometry and the Law of Combining Volumes of Gases

The empirical preconditions of Berzelian formulas were the so-called stoichiometric laws, in particular the law of multiple proportions and Joseph Louis Gay-Lussac's law of combining volumes of gases.[13] Quantitative analyses of inorganic compounds consisting of the same elements A and B, such as nitrous acid, nitric acid, and nitrous gas, had shown that, taking the weight of one element as a constant referent, the other element forms a series of combining weights in ratios of small integers. John Dalton was one of the first scholars to explain this discontinuity of macroscopic combining weights as strong evidence for the assumption of an underlying discontinuity of hidden chemical entities, which he conceived as submicroscopically small atoms. "I remember the strong impression," he wrote in 1811, "which at a very early period of these inquiries was made by observing the proportion of oxygen to azote as 1, 2, and 3, in nitrous oxide, nitrous gas, and nitric acid, according to the experiments of Davy."[14] As early as 1808, Thomas Thomson published an experimental report on multiple proportions in which he states that in the normal and acid oxalates of potassium and strontium, one salt, for the same amount of acid, contains exactly double the proportion of base contained in the second.[15] Shortly thereafter, William Hyde Wollaston published experimental results showing that, when combined with equal weights of potash, the amount of carbonic acid in the carbonate of potash stood to that in the subcarbonate of potash in a ratio of 2:1. Wollaston referred to the earlier paper by Thomson and to Dalton's theory of chemical combination by remarking that his results were a particular instance of that theory.[16] In 1810, Berzelius presented a similar law of proportions in the composition of salts. Early nineteenth-century chemists conceived of salts as binary compounds consisting of an acid and a base, both oxides. Berzelius's law of oxides stated that the quantity of oxygen in one component of a salt must be a small integral multiple of the quantity of oxygen in its other component.[17]

In light of the law of multiple proportions, the other stoichiometric laws added up to a bedrock for Daltonian atomism or similar theories, such as Berzelius's theory of chemical proportions or portions (see later), as well as for chemical formulas. The law of definite proportions assumed that all chemical compounds have a fixed quantitative composition. Two years before the French chemist Louis Joseph Proust stated this law, the Prussian chemist Jeremias Benjamin Richter had coined the term *stoichiometry*.[18] In his experiments, Richter found that the weights of different acids saturating a given weight of any base and forming a neutral salt (or of different bases saturating a given weight of any acid) were in a constant ratio. If the saturating weights of different acids were related to a conventional standard combining weight of a base (or the bases to a conventional standard combining weight of an acid), numbers representing invariant and dimensionless "equivalent" weights could be ascribed to each acid and each base. By comparing the quantitative composition of oxide series and other simple inorganic compounds like sulfides, chemists also found equivalent weights, or simply *equivalents*, as they were called; for example, 2703 parts (by weight) of silver or 1713 parts of barium or 791 parts of copper combined with 200 parts of oxygen to form oxides and with 400 parts of sulfur to form sulfides.[19] The numbers representing the combining weights of the three metals were "equivalent" with respect to their saturating effects when combined with 200 parts of oxygen or 400 parts of sulfur, respectively. Likewise, sulfur and oxygen were equivalents with respect to the fixed quantities of metals. Furthermore, 200 parts of oxygen and 400 parts of sulfur (or 100 parts of oxygen and 200 parts of sulfur) combined to form hyposulfuric acid.[20] When all numbers were related to the same unit of combining weight, the combination of the laws of multiple proportions and of equivalents made it possible to ascribe to each element a series of numbers representing relative combining weights in small integer ratios.[21]

In 1808, the French chemist Joseph Louis Gay-Lussac reported the results of new experiments together with a generalization known today as *Gay-Lussac's law of combining gases*.[22] This law represented a further discontinuity in the quantitative relations of combining substances: When any two kinds of gaseous substances combine chemically, the ratio of their volumes is 1:1, 1:2, or

1:3.[23] Gay-Lussac also stated that the "contraction" of the volume of the two combining gases was related to the volumes of one of the two initial substances by a simple ratio.[24] For example, one part by volume of oxygen and two parts by volume of hydrogen form two parts by volume of water vapor. Furthermore, when two gases formed different compounds, the parts by volume stood to each other in a small whole number ratio. For instance, one part by volume of oxygen united with two parts by volume of nitrous gas to form nitric acid and with three parts by volume to form nitrous acid.

It should be mentioned that in their application as an empirical foundation of Berzelian formulas and of explanatory theories, such as Dalton's atomic theory and Berzelius's theory of chemical proportions (or portions), both the original law of multiple proportions and Gay-Lussac's law were tacitly transformed in an important aspect. The law of multiple proportions assumed small integer multiples of proportions, *small* meaning one, two, or three. Accordingly, Gay-Lussac's law stated that the relation of combining volumes of gases was always 1:1, 1:2, or 1:3. Yet, when applied in association with Berzelian formulas, the original numeric restrictions were given up and increasingly larger numbers were allowed, in particular in the context of organic chemistry.[25]

THE VARIOUS MEANINGS OF BERZELIAN FORMULAS

In the historiography of chemistry, Berzelian formulas have been interpreted either as mere abbreviations for names, thus denoting macroscopic chemical compounds and their composition from chemical elements,[26] or as a notational system that refers to empirical, stoichiometric relations,[27] or as unequivocal signs for atomic weights and "atoms" in the sense of submicroscopically small particles.[28] I propose that these are not alternatives and that Berzelian formulas simultaneously displayed all these meanings. As the following chapters will demonstrate, the reference and meaning of Berzelian formulas depended on their context of application. In the various practical contexts, it was possible that a particular layer of reference and meaning came to the foreground while others receded to the background. For example, because it was possible to calculate the outcome of a quantitative analysis from the chemical formula of a substance, formulas often were applied merely as signifiers for the quantitative stoichiometric or volumetric relations of chemical compounds. In contrast, in the context of an atomic theory designed to solve fundamental explanatory problems and that defined atoms as submicroscopically small particles, they could unequivocally signify atoms and their characteristic atomic weight. In most of their applications, however, Berzelian formulas had a theoretical meaning that differed from "atomic" composition. In this section, I argue in more detail for the latter proposition.

Berzelius's Theory of Chemical Proportions or Portions

In the first two decades of the nineteenth century, leading European chemists were fascinated by the idea of mathematizing chemistry. Dalton's atomic theory, Berzelius's theory of chemical proportions (or portions), and the associated endeavors to assign a unique and invariant relative combining weight—also called *proportion*, *equivalent*, *atomic weight*, and so on—to each chemical element provided the theoretical framework for this enterprise.[29] In 1813, for example, Thomas Thomson wrote of Dalton's theory: "It puts us in the way of establishing principles of rigid accuracy as the foundation of our reasoning, and to call in the assistance of mathematics to promote the progress of a science which has hitherto eluded the aid of that unrivalled instrument of improvement."[30] Similarly, in a review of John Black's translation of Berzelius's *System of Mineralogy*, the reviewer wrote about Berzelius's theory of proportions: "The doctrine of chemical proportions, though but recently introduced into chemistry, has produced a great reform in the science, and has given birth to a degree of accuracy, both in experimenting and reasoning, which has already placed chemistry on a footing with the mathematical sciences."[31]

The introduction of chemical formulas was intertwined with these events. In the journal articles introducing chemical formulas, Berzelius was concerned mainly with elaboration of what he called "laws of chemical proportions" and a "doctrine" or "theory of proportions." Since 1807, he had been carrying out quantitative analyses of chemical substances, in particular of inorganic compounds. These experiments, together with the ongoing attempt to assign a unique, invariant relative combining weight to each chemical element, contributed to his growing conviction that there are general laws that determine the number of possible stoichiometric proportions, although many chemists were still skeptical because the concept of "proportions" contradicted Claude Louis Berthollet's theory of affinity, which was widely accepted by European chemists.[32] Berzelius wrote the following:[33] "I do not know how far chemical philosophers will allow them [the laws of chemical proportions] to be well founded; but, in hopes that the laws of chemical proportions which I have endeavored to establish will be one day examined and admitted, I will continue in this paper my researches. . . ." To achieve his goal, Berzelius followed a double strategy. He continued experimentation and at the same time attempted to elaborate a chemical theory from which the general laws of proportion could be deduced. John Dalton's atomic theory offered one possible route. Yet some "anomalies" occurring in this theoretical framework[34] and speculations about mechanical qualities of the atoms that were far from being the subject of experiments at the time prevented Berzelius from accepting it wholeheartedly. Instead, he tried to go his own way by developing a chemical theory, which he called the *theory of chemical proportions*.[35] His new formulaic notation became not only a suitable medium for expressing this theory but also a tool for forging its distinction from Dalton's atomic theory.

In his attempt to elaborate a theoretical alternative to Daltonian atomism, Berzelius made a particular experience: He encountered the limits of the available sign systems. Ordinary language, which was the most common sign system chemists used for theoretical purposes, was also the language of philosophers and of atomic theories in the natural philosophical tradition. Speaking of "atoms" would therefore immediately invoke the idea of invisibly small bodies, defined by their orientation in space, size, shape, and other mechanical properties, which chemists could not link to laboratory practice. Daltonian diagrams, which represented simple atoms by circles and composed atoms (later "molecules") by juxtaposing them, had the same effect.

Berzelius's confusion about language can hardly be overlooked in his writings. In the two journal articles in which he introduced chemical formulas, he gave different, apparently contradicting explanations of the meaning of these signs. In his 1813 article he wrote:[36]

> To avoid long circumlocutions, I shall here employ a simple and short method of *expressing determinate combinations*, which I always use in my annotations. . . . I shall take the liberty of giving a short explanation here of this method, *which is founded on something very analogous to the corpuscular hypothesis of Dalton*. It is known that bodies in their gaseous state either unite in equal volumes, or one volume of one combines with 2, 3, &c. volumes of the other. Let us express by the initial letters of the name of each substance *a determinate quantity of that substance*; and let us determine that quantity from its relation in weight to oxygen, both taken in the gaseous state, and in *equal volumes*; that is to say, *the specific gravity of the substances in their gaseous state, that of oxygen being considered as unity*. . . . It is obvious that this comes to the same thing as Mr. Dalton's *weights of atoms*; but I have here the advantage over him, of not founding my numbers on an hypothesis, but upon a fact well known and proved.

Berzelius here states that the letters in his formulas refer to a "determinate quantity" of an element and that this quantity is identical to the specific gravity of the denoted element, the specific gravity of oxygen taken as the unit. He adds something that sounds strange to today's readers, however. On the one hand, he asserts that his signs denote the specific gravity of combining substances, that is, measurable magnitudes that are not founded on a hypothesis. On the other hand, he insists that his "method" is founded on a theory similar to Dalton's "corpuscular" or atomic theory, which assumed "atoms" in the sense of submicroscopically small bodies. He also adds that the specific gravity denoted by the letters "comes to the same thing" as Dalton's atomic weight.

A year later, Berzelius emphasized that from the quantitative information given by chemical formulas, the numeric result of a quantitative analysis could be calculated. Here he unequivocally defined his formulas as signifiers for empirical, stoichiometric relations:[37]

> When we endeavour to express chemical proportions, we find the necessity of signs. . . . they are destined solely to facilitate the expression of chemical proportions, and to enable us to indicate, without long periphrases, the relative number of volumes of the different constituents contained in each body. By determining the *weight of the elementary volumes, these figures will enable us to express the numeric result of an analysis* as simply, and in a manner as easily remembered, as algebraic formulas in mechanical philosophy.

A short time later, he wrote, "In order to express without a multitude of words the *composition of the body with respect to chemical proportions, I avail myself of formulae* in which each body is denoted by the letter which is placed opposite in table I."[38]

This definition says that formulas signify "chemical proportions" rather than "volumes" or "specific gravity." Yet it is by no means evident that Berzelius unambiguously meant "proportions" in the sense of numeric relations of combining weights, for he gives the following example: "for example, $2\ SO^3 + CuO^2$ = persulphate of copper; The number 2 denotes that the acid in the salt contains not only *two particles* of sulphur but six of oxygen, &c."[39] In this example, Berzelius explains that the formula for copper persulfate denotes first the substance and second the number of "particles" contained in it.

Specific gravity or weight of volumes of elements and entire compounds, atomic weight of atoms or particles, chemical proportions—according to Berzelius, his chemical formulas signified all these entities. Why did he allow talk of numeric quantitative relations of the components of a compound, which can be determined by experiment, to run seamlessly into talk of unobservable particles? And how was it possible that Berzelian formulas signified both observable laboratory substances and unobservable Daltonian atoms, given the fact that physicists and philosophers of the time clearly distinguished between macroscopic and submicroscopic entities? To answer these questions, we must study in more detail what Berzelius meant when he spoke about *atoms*, *vol-*

umes, and *proportions* in connection with his introduction of chemical formulas and his theory of proportions.

Signs for Scale-independent Chemical Portions

Whereas in the context of stoichiometric laws *proportion* meant a relative proportion by weight of one substance to another—that is, the numeric relation of weights of combining substances—in the new theoretical context *proportion* meant a portion or a bit of a substance defined by its unique and invariant relative combining weight. A measurable *attribute* of things was transformed into a new, unobservable *thing*, a carrier of weight, the weight being empirically determined in a relative way by referring it to a conventional standard. Alan Rocke coined the term *chemical atoms* for these entities, which he described as "building blocks" that were "shorn of all physical aspects."[40] In accordance with this, he made a sharp distinction between *chemical* atoms and *physical* atoms. Although the following discussion draws heavily on Rocke's former analysis, I prefer the term *portion*, which has the dual advantage of being an actor's term[41] that additionally brings to mind Berzelius's theory of proportions and avoids possible anachronistic misunderstandings.[42]

In analogy to the transformation of the meaning of *proportion* within his theory of proportions, Berzelius also gave a new theoretical meaning to the term *volume*. His theory of volumes was based on Gay-Lussac's law of combining volumes of gases, which previously had been rejected by John Dalton.[43] This law stated that, on the assumption that all chemical substances can be transformed into gases, *one* volume of a substance always combines with one, two, or three volumes of a second substance. Berzelius's theory of volumes modified Gay-Lussac's law, first, in its speculative generalization, including also organic compounds and, second, in raising the law's numeric restriction, thus allowing volume ratios larger than 1:1, 1:2, and 1:3. Moreover, Berzelius played with the idea of a shift in the scale of "volumes" down to immeasurably small units and vice versa. This is implicitly expressed by suggestions such as the following: "there is no other difference between the theory of atoms and that of volumes, than that the one represents bodies in a solid form, the other in a gaseous form. It is clear, that what in the one theory is called an atom, is in the other theory a volume."[44]

How can the only difference between the "volume" of a body and an "atom" consist of the gaseous form of the former and the solidity of the latter? In keeping with a traditional mode of reasoning about the existence of "atoms" as the smallest (solid) parts of a solid body that shifts in scale, one can imagine an analogous shift in scale of a gaseous body resulting in its smallest gaseous parts, or smallest "volumes." What remains unchanged in this shifting of scale is the idea of a scale-independent unit, bit, or portion of a substance that bears all the macroscopic substance's qualities and has a relative, invariant combining weight.

The theoretical concept of a chemical portion, either in its solid or gaseous state, overlapped with Dalton's concept of an elemental atom without being identical with it. Both the concept of chemical portion and the Daltonian "atom" were superimposed on the concept of a chemical element. Both concepts postulated unobservable bits of elements identified by their unique and invariant relative combining weight. Whereas Dalton's atoms were defined as unobservable small particles or "bodies" of the microworld having a certain shape, size, and orientation in space, the chemical portion was a scale-independent entity without mechanical properties. This distinction between scale-independent portions of chemical substances and submicroscopically small atoms may be strange for today's readers and even for today's chemists, who are accustomed to thinking in terms of submicroscopic particles. It was different in the first half of the nineteenth century, however, when the most fundamental category of the chemical practice—that is, of experimentation and classification—was that of a chemical substance. Chemists spoke, for example, of copper, sulfuric acid, and nitrous gas, without considering the shape of the body made up of copper or the size of the vessels containing sulfuric acid or nitrous gas. They were interested in the qualities of the substances, what colors they had, how they smelled, what characteristic effects they exhibited when mixed with reagents, what kind of reactions they underwent, what reaction products were produced from them, and what relative combining weights they had. In sum, *chemical substance* was (and is) a category that abstracted from the mechanical properties of bodies, such as shape, spatial orientation, and size. It was as scale independent as the superimposed concept of a quantitative unit of substance, the chemical portion.

Given the fact that Berzelius's introduction of chemical formulas was embedded in his struggle to introduce a new chemical theory that differed from Dalton's atomic theory, his formulaic sign system appears in a new light. Letters and numbers were perfectly suited to denoting units of chemical substances without simultaneously invoking ideas about atoms in the natural philosophical tradition. It is no exaggeration to say that chemical formulas enabled Berzelius to elaborate clearly the difference between a submicroscopically small atom and a discrete, scale-independent chemical portion. The newly introduced mode of representation was not an exterior medium for a preexisting conceptual referent, a kind of receptacle for some content, but a constituent of meaning. In contrast to chemical formulas, verbal language was too ambiguous for forging and clearly denoting the theoretical specificity of "chemical portions." Accordingly, in the decades following the introduction of Berzelian formulas, chemists used a broad variety of different terms to refer to chemical portions. Some were eager to emphasize that the concept of chemical portion was not identical to that of an atom in the philosophical tradition and therefore preferred terms like *equivalent* (Wollaston), *proportion* (Davy), *combining weight* (Young), *portion* (Thomson), and *parcel* (Whewell). Others, however, paid less attention to theoretical and terminological boundaries, thus speaking of "atoms" when scale-independent chemical portions were meant.

That Berzelian formulas flourished in nineteenth-century chemistry was due to the fact that they had different layers of reference and meaning and did not commit chemists to elaborate, foundational theories, such as Dalton's atomic theory and Berzelius's atomistic foundation of his electrochemical theory. Dagognet pointed out that "scientific symbolism" is characterized by an inverse quantitative relation: *la minceur d'un signifiant, l'universalité et les capacités du signifié.*[45] Whereas Dagognet believed that exclusively graphical formulas introduced in the second half of the nineteenth century were semantically "dense" scientific signs,[46] I assert that this holds also for Berzelian formulas. Berzelian formulas denoted simultaneously pure chemical compounds made up of elements, the stoichiometric or volumetric quantitative relations of these elemental constituents, unobservable scale-independent portions of elements and of the entire compound, and sometimes, that is, in specific theoretical contexts, also atoms in the sense of submicroscopically small particles. Moreover, as I argue in the next section, in the 1830s, Berzelian formulas often were used to denote the binary "constitution" of organic compounds. In sum, their large "capacity of inscription"[47] made these signs applicable in various practical contexts.

Berzelian Formulas and Organic Compounds

There is a certain irony in the fact that Berzelian formulas were first widely accepted and applied in a chemical domain where their applicability was questioned at the time when this sign system was introduced. "As to the organic volumes," Berzelius wrote in 1814, "it is at present very uncertain how far figures can be successfully employed to express their composition."[48] Berzelius's doubts shed some additional light on the theoretical contents of chemical formulas.

What was the problem? The law of multiple proportions, which was an important pillar of Berzelius's theory of chemical portions and of his formulas, states that the ratio of the combining weights of a substance A that forms different compounds with an invariant weight of a substance B will always be a ratio of small integers. What is easily overlooked here is the fact that the law requires not only integers but also small ones. Berzelius himself repeatedly pointed out how important this aspect was for an atomic theory and for his theory of chemical portions because it followed from this constraint that the number of combining atoms or portions had to be a small one as well. Accordingly, Berzelius wrote:[49]

> If we suppose a combination of 2 atoms of one body with 3 of another, we have equal reason to suppose the possibility of a combination of 3 atoms with 4, 5 with 6, &c. By thus increasing the number of atoms of each body united as much as the possibility of mutual contact will permit, we extend the bounds of chemical proportions indeed but almost entirely destroy the doctrine of determinate proportions.

A few years later, he wrote in his monograph on the theory of proportions that "if it were to be possible that an indeterminate number of atoms of an element could combine with an indeterminate number of a second one a countless number of compounds would be created in which the differences of the relative quantities of their components would be so insignificant that they could not be estimated even in the most careful examinations."[50]

Berzelius applied these principles of numeric restriction to the domain of inorganic compounds, where stoichiometry was first introduced and well elaborated. In organic chemistry, however, things were different. Here chemists were confronted with results that, on the basis of experiments with inorganic compounds, should have been excluded in principle. There were, in fact, countless compounds made up of the same elements, the elemental components having very small quantitative differences, and the formulas denoting these compounds included numbers much greater than those allowed for by the theoretical principles established in inorganic chemistry.

Chemists were well aware of the precarious status of the stoichiometric laws in organic chemistry. For example, Berzelius wrote in 1814:[51]

> I suppose that other chemists, as well as myself, have found it difficult to conceive how such an immense variety of compounds among three or four elementary bodies could exist conformably to the laws of chemical proportions; the different species of vegetable oils, for example, of tannin &c.; for it is to be supposed that the difference in their composition is infinitely small; but this difficulty is obviated

> by the single circumstance that in these ternary, &c. oxides, none of the elements is necessarily an unit; of consequence the number of possible combinations becomes almost infinite.

Two decades later, Liebig wrote in a similarly skeptical mood: "The study and analysis of organic bodies has in my eyes proved irrefutably that bodies are able to combine in all proportions."[52]

In the thriving field of organic chemistry, the law of multiple proportions could not be confirmed, at least not for organic compounds in the proper sense. The only experimental confirmation that Berzelius and other chemists could present came from experiments with mixed organic and inorganic compounds consisting of an organic compound and lead oxide. In these mixed compounds, the ratio of oxygen contained in the inorganic component and in the organic one was in small integers. "It is in this manner, by consulting not merely analysis, but by examining the compounds of which the substance is capable, that we are enabled to unfold the laws of chemical proportions in organic nature,"[53] Berzelius asserted with some satisfaction.

The lack of empirical confirmation did not prevent Berzelius from extending the laws of proportion, the law of combining gases, and the theory of chemical portions to organic chemistry. In doing so, however, he made an important modification:[54]

> The laws of proportions in organic nature may be comprehended under the two following general rules: 1. When three or more elementary bodies, of which oxygen is always one, combine so as to produce a ternary, quaternary etc. oxide, *a certain number* of atoms or volumes of one of the elements combines with *a certain number* of atoms or volumes of each of the others; *but it is not necessary that any one of these elements should be considered as unity*. 2. When these oxides combine with each other or with binary oxides, the oxygen in the one is always a multiple by a whole number of that in the other.

With respect to organic compounds, Berzelius no longer restricted the laws of proportions to small integers. He now spoke merely of "a certain number" of "atoms" or "volumes," and he explicitly revoked the principle that one component is always considered as unity, that is, that it is represented in the formula by one "atom" or portion. For example, as early as 1814, Berzelius asserted that oxalic acid was made up of eighteen parts by volume of oxygen, twelve of carbon, and one of hydrogen.[55] It is obvious that in this case the term *volume* no longer denoted a measurable magnitude but had a new theoretical status.

Yet, in many cases, even this important modification did not provide a safe ground for the theory of chemical portions in their application to organic compounds. Even if greater numbers of chemical portions were allowed, often it was not possible to obtain integers by dividing the data of quantitative analysis by the theoretical combining weight (or atomic weight) of the elements.[56] In an article published in 1815, Berzelius reported the quantitative analyses of thirteen organic compounds and their transformation into formulas. An examination of Berzelius's calculation reveals that in only eight cases did he succeed in transforming the analytical results into integral numbers of "atoms."[57] Of these eight compounds, three formulas contained relatively large numbers of portions: oxalic acid, 1 H + 12 C + 18 O;[58] cane sugar, 10 O + 12 C + 21 H;[59] and gum arabic, 12 O + 13 C + 24 H.[60] In the following five cases, there was a considerable deviation from integers:

tannic acid: O = 3, C = 3,94 (rounded up to 4), H = 4,25 (rounded down to 4)

tartaric acid: O = 5, C = 3,87 (rounded up to 4), H = 4,67 (rounded up to 5)

benzoic acid: O = 1, C = 4,85 (rounded up to 5), H = 3,81 (rounded up to 4)

tannic acid: O = 4, C = 6,10 (rounded down to 6), H = 5,65 (rounded up to 6)

sugar: O = 10, C = 12,00, H = 20,86 (rounded up to 21)

These examples demonstrate that the theory of chemical portions that required integral numbers of portions were far from being clearly confirmed by the quantitative analysis of organic compounds. They thus emphasize the speculative boldness of chemists when they began to apply Berzelian formulas to organic compounds.

IMAGE AND LANGUAGE: THE SYNTAX OF BERZELIAN FORMULAS

Throughout the literature about chemical formulas, a sharp distinction has been made between Berzelian formulas on the one hand and structural and stereochemical formulas on the other. Berzelian formulas have been viewed as algebraic, linguistic, or symbolic signs, the other two as graphical, pictorial, or iconic. This dichotomy fits a popular distinction made by semioticians between language-like or logical types of signs, which are arbitrarily related to their objects and have a compositional semantics and image-like or iconic signs, which represent their objects by virtue of being similar, isomorphic, or analogous to them.[61] The distinction, which goes back to Charles S. Peirce's famous triad of "symbols" (arbitrarily linked with their objects), "icons" (similar to their objects), and "indices" (physically connected with their objects),[62] has been productively applied to twentieth-century particle physics by Peter Galison.[63] As applied to chemical formulas, however, this distinction raises some questions that will be posed in the following section.

The Algebraic Form of Berzelian Formulas and Iconicity

Berzelian formulas consist mainly of letters and numbers, the numbers preceding the letters or being superscripts of them.[64] The letters, such as O (oxygenium), S (sulphur), or Cu (cuprum) were taken from the Latin names of the substances. In addition, Berzelius often used dots instead of O for oxygen, letters combined with dashes to denote two "atoms" of an element, as well as the plus sign and parentheses to combine letters and numbers. For example, in his introductory article of 1814, Berzelius represents the composition of sulfuric acid by the formula S + 3 O, but also by SO^3, which he viewed as identical in meaning with the former formula (Fig. 1.1). The plus sign could be omitted in the case of the simplest compounds consisting only of two elements, such as SO^3, but Berzelius always used it for denoting compounds consisting of more than two elements. For example, the formula for copper sulfate was SO^3 + CuO and that for alum 3 (AlO^2 + 2 SO^3) + (Po^2 + 2 SO^3). The reason for this was that the formulas represented not only the elemental "composition" of compounds but also their binary "constitution," that is, the internal association of the elements to form two composed "immediate constituents."[65] Thus, the formula for copper sulfate was not $CuSO^4$ as we would be inclined to write today; it had to be SO^3 + CuO. Partition of the formula into two partial formulas linked by the plus sign signified that copper sulfate was not immediately composed from the three elements sulfur, oxygen, and copper but from sulfuric acid and copper oxide and that these two "immediate constituents" were stable enough to be isolated experimentally in chemical analyses. In the case of more complicated compounds, such as alum, parentheses were necessary to denote the binary constitution. The two parentheses in the formula for alum signified that its two immediate constituents, obtained through analysis, were aluminium sulfate (AlO^2 + 2 SO^3), itself consisting of the two components aluminium oxide and sulfuric acid, and sulfate of potash (Po^2 + 2 SO^3), which again is shown as being made up of the two components potash and sulfuric acid.[66]

Berzelius explicitly mentioned that he used parentheses "as is done in algebraic formulas" (Fig. 1.1). Thus, a number before the parenthesis referred to all letters included by it; however, Berzelian formulas did not strictly follow algebraic notation. The plus sign always denoted additivity, but additivity was not always denoted by the plus sign. For example, the formula SO^3 meant $S + 3\ O$, whereas according to algebraic notation, it would have signified multiplication. Although many scientists criticized this inconsistent application of algebraic notation,[67] Berzelius was little impressed. He insisted that his notation was algebraic, and he viewed this as one of its major advantages: "These figures will enable us to express the numeric result of an analysis as simply, and in a manner as easily remembered, as the algebraic formulas in mechanical philosophy."[68] Like many other chemists of the time, mathematization of chemistry, and thereby enhancement of its reputation, was one of Berzelius's main goals.

Letters, numbers, superscripts, the plus sign, and parentheses were inscriptions of an algebraic mode of representation par excellence. Furthermore, Berzelian formulas were read progressively, as algebraic notation is read. Yet the algebraic form of Berzelian formulas was not exhausted by its mathematical significance, and neither were Berzelian formulas completely lacking in imagery. Rather than being a purely mathematical token signifying the additivity of numbers (i.e., the theoretical combining weights of the elements), the plus sign also signified the chemical additivity or chemical combination of portions of elements constituting a chemical compound. This becomes particularly clear when we compare the Berzelian formula models of binary constitution, for which Berzelius coined the term *rational formulas*,[69] with raw formulas or empirical formulas.

Roughly between 1833 and 1840, European chemists shared the assumption that organic substances have a binary constitution in analogy to inorganic compounds. Chemists often constructed different formula models of the binary constitution of an organic compound. Between 1833 and 1840, these alternative formula models became the subject of a prominent controversy in European chemistry. For example, if the Berzelian raw formula for alcohol was $2\ C + 6\ H + O$ or C^2H^6O, several formula models of its binary constitution were possible, such as $(2\ C + 6\ H) + O$ or $C^2H^6 + O$ (Berzelius) on the one hand or, on the other, $(2\ C + 4\ H) + (2\ H + O)$ or $C^2H^4 + H^2O$ (Dumas; all formulas refer to the "atomic weight" $C = 12$). Because there is no mathematical difference between these formulas, the controversy remains inexplicable as long as we fail to recognize that the plus sign had an additional chemical meaning. In these differing formula models, the modes of distribution of the letters and the plus sign show how the elements of an organic compound are chemically linked together to form its composed immediate constituents. On Berzelius's model, alcohol was constituted of a hydrocarbon denoted by C^2H^6 and oxygen; on Dumas's alternative model, it was constituted of the hydrocarbon C^2H^4 and water. In both cases, the plus sign also signified a chemical combination and thus denoted the different ways in which the elements of alcohol chemical associate to form its two immediate constituents.

Just as the plus sign signified chemical additivity, a letter was not merely shorthand for the name of an element and a sign for its theoretical combining weight ("atomic weight"). It also stood for the unobservable chemical entity postulated by the theory of chemical portions: the portion of an element, which was identified by its relative combining weight or its specific gravity. Letters had a certain "graphic suggestiveness"[70] owing to the one-to-one correspondence between a letter and the denoted chemical portion. Hence, a Berzelian formula conveyed a building-block image of the denoted chemical compound. With respect to algebraic notation in general, Rudolf Arnheim gave cogent expression to what is meant here:[71]

> In the strictest sense it is perhaps impossible for a visual thing to be nothing but a sign. Portrayal tends to slip in. The letters of the alphabet used in algebra come close to being pure signs. *But even they stand for discrete entities by being discrete entities*: a and b portray twoness. Otherwise, however, they do not resemble the things they represent in any way, because further specification would distract from the generality of the proposition.

The letters of the alphabet in algebraic notations, Arnheim says, come close to "pure" or arbitrary signs because they do not resemble in any way the signified objects. For example, the letters of Berzelian formulas did not resemble the mental image of an elemental portion—if there was such a mental image at all. Yet the fact that a letter is a visible, discrete, and indivisible thing (unlike a written name) constitutes a minimal isomorphy with the postulated object it stands for, namely, the indivisible unit or portion of chemical elements. Moreover, the partition of a Berzelian raw formula into the two parts of a Berzelian formula model of binary constitution, such as (2 C + 6 H) + O, conveyed twoness, that is, the binary constitution of a compound. Figure 1.2 presents a more pictorial form of this implicit building-block image of both Berzelian raw formulas and the formula models of binary constitution constructed from them.[72] Albeit for today's readers the building-block image of Berzelian formulas may be far from obvious, nineteenth-century chemical practitioners explicitly appreciated it. The American chemist Campbell Morfit highlighted the iconic aspects of Berzelian formulas as follows:

> There are many advantages attending the employment of formulae, and nothing has tended to advance the science of chemistry further and more rapidly than their use. They convey to the eye, like pictures, a far clearer view of the nature

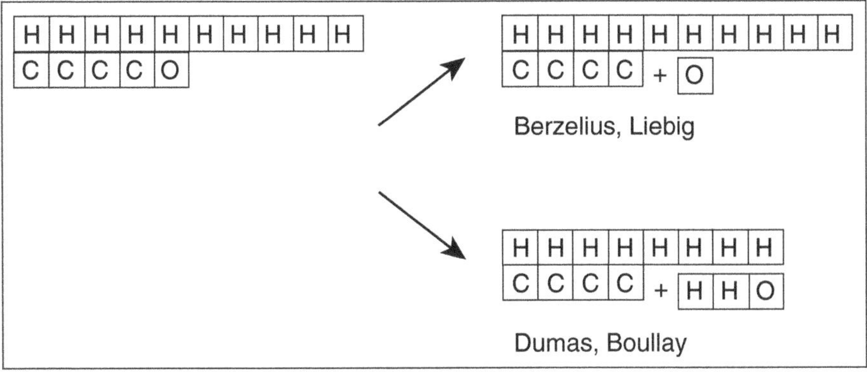

Figure 1-2 Building-block models of the composition and constitution of ordinary ether (drawn after Berthelot 1864; the "atomic weight" of carbon is 6)

> of a compound than the most labored description could effect. . . . As they are pictorial representations, the memory may retain the composition of thousands of compounds, and yet not be overburdened. (Morfit 1850, 453)

The War of Signs

"Words and images seem inevitably to become implicated in a 'war of signs' (what Leonardo called a *paragone*)," wrote W. J. T. Mitchell in his *Iconology.*[73] With respect to the different types of chemical formulas, the French philosopher and semiotician François Dagognet has been the most sophisticated proponent of the idea of such a war. He not only views Berzelian formulas on the one hand and structural and stereochemical formulas on the other as two irreconcilable opponents, but he goes on to "spice up" the dichotomy with normative overtones. Whereas he views the structural and stereochemical formulas as guarantor of the future, he characterizes Berzelian formulas as having hardly any advantages:[74]

> The first mode of writing, which merely translated speech by applying letters and vocal symbols, hardly offers any advantage (in comparison with spoken chemistry, which it perpetuates). . . . This stenography will occupy or invade chemistry during the first half of the nineteenth century until that moment (rather near) when its insufficiencies will become obvious. At the beginning of the nineteenth century it was mainly preached by Berzelius, who established the rules of its application.

According to Dagognet, Berzelian formulas are a kind of stenography that transcribes only the spoken word.[75] They are a shorthand for names whose function in no way goes beyond that of verbal language: "*la chimie écrite ne déborde pas la parlée.*"[76] The "voco-structural," as he terms both ordinary language and Berzelian formulas, is only a means of memory, transport, and communication that conveys information about the composition of reality.[77] In contrast, graphical formulas are portrayed as a "figurative and geometric instrument" possessing the generative function of a true "tool."[78] "The symbol ceases to be a means of fixation, recall or doubling: It increasingly becomes an 'ideal body' that can be manipulated directly," Dagognet points out with respect to graphical formulas, thus ascribing the passive function of doubling to the "voco-structural."[79]

What are Dagognet's criteria for describing Berzelian formulas as unequivocal linguistic or logical signs, which can be sharply distinguished from graphical formulas? The mere fact that the letters of Berzelian formulas are taken from the Latin names of the substances is sufficient to characterize them as languagelike.[80] Dagognet offers neither an additional semiotic analysis of the syntax and semantics of Berzelian formulas nor an analysis of their practical application by the historical actors. He takes no notice of the historical existence of binary formula models in the 1830s, nor does he take into account that Berzelian formulas also denoted unobservable portions of elements, whereas the names denoted only observable laboratory substances. As to his criteria for defining stereochemical formulas as "graphical," the following statements make them quite clear. Referring to a perspective drawing of methane (Fig. 1.3)[81] he writes, "The evolution of graphs is evidently *not a question of arbitrary notation*, but rather the quest for a conventional projective system which makes better representations of the articulations or *real arrangements possible.*"[82] He adds that the graph accords better with chemical isomerism than with Berzelian formulas.[83] This view largely agrees with Peirce's assertion that the icon "may represent its object mainly by its similarity."[84]

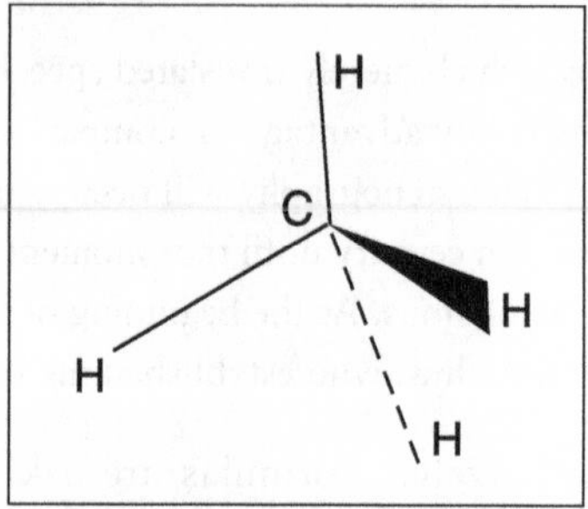

Figure 1-3 Dagognet's perspective drawing of methane (Dagognet 1969, 195)

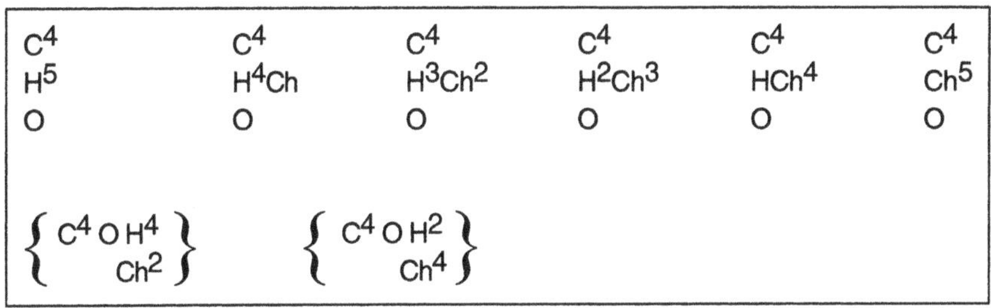

Figure 1-4 Type formulas of chlorinated "methylether" by Dumas (Dumas 1840a)

It is well known that Peirce's notion of similarity as a criterion for iconicity is highly contestable (see later discussion). This criterion nevertheless presupposed, the question arises as to whether it justifies a rigid dichotomy when applied to Berzelian, structural, and stereochemical formulas. The first objection that comes to mind is the continuous historical evolution of chemical formulas. Starting with Berzelian formulas, each successive formula—first type formulas, then structural formulas, and later stereochemical formulas—added new elements to the older sign system while conserving some elements of the original syntax. In particular, all the later types of formulas retained the letters introduced by Berzelius. Letters were just as arbitrary when used in Berzelian formulas to denote chemical portions as they were when used in stereochemical formulas to denote atoms oriented in space. All types of chemical formulas used arbitrary signs of one kind or another. In the case of type formulas, introduced by Jean Dumas in 1840 and further developed after this, the letters and numbers of the Berzelian formulas were preserved, but two-dimensionality and brackets were added (Figs. 1.4 and 1.5). Two-dimensionality and brackets were only convenient ways to represent the "type" of a class of substances, without "corresponding" in any way to the properties of the represented substances. The dots used in type formulas to denote units of affinity were arbitrary signs in their turn as well.

Structural formulas, introduced by Archibald Scott Couper in 1858, added lines that signified units of force or valences (Figs. 1.6 and 1.7). Despite their graphic suggestiveness, enhanced when numbers were omitted completely, structural formulas were largely arbitrary signs. The lines, dotted or fully drawn, did not "correspond" to units of force (valences), and two-dimensionality did not signify the spatial orientation of the atoms but only the sequence of their chemical combination. Because many chemists believed what they saw—namely, a certain spatial orientation of the structural formulas—

PRIMARY TYPES OF DOUBLE DECOMPOSITION				
H . Cl	Chloride or Hydride	Cl . Cl	Na . Cl	Et . Cl
H H } O	Oxide or Hydrate	Cl H } O Cl Cl } O	Na H } O Na Na } O	Et H } O Et Et } O
H H H } N	Nitride or Amide	H H } N I I H } N Cl Cl Cl } N	Na H H } N H } N Na Na Na } N	Et H H } N Et Et H } N Et Et Et } N
H H H H } C	Carbide or Methide	Cl H H H } C Cl_2H_2C Cl_3H C Cl_4C	Na H H H } C	Et H H H } C?

Figure 1-5 Odling's type formulas (Odling 1864)

structuralists had to go to considerable argumentative lengths to convince their opponents that they did not intend to represent spatial orientation. In turn, the stereochemical formulas introduced by Jacobus Henricus van't Hoff in 1874, which did represent the spatial orientation of atoms, depicted tetrahedrons (Figs. 1.8 and 1.9). For most chemists, the corners and edges of the tetrahedron had no realistic chemical significance whatsoever. Moreover, it took some time until chemists were convinced that tetrahedral diagrams were suitable representations of molecular spatial relations. For the untrained eye, it is not natural to read the diagrams this way because the carbon atom (or the skeleton of several carbon atoms), imagined as placed at the center of the tetrahedron, and its four valences, imagined as directed to the four corners of the tetrahedron, usually were not depicted. Given the fact that the earlier structural formulas signified valence bonds, representing them by lines, and did not signify

Methylic alcohol:	$C^2 \begin{matrix} \cdots O \cdots OH \\ \cdots H^3 \end{matrix}$	$C^2 \begin{Bmatrix} O - OH \\ H^3 \end{Bmatrix}$
Æthylic alcohol:	$\begin{matrix} C^2 \begin{matrix} \cdots O \cdots OH \\ \cdots H^2 \end{matrix} \\ \vdots \\ C^2 \cdots H^3 \end{matrix}$	$\begin{matrix} C^2 \begin{Bmatrix} O - OH \\ H^2 \end{Bmatrix} \\ \mid \\ C^2 - H^3 \end{matrix}$

Figure 1-6 Couper's structural formulas (Couper 1858a, b)

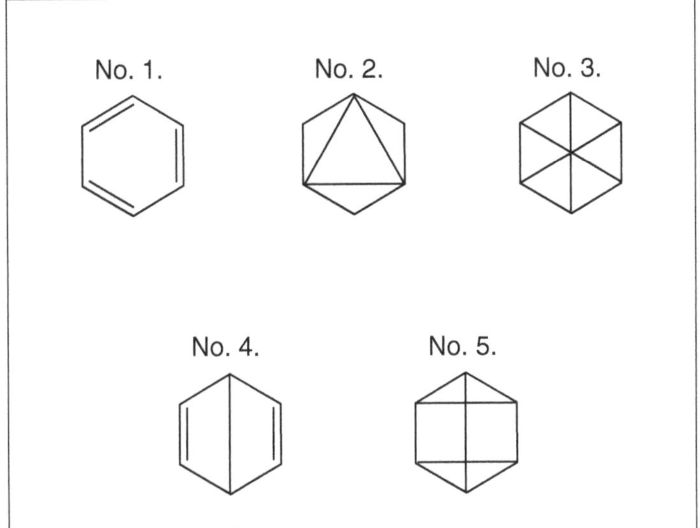

Figure 1-7 Kekulé's structural formulas of benzene (Kekulé 1869)

orientation in space, it must have seemed strange to the contemporary reader that it was exactly the other way around with stereochemical formulas; chemical valence bonds were now absent, and the focus was on spatial orientation.

As to Berzelian formulas, they can appear only as unequivocal logical signs when their theoretical meaning and their context of application are ignored. For chemists of the time, however, they also conveyed a building-block image of their referents. In sum, all types of formulas were to some degree arbitrary, and all displayed some imagery. They were mixed logical and iconic signs with differences of degree only. In his *Picture Theory*, Mitchell comes to a similar conclusion, that "all media are mixed media, and all representations are heterogeneous; there are no 'purely' visual or verbal arts, though the impulse

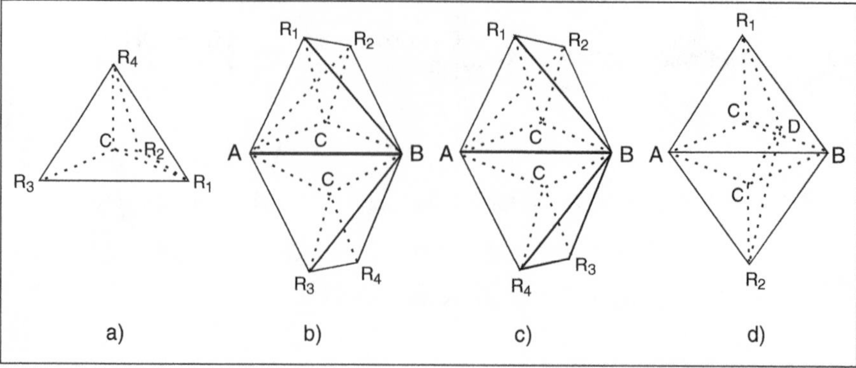

Figure 1-8 van't Hoff's stereochemical formulas (a: single bond; b and c: isomers with a double bond; d: triple bond; van't Hoff 1874)

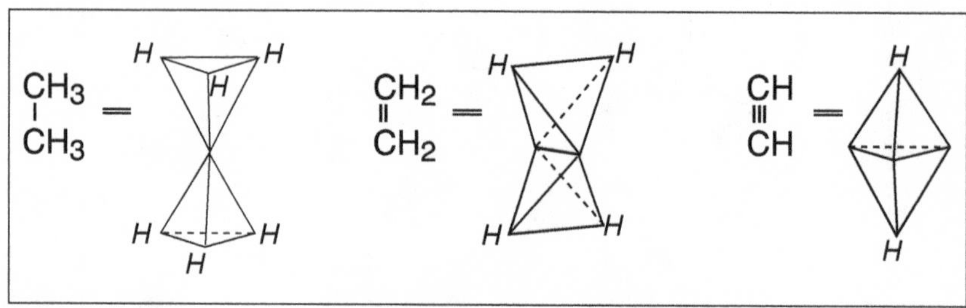

Figure 1-9 Wislicenus's stereochemical formulas (Wislicenus 1887)

to purify media is one of the central utopian gestures of modernism."[85] This view is confirmed if we follow a different semiotic approach that completely dismisses Peirce's criteria for a typology of signs.

Graphic Suggestiveness and Maneuverability

Peirce's view that iconicity can be explained by the notion of "similarity to objects" has been criticized by semioticians, in particular because it assumes a given object as the distinguishing parameter between types of signs and largely ignores the cultural impact on our perception of resemblance.[86] Nelson Goodman is one of the few philosopher–semioticians who drew out all the consequences of this criticism by rigorously questioning the language–image boundary and rejecting the criterion of similarity. According to Goodman, a language-like symbol system is *digital*, defined as discontinuous and differentiated throughout.[87] A nonlinguistic or pictorial system is *analog* in the sense of syntactic and semantic density.[88] Goodman exemplifies the difference between *digital* and *analog* by an ordinary watch and a simple pressure gauge that lacks figures or other marks on its face. An ordinary watch without a second hand displaying numerals only for hours and minutes is a digital system that is normally used to pick out hours and minutes. By contrast, the pressure gauge is a simple example of an analog system in that it registers pressure in a continuum. If the pointer of the gauge moves smoothly as the pressure increases continuously, and each position of the pointer marks a difference, the report on the pressure lacks any differentiation. Goodman summarizes as follows: "Nonlinguistic systems differ from languages, depiction from description, the representational from the verbal, paintings from poems, primarily through lack of differentiation—indeed through density (and consequent total absence of articulation)—of the symbol system."[89]

In this conceptual framework, Goodman characterizes diagrams of carbohydrates and even three-dimensional molecular models made of ping-pong balls and chopsticks as digital signs.[90] Hence, any kind of chemical formulas will have to be viewed as a digital sign system. This does not mean, however, that Goodman's approach leaves no room for conceptualizing the differences between Berzelian formulas, structural formulas, and stereochemical formulas; nor does it mean that it necessitates the identification of Berzelian formulas with ordinary language. Linguistic and notational systems in general, Goodman pointed out, may differ in several features, for example, in the number of their atomic constituents, mnemonic efficacy, graphic suggestiveness, and maneuverability.[91] The latter two notions, although not explained in detail by Goodman, are particularly useful analytical tools for explaining the difference between Berzelian formulas and verbal language (and also between the types of formulas).

Berzelian formulas have a certain "graphic suggestiveness" compared with verbal language. For example, the raw formula for alcohol suggested by Berzelius in 1833 was C^2H^6O; Berzelius's model of its binary composition was $C^2H^6 + O$ (see preceding). The raw formula allows us to see at a glance both the elemental composition of alcohol and the number of the portions (or "atoms") of the elements. The formula model $C^2H^6 + O$ gives additional visual information about the two immediate constituents. In both cases, a certain "graphic suggestiveness" is present because of the one-to-one correspondence of the letters and the chemical portions and the spatial partition of the constitution model into two parts, separated by the plus sign. This becomes particularly clear if we compare the two formulas with the representation of the same contents by verbal language. The additivity of the names of the elements is the same as the additivity of the letters (which actually were taken from the names). Thus, we might call alcohol *carbon–hydrogen–oxygen.* Also to denote the number of elemental portions, we would have to transform this into dicarbon–hexahydrogen–monoxygen. This name is already a compromise meant to avoid the clumsy alternative of repeating each of the constituent elements, for example, carbon–carbon–hydrogen–hydrogen–hydrogen–hydrogen–hydrogen–hydrogen–oxygen. Whereas the syntax of the formula C^2H^6O visually suggests a composition from units or portions of the elements, the visual perception of the verbal name does not display this kind of information. The same is the case for the representation of the binary constitution of alcohol. Using verbal language, the constitution might be distinguished from the composition by different end syllables of the names of the two immediate constituents, such as *dicarbon–hexahydride–monoxid.* If the reader knows that the suffix *ide* is used for substance components within a binary compound, he or she will understand easily that this name denotes a binary constitution of alcohol from dicarbonhexahydrogen and oxygen; however, the graphic suggestiveness of the formula model $C^4H^6 + O$, which shows binarity by its partition into two separate partial formulas, is absent in the verbal form.

The particular advantage of Berzelian formulas in visually representing the binary constitution of compounds was emphasized by William Whewell, who was a decided defender of algebraic notation in chemistry at a time when British chemists were still skeptical toward it. He wrote: "I can hardly perceive how any person, at all acquainted with mathematical symbols, can adopt any other modes of notations than this, inasmuch *as no other can assist us in reasoning on the constitution of chemical compounds.*"[92] Whewell explained his statement by way of the example of the constitution formula of stilbite, whose raw formula was 15 S + 4 A + C + 6 q (S standing for silica, A for alumina, C for lime, and q for water):[93]

> If 12 atoms of the silica go with the alumina, and 3 with the lime, the symbol may stand thus: (12 S + 4 A) + (3 S + C) + 6 q; or, what is the same thing, 4 (3 S + A) + (3 S + C) + 6 q; *in which form it is clearly seen* that we have 3 atoms of S with 1 of A, also 3 of S with 1 of C, and that 4 of the former parcels are combined with 1 of the latter.

We will see that not only verbal language, used as a means for representing the binary constitution of compounds, lacked graphic suggestiveness but Daltonian diagrams as well. This is the reason behind Whewell's assertion that there is "no other" notation that can assist in clearly representing the constitution of compounds.

There is also a difference in the "maneuverability" of Berzelian formulas and verbal language. With a view to the performative aspects of constructing models, the letters of Berzelian formulas were much easier to maneuver than names. They can be written down quickly without occupying much space, thus facilitating the tinkering on paper for exploring different possibilities of grouping the letters. Moreover, the additivity of letters was the only syntactic rule constraining the work on paper. In this respect, letters had no advantage over names but over another representational alternative of the time: Daltonian diagrams. It has been suggested that chemists preferred Berzelian formulas to Daltonian diagrams because of their iconoclastic attitude.[94] I do not share this view, first, because European chemists in fact did depict objects in the eighteenth century and the first half of the nineteenth century, such as laboratory apparatus and laboratories, and, second, because I think they selected Berzelian formulas for other reasons. These reasons have to do, first, with the fact that Berzelian formulas were enormously productive paper tools that could be used for a variety of practical purposes, which will be discussed later, and, second, with the semiotics of Berzelian formulas. The fact that they had different layers of meaning and conveyed a building-block image of chemical portions without simultaneously requiring an investment in atomic theories, together with the simplicity of their maneuverability on paper, rendered them much more attractive for early nineteenth-century chemists than Daltonian diagrams.

Dalton's Diagrams and Their Historical Disadvantages

In 1837, soon after the British Association for the Advancement of Science recommended that its members adopt Berzelian formulas, John Dalton wrote a harsh criticism of them: "Berzelius's symbols are horrifying: a young student in chemistry might as soon learn Hebrew as make himself acquainted with them. They appear like a chaos of atoms. Why not put them together in some sort of order? . . . [They] equally perplex the adepts of science, discourage the learner, *as well as to cloud the beauty and simplicity of the Atomic Theory.*"[95]

Of course, Dalton was firmly convinced that his own diagrams of simple and composed atoms (later "molecules") avoided the alleged shortcomings of Berzelius's notation (Fig. 1.10). It was in particular the fact that algebraic notation did not unequivocally represent that made Dalton confident that his own sign system was superior to that of Berzelius.

Dalton presented his diagrams in his *New System of Chemical Philosophy*,[96] in which he elaborated the chemical aspects of his theory of atoms. This theory occupied him for more than a decade in connection with his research

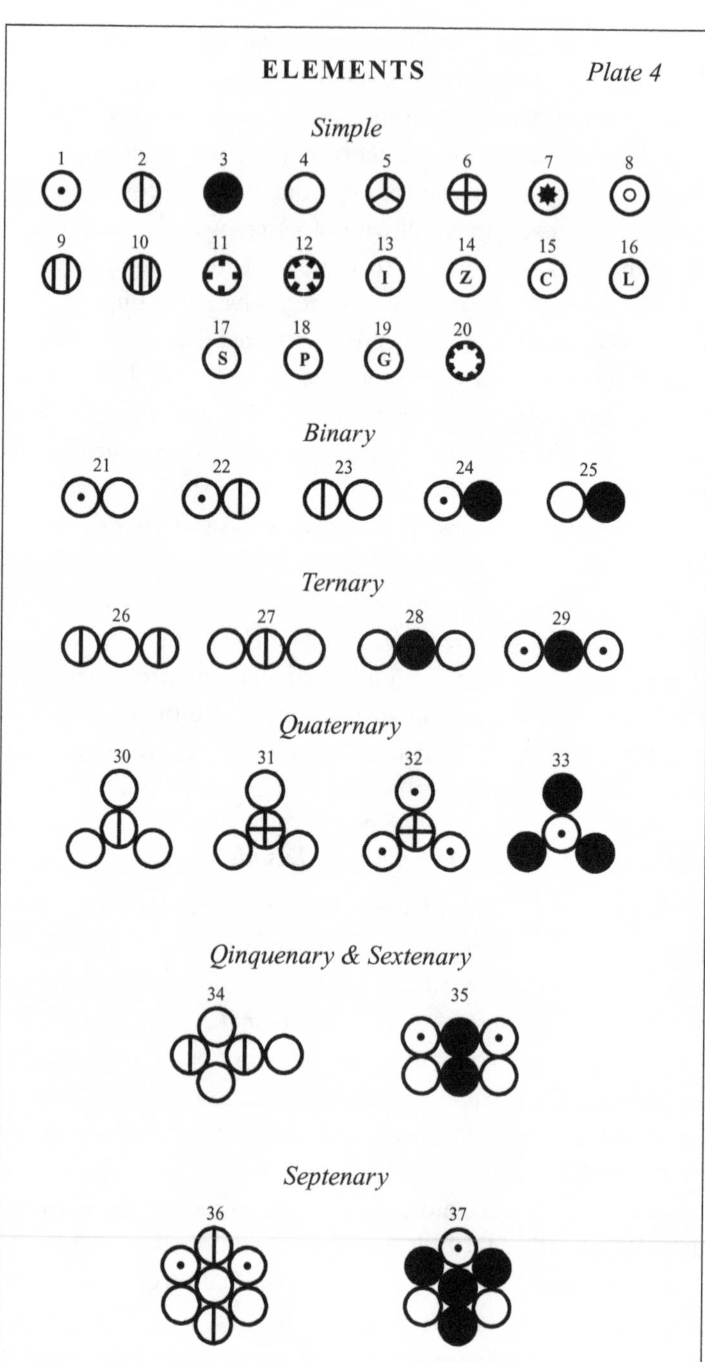

Figure 1-10a Diagrams of atoms (after Dalton 1808–27)

PLATE IV. This plate contains the arbitrary marks or signs chosen to represent the several chemical elements or ultimate particles.

Fig.			Fig.		
1	Hydrog. its rel. weight	1	11	Strontites	46
2	Azote	5	12	Barytes	68
3	Carbone or charcoal	5	13	Iron	38
4	Oxygen	7	14	Zinc	56
5	Phosphorus	9	15	Copper	56
6	Sulphur	13	16	Lead	95
7	Magnesia	20	17	Silver	100
8	Lime	23	18	Platina	100
9	Soda	28	19	Gold	140
10	Potash	42	20	Mercury	167

21. An atom of water or steam, composed of 1 of oxygen and 1 of hydrogen, retained in physical contact by a strong affinity, and supposed to be surrounded by a common atmosphere of heat; its relative weight = 8
22. An atom of ammonia, composed of 1 of azote and 1 of hydrogen 6
23. An atom of nitrous gas, composed of 1 of azote and 1 of oxygen 12
24. An atom of olefiant gas, composed of 1 of carbone and 1 of hydrogen 6
25. An atom of carbonic oxide composed of 1 of carbone and 1 of oxygen 12
26. An atom of nitrous oxide, 2 azote + 1 oxygen 17
27. An atom of nitric acid, 1 azote + 2 oxygen 19
28. An atom of carbonic acid, 1 carbone + 2 oxygen 19
29. An atom of carburetted hydrogen, 1 carbone + 2 hydrogen 7
30. An atom of oxynitric acid, 1 azote + 3 oxygen 26
31. An atom of sulphuric acid, 1 sulphur + 3 oxygen 34
32. An atom of sulphuretted hydrogen, 1 sulphur + 3 hydrogen 16
33. An atom of alcohol, 3 carbone + 1 hydrogen 16
34. An atom of nitrous acid, 1 nitric acid + 1 nitrous gas 31
35. An atom of acetous acid, 2 carbone + 2 water 26
36. An atom of nitrate of ammonia, 1 nitric acid + 1 ammonia + 1 water 33
37. An atom of sugar, 1 alcohol + 1 carbonic acid 35

Figure 1-10b Dalton's explanation of the diagrams (after Dalton 1808–27)

on meteorology and on the mixture of different gases.[97] It postulated atoms in the sense of invisibly small bodies with mechanical qualities, such as shape, size, and orientation in space. To explain the properties of mixed gases, Dalton also assumed that these particles were surrounded by an atmosphere of heat, which exerts a repulsive force, and that they had a globular form because of the heat atmosphere. In his elaboration of the impact of his atomic theory on chemistry, Dalton adopted Lavoisier's concept of chemical elements in 1808 and gave it a new ontological dimension: There were as many kinds of simple atoms as chemical elements, differing in their weight, size, and chemical qualities. The atoms of chemical elements were their smallest, indivisible parts, which could not be created or destroyed in any chemical experiment.[98] In contrast to elemental atoms, the smallest particles of chemical compounds, which Dalton named *atoms* as well, could be decomposed and recomposed in experiments. According to Dalton, the juxtaposition of heterogeneous atoms within the compound atom was caused by chemical affinity and creation of a common atmosphere of heat. For instance, he described the compound atom of water or steam as follows: "An atom of water or steam, composed of 1 of oxygen and 1 of hydrogen, retained in physical contact by strong affinity, and supposed to be surrounded by a common atmosphere of heat."[99]

Dalton's diagrams of atoms, drawing elementary atoms as circles distinguished by arbitrary marks and compound atoms as regular geometric figures consisting of different juxtaposed circles, were designed to illustrate these ideas. In doing so, Dalton drew on an old philosophical tradition, going back to the seventeenth century and perhaps earlier,[100] in which atoms were represented by circles. In light of this tradition, Dalton's circles and composed diagrams were read as unequivocal signs for submicroscopically small atoms that have a certain shape, size, orientation in space and additional chemical properties, such as their relative combining weight. This raises the question of the extent to which Dalton viewed his diagrams as realistic images of atoms. Although there is no doubt that in the case of the diagrams' shape he assumed a realistic depiction of globular atoms, surrounded by the heat atmosphere, with respect to size, he was less decided. The diagrams indeed show that circles of the same size denote different kinds of elemental atoms. Yet, in his comment in the text, Dalton was silent about the different atoms' sizes. A few years later, he even rejected Berzelius's hypothesis, published shortly before,[101] that the atoms of the different elements had the same size: "Dr. B. [Berzelius] seems to hold it necessary that all atoms should be of *the same size*. This, he thinks, is required in order to form bodies into regular figures. Now this is not part of my doctrine."[102]

Why, then, did Dalton nevertheless draw equally sized circles in his diagrams? I propose that this was due to constraints in drawing the diagrams.[103] The representations of compound atoms had to be regular geometric figures that approximated the globular shape because of the assumption that compound atoms had a common heat atmosphere. This presupposed, circles of the same size offered the simplest solution for drawing such figures.

Additional problems of drawing were caused by the fact that Dalton also designed his diagrams to represent orientation in space in a realistic way, that is, according to the requirements of his atomic theory. Thus, the construction of the composed diagrams also was constrained by the necessity of taking the affinity and repulsion of atoms into account. Because there was affinity only between heterogeneous atoms, juxtaposition of the same circles had to be avoided. This was even more the case in that Dalton believed that "particles of the same kind repel each other, and therefore take their stations accordingly."[104] We see in the diagrams 21 through 34 in Figure 1.10a that only different circles are juxtaposed. The more atoms of the same element contained in a compound atom, however, the more difficult it became not to violate the theoretical constraints. Indeed, diagrams 35, 36, and 37, where equal circles are next to each other, fail to conform to the theoretical constraints. For example, in the case of diagram 37 for sugar, which is shown as consisting of four atoms of carbon, two of oxygen, and one of hydrogen, there was no way to avoid the juxtaposition of the four black circles. By 1830, this kind of problem led Dalton to concede the following: "My method of denoting atomic combinations cannot be so advantageously applied where the atoms are numerous as where they are few. The arrangement cannot be made upon a plane such as it would be by nature. In such case a method somewhat like that of Berzelius might be advantageously adopted."[105]

Affinity and repulsion were not the only constraints on the spatial arrangement of the diagrams. Like Berzelian formulas, the diagrams also had to represent the binary constitution of the compounds. Thus we read in Dalton's "explanation of the plates," point 37: "an atom of sugar, 1 alcohol + 1 carbonic acid." Accordingly, the upper part of the diagram for sugar, consisting of one circle with a point (for hydrogen) and three black circles (for carbon), signifies one atom of alcohol, and the lower part, consisting of two white circles (for oxygen) and one black one (for carbon), signifies carbonic acid. The same principle can be observed in diagrams 34, 35, and 36. At this point, I would like to come back to Whewell's claim that "no other [sign system] can assist us in reasoning on the constitution of chemical compounds" than Berzelian formulas, "in which form it is clearly seen" how the compounds are constituted (see preceding). Daltonian diagrams really did *not* allow chemists to see the binary constitution at a glance. Their graphic suggestiveness lay in a different direction. Instead of focusing chemists' perception on two constituents, as the Berzelian binary formula models of constitution did, the individual atoms and their orientation in space captured the chemists' attention. Only with the additional verbal information contained in the legend does one begin to see the "*Gestalt*" of binarity.

When the British Association for the Advancement of Science met in 1835 to discuss chemical notation, Dalton again compiled diagrams, but this time he carefully selected his examples to avoid juxtapositions of equal circles. He also omitted the representation of the binary constitution of the compounds.[106] His colleagues remained skeptical, however: "The method of symbols proposed by Dr. Dalton *might, perhaps apply to the simple cases of combination* that had been selected by that philosopher as examples, but when they come to be applied to groups consisting of 40, or 60, or 100 molecules, *how could the eye recognize at a glance their method of arrangement.*"[107] In this statement, an additional advantage of Berzelian formulas is highlighted. In the case of large organic compounds, numbers were much easier to handle than the repetitive elements of Daltonian diagrams. Accurate drawings of the spatial arrangement of large molecules were indeed a difficult enterprise. Even more difficult was depiction of the arrangement of chemical portions into the two immediate components of a binary organic compound, which was the shared goal of European chemists of the time. In the case of large molecules, it became entirely impossible to recognize the binary arrangement at a glance.

CHAPTER TWO

Two Cultures of Organic Chemistry in the Nineteenth Century: A Structural Comparison

In the late 1820s, the chemical formulas that Berzelius had introduced in 1813 began to develop their productivity in organic chemistry. In an unforeseen way, their application became a decisive precondition of a deep structural transformation of organic chemistry, which took place between roughly 1827 and 1840. This structural transformation can be described as the decline and disintegration of the pluricentered culture of plant and animal chemistry and the emergence of the experimental culture of carbon chemistry. In the new experimental culture of carbon chemistry that took shape during the 1830s and thrived in the 1850s, chemists applied Berzelian formulas as paper tools for the experimental investigation of the chemical reactions and the invisible constitution of organic substances and for their classification. In doing so, chemical formulas became a sign system that contributed to reference and mediated between different practices, such as experimenting, modeling, classifying, and theory formation.

The purpose of this chapter is to provide a survey, or a kind of orienting historical map, for the subsequent analyses and reconstructions in Chapters 4 through 8 of the transformations of organic chemistry during the 1830s and the new role played by Berzelian formulas in this process. To this end, I compare plant and animal chemistry before the late 1820s and the experimental culture of carbon chemistry after 1840. This comparison of before and after is not a dense historical description but a rough sketch concentrating on structural features of the two forms of organic chemistry. If it treats both scientific cultures as though they were neatly separated, static systems, it is only for the sake of clarity; all aspects of the process of transformation and of its dynamics are discussed later and are summarized in Chapter 8. Based on the historical specificity of the topic under consideration, the following five criteria have proved the most suitable for such a structural comparison: the entire spectrum of research objects in both cultures; the concept of organic substance and its material referents as the predominant research objects; the way in which these things were ordered, that is, the classification of substances; the type of experiments performed; and the semiotic forms for representing the invisible scientific objects. The last of these criteria applies to the use of Berzelian formulas. Because this is a central issue in the remaining chapters of this book, I will not address this issue in the following overview. Furthermore, these five criteria are not exhaustive. For example, aspects of chemists' beliefs and self-image are mentioned in the following but not discussed in any detail; in particular, a scrutiny of the social institutions and relationships in both cultures of organic chemistry still needs to be done.

My main argument, that organic chemistry before the late 1820s and after 1840 is structurally so different that we should speak of two different scientific cultures, is a sweeping claim that requires a great deal of evidence. Historians of chemistry have not ignored these processes, but they have conceptualized them in ways that differ from my assertion that there was a deep structural transformation.[1] Furthermore, there are also differences of emphasis concerning our understanding of the scope and impact of the transformation of organic chemistry between the late 1820s and the early 1840s and the subsequent transformations until the 1860s. Alan Rocke characterized the changes of organic chemistry during the 1850s as a "quiet revolution," and he considered the events before a "preparation" for these changes. His points of comparison have differed from mine. He has named in particular the reforms of atomic weights; the introduction of the concept of atomicity in the 1850s, which led to what became known as *valence and structure theory*; the refinement of organic synthesis; the acceleration of compounds produced; institutional indicators, such as the formation of new institutional boundaries of organic chemistry and its entrenchment as a "subdiscipline"; and an increase in the number of chemists, the number of journals, the number of papers published, and so on.[2] From the perspective of these criteria, Rocke's assertion that profound transformations took place in the 1850s is correct. From the perspective of the criteria that I choose, however, the events in the 1830s were not mere "preparations" for these processes but decisive preconditions on the deep structural level of what I call a "scientific culture." Based on my criteria, the processes during the 1850s appear as developments within the new framework of the experimental culture of carbon chemistry established between the late 1820s and the early 1840s. In the last section of this chapter, I argue in more detail for my assertion that pre-1830 plant and animal chemistry and post-1840 carbon chemistry were two different "scientific cultures." It should be mentioned that the plausibility of this thesis, as well as the plausibility of my points of comparison, depends heavily on the detailed analyses and discussions presented in Chapters 3 through 8. At issue here is a problem of presentation: criteria of comparison and judgments that are not formal but rather are elaborated during the course of the research process cannot be sufficiently justified beforehand. Hence, the question of the plausibility of my comparison in this chapter depends to some extent on the entire book.

THE AREA OF RESEARCH OBJECTS IN EARLY NINETEENTH-CENTURY PLANT AND ANIMAL CHEMISTRY

> All that our research in this secretive area of chemistry can accomplish, is to observe chemical changes that are produced in living bodies . . . to follow the phenomena which accompany the life process as far as possible, and then to separate the organic products from one another, study their properties, and determine their composition.
>
> —BERZELIUS, 1833–41, 6:27 f.

Plant and animal chemistry of the late eighteenth and early nineteenth centuries, or *organic chemistry* as it was sometimes termed, was a subculture of chemistry that, terminologically as well, the historical actors considered a coherent area separate from inorganic chemistry. When considered alongside contemporary inorganic chemistry, several characteristic features stand out. Plant and animal chemistry was strongly oriented toward practical applications, and it had a particularly close relationship with pharmacy. Foundational chemical theories influenced by natural philosophy, such as the Newtonian atomic theory of affinity, were scarcely applicable to it. Furthermore, experimental investigations of chemical reactions, which were the dominant kind of experiments in inorganic chemistry, were extremely rare in early organic chemistry. Consequently, classification of plant and animal substances was based on natural historical criteria rather than on knowledge about the composition and constitution of substances gained by studying chemical reactions.

Plant and animal chemistry was a remarkably stable scientific culture, left untouched by the "chemical revolution" in the last third of the eighteenth century.[3] In the course of the eighteenth century, certain aspects of experimentation in animal and plant chemistry had slowly transformed. In particular, the method for obtaining substances from plants and animals had shifted from distillation to extraction with solvents, the latter having undergone refinement during this period.[4] Furthermore, Antoine-Laurent Lavoisier's analyses of plant and animal substances showed that these substances were made up of a few elements, such as carbon, hydrogen, and oxygen.[5] The quantitative analysis of these substances, which was also begun by Lavoisier, was resumed in the first decade of the nineteenth century; it led to problems concerning the identification and classification of plant and animal substances. In terms of the overall character of plant and animal chemistry, however, little had changed until the late 1820s.

In the historiography of nineteenth-century chemistry, plant and animal chemistry has been presented almost uniformly as a scientific culture concerned with the same type of research objects as inorganic chemistry of the time and the later experimental culture of carbon chemistry; hence, it has been portrayed as the study of organic substances.[6] The following analysis demonstrates that the scientific objects of plant and animal chemistry can be understood only by viewing it within a specific historical constellation of practices when chemical experimentation found itself grouped with natural historical investigations and commercial practice. Plant and animal chemistry was concerned with substances extracted from plants and animals, especially those with pharmaceutical value, as well as with many aspects of animal and plant life. It was not only the chemistry of substances but also the chemistry of living natural bodies. Thus, it was also concerned with complex organic bodies (for example, leaves, roots, barks), the external structure and anatomy of plants and animals, and physiological processes. Everything identified as animal or plant "economy," such as digestion and breathing, assimilation and vegetation, fermentation and putrefaction, belonged to its area of study. The issue of physiological processes entailed the question of their anatomic location and therefore the study of the external and internal construction of plants and animals, which naturally intersected with botany and zoology. Moreover, a significant number of chemists active in plant and animal chemistry, first and foremost Jacob Berzelius, reflected on "life," which became a basic concept of the then developing field of "biology." A historically appropriate definition of the research area of plant and animal chemistry thus would read as follows: It included all aspects of the plant and animal kingdoms that could be studied with the conceptual and experimental resources of the chemist. Because many of its objects could not be systematically studied in the chemical laboratory, but rather only by ordinary or systematic observation in "free nature," in its early form organic chemistry was both an experimental and an observational science.

Antoine François Fourcroy's *Système des Connaissances Chimiques* (1801–02) was one of the most extensive and widely read textbooks at the turn of the century in which plant and animal chemistry is extensively treated. Fourcroy describes animal chemistry's (*chimie animale*) area of study in the following way:[7]

> Today, animal chemistry is in the same fortunate condition as plant chemistry. Like it, it is not limited to the analysis of materials from animal bodies, and the advancement of the art of preparing these materials for our needs. It is destined to become something much higher, and its operations extend much further. Armed with exact instruments and ingenious methods for determining the true differences that exist between plant and animal materials, it shows what happens to the state of the former when they are transformed into the latter in animal organs; it explains the effects of digestion, respiration, transpiration, and many other functions of the animal economy; it opens a new route for the physics of organized and sensible bodies; its task is to come to the aid of anatomy, in order to base physiology on certain knowledge.

According to Fourcroy, animal chemistry was not confined to the study of substances obtained from animals, but it also explored the state of these substances in animal organs, physiological processes, and the general "physics" of organized and sensible bodies. Similarly, plant chemistry was not limited to the study of substances obtained from plants. Instead, Fourcroy saw its ultimate goal as the discovery of the "laws of plant physics."[8]

Fourcroy distinguished six classes of "facts" as belonging to plant chemistry:[9]

> The six orders of facts, in which I divide and subsume everything which currently belongs under plant chemistry, are based on the following six main points: the first relates to the *structure of plants* and the difference between this structure and the grain or entire masses of mineral materials. It would be impossible to understand their chemical properties, if one did not have an exact or precise notion of its *organization*. . . .
>
> I count among the second class of facts those which belong to *nature or plant composition* in general. . . .
>
> In the third class are the kind of actions which the already known major substances carry out on the plants, that is, the *chemical characters* which they represent. . . .
>
> The fourth class of facts includes the examination of *all materials extracted from plants*, which truly constitute them and which one can therefore call the principles of the plants, or immediate materials. . . .
>
> In the fifth, I place the report and *study of the different natural alterations to which plants are susceptible*. . . .
>
> Finally, I count in the sixth class of facts everything which I call the chemical phenomenon of *vegetable life*, or the application of all the facts preliminarily

> presented in the preceding orders to *plant physics*. The last, the fulfillment of all and the object to which they all flow, is one of the most beautiful results of modern chemistry.

Fourcroy divided the section of his textbook covering plant chemistry according to these six "facts." Only the second, third, and fourth items are objects of knowledge in the sense of substance chemistry; these "facts" study the procurement of substances from plants,[10] the experimental test of its chemical properties,[11] and the special characteristics of individual plant substances.[12] In contrast, the first group of research topics refers to the external structure and anatomy of plants, their principal life expressions, as well as their role on the earth's surface and in the "economy" of nature as a whole.[13] In the fifth group, Fourcroy includes all those research topics closely or more distantly related to fermentation as well as the formation of fossils, petrifaction, and other products known from mining.[14] Finally, a large part of his textbook is dedicated to an extensive treatment of the sixth group, that is, physiological processes such as plant nutrition, vegetation, germination, fruit formation, and others.[15] It is worth noting how Fourcroy justifies the connection between these diverse research objects. It is not the study of substances that unites them but rather the "phenomena of vegetable life" or "plant physics."

A typical problem for plant chemistry was the natural development of substances in the organs of plants:[16]

> Since the most exact analyses have all shown, without exception, that plant materials are made up in their first principles from carbon, hydrogen and oxygen . . . it is evident that the problem of the natural formation of these compounds consists in knowing from where the plants obtain these simple substances, how they appropriate them and combine them in groups of three with one another. This problem in fact encompasses all of plant physics.

Fourcroy's inquiry into the natural formation of the substances that constitute plants took place within the context of a historically specific state of collectively shared knowledge. The chemical analyses in the last third of the eighteenth century, especially those of Lavoisier, had shown that the substances extracted from plants as a rule consisted of only three elements: carbon, hydrogen, and oxygen. This raised the question of how the various plant substances could be formed from only three elements. For Fourcroy, it was obvious that plant formation was to be understood in principle as *chemical synthesis*. Plant organs were accordingly instruments or vessels in which analyses and syntheses took place under the influence of chemical attraction.[17] Despite this adaptation of central concepts of inorganic chemistry for the theoretical understanding of physiological processes, Fourcroy repeatedly emphasized that the methods of studying the processes of formation of plant and animal substances must be distinguished from those of inorganic chemistry.[18] His reasoning for the difference in research methods was linked to an ontological distinction between the bodies of the mineral kingdom and those of the plant and animal kingdom. Similar distinctions were made by other chemists, but they had different opinions about the extent to which the bodies of the three natural kingdoms were different and precisely where the boundaries between all three should be drawn.

Fourcroy's discussion of the scientific objects of plant and animal chemistry makes clear that this domain of early nineteenth-century chemistry cannot be reconstructed solely on the basis of its experimental practice. Just as the external structure and anatomy of plants was studied by way of observation and dissection, according to Fourcroy, observation was also sufficient for investigation of the most important functions of vegetable life.[19] Merely by observation, for example, it could be established that plants take up fluids and transport them, that their organs develop, that they excrete substances and reproduce them, and so on. Fourcroy added that "*Observation alone also proves* that in different stages of plant life, and for execution of their functions multiple parts undergo movements, which in certain ways are related to the movements one can observe in animals."[20]

Fourcroy's definition of the scope of plant and animal chemistry was not idiosyncratic. This is shown, for example, by the section on plant chemistry in one of the leading British chemistry textbooks, Thomas Thomson's *System of Chemistry*. Thomson described plant chemistry in the following terms:[21]

> We have now seen the different *substances which are contained in plants*; but we have still to *examine the manner in which these substances are produced*, and to endeavour to trace the different *processes which constitute vegetation*. But I will warn the reader not to expect complete information in this chapter. The wonders of the vegetable kingdom are still but very imperfectly explored; many of the *organs of plants* are too minute for our senses, and scarcely a single process can be completely traced. The multiplicity of *operations continually going on in vegetables* at the same time, and the variety of different, and even opposite substances, formed out of the same ingredients, and most in the same place, astonish and confound us. The orders, too, and the skill with which every thing is conducted, are no less surprising.

According to Thomson, plant chemistry studied substances contained in or extracted from plants, the production of these substances in living plants, plant growth (vegetation), plant organs (as the natural sites of substance production and the processes of vegetation), as well as all other types of chemically explicable transformation processes in plants. The individual chapters of the section on plant chemistry in Thomson's textbook thus include topics such as plant nutrition,[22] the transformation of fluid nutrients in plant organs,[23] the function of the leaves,[24] the different plant juices,[25] and plant decay.[26]

Thomson introduced animal chemistry in a similar way:[27]

> This part of the subject [the chemistry of animals] naturally divides itself into four Chapters. In the First Chapter, I shall give an account of the *different ingredients hitherto found in animals*, such of them at least as have been examined with any degree of accuracy; in the Second, I shall treat of the *different members of which animal bodies are composed*; which must consist each of various combinations of the ingredients described in the First Chapter; in the Third, I shall treat of *those animal functions* which may be elucidated by chemistry; and in the Fourth, of *the changes which animal bodies undergo after death*.

Like his plant chemistry, Thomson's animal chemistry included all aspects of the physiological functions of animals. Although the simple animal substances that could be extracted from animal bodies formed the center of research in animal chemistry, research objects which at the start of the nineteenth century could not be readily studied in experiment also were part of it.

The Chemical Study of "Life"

Jacob Berzelius, who was the most influential chemist of the first third of the nineteenth century, combined plant and animal chemistry under the phrase *organic chemistry* and used the term *organic* more systematically than his contemporaries to contrast it with *inorganic* chemistry. His "chemistry of organic bodies" included the entire spectrum of scientific objects described here. Furthermore, the section on plant chemistry included a detailed natural history—hundreds of pages long—of various plant materials, mostly relevant to pharmaceutical uses, such as barks, roots, woods, herbs, leaves, fruits, and seeds.[28] At the same time, Berzelius—more than any other chemist—was concerned with "life," the fundamental concept of the emerging life sciences. Even if at various times Berzelius expressed different conjectures about the modifications of chemical affinity in living beings or about a special "life force," life in his eyes was to a large degree a chemical phenomenon and thus a subject of research for organic chemistry. The section on plant chemistry in the third edition of his textbook begins with the following reflections on life, which use the workshop metaphor to characterize it as a chemical process:[29]

> In living nature, the elements seem to follow entirely different laws as in dead nature, the products of their mutual effects are thus totally different from the area

> of inorganic nature. To discover the cause for this difference between the behavior of the elements in dead nature and in living bodies would be *the key to the theory of organic chemistry*. This cause is hidden in such a way that we, at least at present, have no hope of discovering it. . . . A *living body, considered as object of a chemical study, is a workshop in which a great many chemical processes take place, the end result of which is to cause all those appearances which in their totality we call life.*

According to Berzelius, the study of life processes was so central to organic chemistry that he called it the "key" to its theory. In this context, it is important to note that Berzelius's and other chemists' notion of life and the attempt to explain it with a life force was not a foreign "belief" laminated to "correct" objects of organic chemical research, but rather was part of the research area of organic chemistry in the early nineteenth century, whose scope was not identical with that of organic chemistry after 1840.[30]

PLANT AND ANIMAL SUBSTANCES: A CROSS BETWEEN NATURAL HISTORY AND CHEMISTRY

> I mentioned that, seen from a chemical perspective, the living body is a workshop of chemical processes carried out by its own instruments, designed to match the properties of the products which are to be produced. Each of these instruments is called an organ. Thus, living nature is called organic, a term which we can extend from the remains and products of living bodies through to the point where the elements are entirely and only united in the fashion of inorganic nature.
>
> —BERZELIUS, 1833–41, 6:3

Although the study of plant and animal substances was not the only research object of plant and animal chemistry, it was the most extensive area, which, in Fourcroy's words, was "the most detailed" and "the richest in facts."[31] The substances studied in plant and animal chemistry were exclusively materials obtained from plant and animal tissues by means of various extraction techniques. The extraction, or "first analysis," of plant and animal substances answered questions about the simplest material constituents of plants and animals that were still "organic" in character. To the extent that these materials were extracted from plants and animals through chemical operations, they were chemical objects. As constituents of plant and animal bodies, however, they were entities that also belonged to the domain of natural history. This dual character of plant and animal substances also informed the meaning of the term *immediate components*, or *immediate principles* of plant or animal bodies.

Fourcroy defined the immediate principles (*principes immédiats*) as "combined materials which make up the tissue" and as "stable materials, which are contained preformed in the vessels and reservoirs of the plants."[32] In the context of inorganic chemistry of the eighteenth and early nineteenth centuries, the term *immediate constituent* or *immediate principle* was used for a complex component of chemical compounds that behaved like a stable building block in chemical reactions, such as an acid and an alkali, both viewed as oxides, constituting salts.[33] On the one hand, in reference to plant and animal bodies, *immediate principle* meant an analogously preformed and stable component of the various tissues and organs of plants and animals.[34] On the other hand, in organic chemistry, the term *immediate principle* did not refer exclusively to pure chemical compounds as was the case in inorganic chemistry.[35] This resulted from the fact that the term *immediate principle* was not embedded in the same laboratory practice in plant and animal chemistry as it was in mineral or inorganic chemistry.[36] From a later perspective, most plant and animal substances of the early nineteenth century prove to be mixtures or organized substance complexes rather than pure chemical compounds. For example, wood fibers, cork, or animal membranes were substance complexes of a higher level of organization. Most volatile oils, fats, resins, and so on were mixtures that were separable into simpler chemical compounds. In accordance with this, Fourcroy, for example, used the terms *organic compounds*[37] and *organized materials*[38] synonymously and included in his classification organized substances.

The Identification of Plant and Animal Substances

In the eighteenth century and at the beginning of the nineteenth century, the identification of plant and animal substances was based primarily on natural historical criteria rather than on experimental knowledge about the composition and constitution of substances as in inorganic chemistry. This included specification of the animal or plant species from which the substance could be obtained and the observation of both the immediately observable properties of the substance and its so-called chemical properties, tested by means of chemical reagents. Whether a substance was sweet or sour, salty or bitter, colorless or colored, transparent or opaque, crystalline or amorphous, elastic or rigid, hard or soft, intensively odorous or odorless, flammable or heat-resistant, soluble in water or alcohol or insoluble—each of these items was used to identify and classify plant and animal substances. The kind and number of properties used to identify substances varied from case to case. For example, Thomson identified the so-called bitter principle by the following properties: "dissolvable in water and alcohol, precipitates through the addition of silver nitrate, usually yellow."[39]

The description of methods of obtaining plant and animal substances offered an additional possibility for their identification, as this following example for the "bitter principle" shows:[40]

> When water is digested over quassia [a wood] for some time, it acquires an *intensely bitter taste and a yellow color, but no smell.* When water thus impregnated is evaporated to dryness in a low heat, it leaves a *brownish-yellow substance which retains a certain degree of transparency.* It continues *ductile* for some time, but at last becomes *brittle. This substance I shall consider as the bitter principle in a state of purity.* If it contains any foreign body, it must be in a very minute proportion.

By specifying the method of their extraction and their properties, Thomson distinguished between twenty-three plant substances, which he termed, like Fourcroy, *plant principles.*[41] These were "sugar, gum, jelly, sarcocoll, tan, bitter principle, narcotic principle, acids, starch, indigo, extractive principle, albumen, gluten, fibrina, oils, wax, resins, camphor, caoutchouc, sandracha, gum resins, wood, suber."[42] Three years earlier, Fourcroy had named a total of twenty plant substances, namely, juice, mucus, sugar, vegetable albumen, vegetable acid or vegetable acids, extractive principle, tannin, starch, gluten, coloring matter, fixed oil, plant wax, plant oil, camphor, resin, gum resin, balsam, caoutchouc (rubber), wood fibers, and suber (cork?).[43]

Immediately apparent in these lists is the relatively low number of plant substances compared with the thousands of organic compounds after mid-century. Furthermore, it becomes clear that most of the substance names differ from the names used in later organic chemistry and that they sometimes appear in plural. The same is true for animal substances, of which Thomson listed the following eight: gelatine, albumine, fibrina, mucilage, urea, sugar, oils, resins.[44] John Webster, a professor of chemistry at Harvard University, included the following substances in his list of animal substances: blood, milk, bile, lymph, mucus, cerebral substance, urine and urinary calculi (both were understood as one substance), cuts, skin, membranes, muscles, ligaments, horn and hair, fat, spermatici, shell, and bone.[45] The fact that many terms for plant and animal substances were used in the plural raises the question of their taxonomic status, which I discuss subsequently. At this point, I want to emphasize that, unlike inorganic chemistry of the time and organic chemistry after 1840, plant and animal chemistry of the eighteenth and early nineteenth centuries was concerned nearly exclusively with substances found in nature outside the laboratory. The term *organic substance* referred to materials produced in the *organs* of plants and animals. Organic substances in this sense had a dual natural historical and chemical character. Their *chemical character* was based mainly on the fact that chemists obtained them from plant and animal tissues through chemical operations; additionally, plant and animal substances sometimes were subjected to chemical analysis. From the perspective of *natural history*, these materials were the smallest components of plant and animal bodies, and knowledge about them was "part and parcel" of knowledge about the plant and animal kingdoms. In accordance with this, the identification of plant and animal substances was based on natural historical criteria.[46]

THE CLASSIFICATION OF PLANT AND ANIMAL SUBSTANCES

> The plant substances which we take to be the general building blocks of the vegetables are as different from one another as the plants themselves from which they are obtained.
> —BERZELIUS 1812, 76

Based on the fact that many terms for plant and animal substances were used in the plural, the question of their taxonomic status arises. This question refers to another aspect of the purity problem. In inorganic chemistry, pure chemical substances were both stable experimental entities, singled out by reversible analyses and syntheses and displacement reactions, and the smallest taxonomic entities that could not be further divided into subkinds.[47] In analogy to the natural historical terminology, a pure substance, either a chemical compound or an element, was also called a *chemical species*. In the taxonomy of inorganic substances, which had been reformed by Lavoisier and his collaborators in the last third of the eighteenth century, different species of substances were grouped into *genera*, and genera were grouped into *classes*. The accompanying reform of nomenclature foresaw generic terms only for the taxonomic groups higher than species. The class of salts, for example, was divided into multiple genera, such as the salts of sulfuric acid, nitric acid, or carbonic acid. Whereas the name of generic entities like *salts of sulfuric acid* were used in the plural, the names of a species belonging to this group, like *iron sulfate*, *zinc sulfate*, and *copper sulfate*, were always used in the singular. Speaking of *iron sulfates*, *waters*, or *sulfuric acids* would have been nonsensical in this system.

In contrast, in plant and animal chemistry of the time, this kind of terminological rule was absent. Chemists used names in the plural both for genera and for species of substances, and they also used names in the singular for genera. In most but not all cases, they applied the names to genera. For example, Fourcroy subdivided sixteen of his twenty plant substances into "species" or "varieties." In the case of plant albumen (*l'albumine végétale*), wood fibers (*ligneux*), tannin (*tannin*), and suber (*suber*), such a distinction was missing. This application of substance names indicates a deeper taxonomic problem. It was in fact an open question for chemists of whether their substance names denoted classes or a species of a substance.

Like the identification of plant and animal substances, their classification was based on their observable properties and their natural origins. These natural historical criteria were ambiguous:[48]

> It is necessary therefore to be well acquainted with their [the plant substances'] *essential characters*, that we know the marks by which they are recognised. These unfortunately are *sometimes ambiguous*; so that a good deal of skill and experience are necessary before we can distinguish them readily.

Identifying and distinguishing substances extracted from plants required identification of their characteristic or "essential" properties, but what properties were essential for grouping plant substances extracted from different plant species into one and the same substance species? The practice of identifying and distinguishing between species of organic substances on the basis of a range of observable properties almost inevitably led to intersections between the properties of different kinds of substances. Evan Melhado described the situation quite accurately: Substances with sharply distinguished qualitative characters were readily taken to belong to different species or even higher taxonomic groups, "[b]ut what of small qualitative differences between very similar substances? What about the differences among the sugars or starches, for example, obtained from various plant and animal sources?"[49] If plant substances obtained from different plant species had very similar characteristics, were they one substance species or a genus containing different substance species that varied according to the biological species? If, instead of natural origins the observed properties of the substance were considered the decisive taxonomic criterion, another problem arose. Did minor differences indicate different chemical species or "varieties" of one and the same species? For example, in inorganic chemistry, where the classification of substances was based on experimental chemical criteria (the composition and constitution of a chemical compound), some differences of observable properties, such as the differences between liquid water, steam, and ice, counted as modifications or "varieties" of one and the same species of water rather than essential differences that required ordering into different groups.

In sum, the taxonomic problems chemists faced in plant and animal chemistry were twofold. On the one hand, it was unclear whether the substances could be divided further into subkinds, that is, whether they were genera containing different species. On the other hand, it was also possible that similar substances were divided too far so that substances distinguished as different chemical species were in fact only varieties of the same species. At the beginning of the nineteenth century, there was but one single "plant substance," which most chemists clearly classified as a genus that could be further divided into species: vegetable acids.[50] Lavoisier already had differentiated between thirteen different species of vegetable acids.[51] This was, however, a special case, because the salts made from the vegetable acids could be distinguished by their crystal form. The difficulties confronting chemists can be recognized as late as 1823, when the French chemist Michel Eugène Chevreul found it necessary to complain that the plant and animal substances that in his time were considered immediate principles included many genera that required division into species.[52]

Jacob Berzelius was among the chemists of the early nineteenth century who were convinced that the wealth of variety in nature was not exhausted by some twenty plant substances and roughly the same number of species stemming from the animal realm.[53] In an 1812 paper on chemical nomenclature, he wrote about the necessity of distinguishing species among the substances (genera) known up to that point:[54]

> In certain plant analyses, which I have made known, for example that of Icelandic mosses and other species of lichens that produce starch, as well as of cinchona bark and the inner bark of the pine, *I tried to prove that the plant substances which we take to be the general building blocks of the vegetables are as different from one another as the plants themselves from which they are obtained.* Thus, for example, I have shown that everything which we call tannin, although sharing certain chemical qualities with one another, nevertheless *differs from one another quite dramatically according to the nature and difference of the plants by which it has been produced*, and that the tannin of the oak-apple, the cachou, the uva ursi, the bark of the willow, etc., are many different species of the same substance which we call tannin. The same is true, as everyone knows, of the volatile oils, the fatty oils, the resins, and no less, as I think, of the starch, the sugar, etc. *Thus, in classifying these bodies, one must proceed in the same fashion as in the system of botany,* and devise *genera and species*, but thereby allow each genus to retain the name which it has had up until now.

It should be noted that Berzelius did not justify his claim that plant substances like starch, tannin, volatile and fatty oils, resins, sugar, and others were genera that required division into species by referring to their quantitative composition and constitution explored by chemical experiments. Instead, he continued to argue in terms of natural history. According to Berzelius, the variations between plant substances followed largely the same principle of variation as plants. For example, the starch extracted from lichens varied according to the species of lichen, the plant substance tannin varied according to the species of tree, and so forth. Hence, chemical classification of plant substances should follow botanical classification. Berzelius also used the same natural historical principle for the classification of animal substances. Animal substances, like "blood, gall, muscle, etc.," he wrote, "are genera, and the kind of animal defines the species."[55] At the end of his discussion, he emphasized that "it would not be helpful to classify them according to their composition, that is, according to the amount of oxygen, hydrogen, nitrogen or carbon which they contain."[56]

Three years later, after Berzelius already had carried out numerous quantitative analyses of plant and animal substances, he confirmed his opinion once again that the known plant substances were genera that still required division into species. "We have, then, various species of tannin, gum, sugar, &c.," he wrote, but he immediately added, "I shall not attempt at present to determine the difference of composition of the species. Such experiments appear too delicate to be executed with success before the generic differences are known. To them, therefore, I have directed my first efforts."[57]

In addition to numerous vegetable acids, by 1815, Berzelius also had analyzed gum arabic, cane sugar, milk sugar, and tannin from oak apples and in the process had obtained results about their quantitative composition that satisfied him.[58] Yet he argued that the new quantitative experiments were too "delicate" to make taxonomic distinctions. Why was this the case? The quantitative analyses showed that in some cases the quantitative composition of phenomenologically very different plant substances, such as gum and sugar, which Berzelius viewed as two different genera, agreed with one another so strongly that the differences measured were within the margin of error of quantitative analyses of samples of one and the same inorganic substance.[59] Berzelius describes the problem of the relationship between quantitative analyses and the distinction between species of organic substances with the following words: "We have found that sugar is made up of 6.802 pct. hydrogen, 44.115 pct. carbon, 49.083 pct. oxygen; gum of 6.792 pct. hydrogen, 41.752 pct. carbon, 51.456 pct. oxygen. It is plain to see that very good analyses of inorganic bodies often show much greater discrepancies than these two results."[60] Berzelius's clear conclusion in 1820 was that experimentation alone did not yield certain knowledge for identifying and distinguishing substance species in organic chemistry.

Berzelius's Theoretical Solution of the Taxonomic Problem

Despite these experimental difficulties, Berzelius became convinced by 1820 that it must be possible, at least in theory, not only to draw sharp distinctions between organic genera but also between species on the basis of the difference in the composition of the organic substances. How was this possible if quantitative analyses of different plant substances, even at the higher taxonomic level of the genus, often did not lead to clear demarcations? Berzelius saw the principle of solution in his theory of chemical portions, taken together with the older chemical principle that differences in properties are caused by differences in qualitative and quantitative composition. Chemical formulas now proved to be excellent tools for drawing taxonomic boundaries. For example, the two plant genera gum and sugar were clearly distinguished by the two formulas 12 0 +13 C + 24 H for the former and 10 O +12 C + 21 H for the latter.[61]

If one assumes, Berzelius wrote in 1820, that organic compounds are made up of "atoms," one can distinguish between two large classes of such compounds. First is the class containing organic species like vegetable acids, whose "composed atoms" are made up of very few elementary "atoms." In such cases, the difference of only one "atom" caused clearly observable differences in macroscopic properties, and this was the reason why differentiating between species was possible merely by observation. The second, much more extensive class of organic compounds, in contrast, was made up of "composed atoms" consisting of a far larger number of elementary "atoms." "In this case," Berzelius observed, "the difference in the composition that arises by adding or subtracting one or more atoms of one element is highly unimportant, and the body thus produced is quite similar to the original in its properties but is distinct to the extent that one can no longer regard them as absolutely one and the same body."[62] As examples, he named the "species" of the genera "sugar," "tannin," the volatile and fatty "oils," and so on, all of which differed in only minor characteristics.

Nevertheless, in his taxonomic practice, Berzelius did not span the gap between the natural historical definition of plant and animal substances and his theoretical approach. As Evan Melhado pointed out, Berzelius "conceded that organic chemistry, unlike mineralogy, would have to remain part of natural history."[63] In the various revisions of his textbook, he never abandoned the natural historical classification of plant and animal substances. As late as

1837, at a time when, for example, Jean Dumas had already argued against the identification and classification of organic species based on natural historical principles, Berzelius wrote that a classification of plant species on the basis of their chemical composition and constitution, analogous to the classification of inorganic substances, was "without particular advantages."[64] The revision of his taxonomy by distinguishing between three "major classes" of plant substances—acidic, basic, and neutral substances—made no decisive change in his basic orientation toward botanical classification. The three major classes shared "chemical properties," but not, as in inorganic chemistry, an immediate constituent. Furthermore, this approach was useful only for classification of plant substances into three major classes. The further division of such large classes was still accomplished according to the conventional natural historical criteria. For example, the genus *gum and plant mucus* was classified into the following species: "A. Gum. Gum arabic. Cherry tree gum. Gum from roasted starch. Gum from the spontaneous decomposition of starch paste. Gum from the treatment of linen, wood, starch or gum arabic with sulfuric acid. B. Plant mucus. Gum tragacanth. Plum gum. Linseed mucus. Quince mucus. Salep. Marigold mucus."[65] Even in the fifth edition of his textbook, Berzelius explicitly maintained his natural historical mode of classification:[66]

> We first divide the subject according to the two kingdoms of organic nature into plant and animal substances, thus dividing organic chemistry into plant chemistry and animal chemistry. Just as the individuals in the plant and animal kingdoms can be classified according to sexual similarities in genera and species, this is also the case for the organic substances obtained from these plants or animals; in organic chemistry this situation has been used in order to group under certain divisions those bodies which are most similar to each other in terms of their properties.

In accordance with this natural historical approach, Berzelius was never ready to recognize substances that were produced in the chemical laboratory by the chemical "metamorphosis" of extracted plant and animal substances as "organic." He never grouped these "artificial" organic substances together with the extracted "natural" ones. This had dramatic consequences for the fifth edition of his textbook, in which the crisis and dissolution of traditional plant and animal chemistry is clearly revealed. Whereas in the 1830s Berzelius grouped organic artifacts produced by chemical experiments under the rubric *chemical properties* of natural organic substances, a decade later, this attempt threatened to break the frame of his classificatory order. The space now devoted to descriptions of organic artifacts began to outweigh that of natural substances. To avoid this, Berzelius separated the description of the chemically produced artifacts from that of natural organic substances. For example, following the section on natural vegetable acids, Berzelius placed a chapter of more than a hundred pages called "Special Acids Formed Through the Influence of Acids, Bases or High Temperature."[67] He did the same with the section on vegetable bases, which was now followed by another section of nearly a hundred pages called "Artificially Produced Bases,"[68] and with the neutral substances. In the latter case, the description of the natural substances filled nearly 250 pages, that of the artificially produced compounds nearly 350 pages.[69] For these organic artifacts, their mere separation in a different chapter of course did not provide a classification. In this case, for the first time, Berzelius based his classification on chemical composition and constitution. In doing so, he had created two different classification systems that existed independently from each other.

"NATURE" AND "ART" IN THE EXPERIMENTAL PRACTICE OF PLANT AND ANIMAL CHEMISTRY

> The art of chemistry is certainly allowed to destroy the union of the principles in a plant compound and to isolate these; as soon as it begins to work on this compound, however, it begins to destroy the balance; and no matter how gentle its action might be, it leads to the denaturation of the compound. . . .
>
> —FOURCROY 1801–02, 7:56

Central to the experimental practice of organic chemistry in the early nineteenth century was the extraction of plant and animal substances from juices, tissues, and various other plant and animal parts and the subsequent tests of the chemical properties of the extracted substances. About the extraction techniques, also called the *first analysis*, Fourcroy wrote:[70]

> Dissection, grinding, pressing are particular means of first analyses. A low fire, the use of water, alcohol, oils, without intense heat, that is, if they are applied over an extended period of time, fulfill the same task; they separate the combined materials of the plants to which they are applied, and they produce through this isolation of different materials as a first result the number and relative proportion of these first compounds ("*premiers composans*"), which themselves are combined.

According to Fourcroy, the techniques for obtaining plant and animal substances could be roughly divided into three groups: *mechanical methods* (like pressing and grinding), *distillation*, and *extraction* (in the narrow sense) by way of various solvents. Distillation, which in the first half of the eighteenth century was the most important of these procedures, especially for obtaining plant materials, was pushed to the background during the second half of the eighteenth century in favor of extraction by means of solvents.[71] Thomas Thomson referred to the role pharmacists played for the development of these new techniques: "The older chemists confined their analysis entirely to destructive distillation. By this process they obtained nearly the same products from every vegetable. For every plant when distilled yields water, oil, acid, and carbureted hydrogen and carbonic acid gas; while a residuum of charcoal remains in the retort. For the first introduction of solvents we are indebted to the apothecaries."[72]

Neither in the eighteenth century nor the nineteenth century, however, were experiments for obtaining plant and animal substances marked exclusively by their connection with pharmacy. Extraction with solvents, like plant distillation beforehand, had two sides. It explored improvements in the manufacturing of substances with the intent of possible practical applications, and it also served as "first analysis" of the "immediate principles" of plants and animals. These experiments thus resist the attempt to distinguish between a pure and applied chemistry or internal and external factors in chemistry.

The Dual Nature of Plant Analysis

The concept of plant analysis was on the one hand a chemical concept, which implied the assumption that the substances obtained from plant and animal bodies were preexisting components of these bodies that could be separated from each other without any transformation, that is, like integer building blocks, by means of chemical operations. On the other hand, plant and animal analysis always could be seamlessly embedded in a natural history of plants and animals if it was interpreted as the continuation of plant and animal anatomy by chemical means. Thus, like the plant and animal substances themselves, the analysis of plants and animals had a dual nature: It was both a chemical and a natural historical enterprise. The natural historical character of these experiments also was expressed in chemists' rhetoric, especially when they emphasized that their experiments did not transform the nature of the plant and animal components.

In his *Systême des Connaissances Chimiques*, Fourcroy wrote that the first analysis served to extract from plants the diverse combined materials "without undergoing an alteration and without changing their nature."[73] Similar statements claiming that this chemical operation in fact did not change the nature of plant substances are abundant in Fourcroy's textbook. For example, referring to the mechanical operations for obtaining plant and animal substances (such as rubbing, pressing, and grinding with mortar and pestle), Fourcroy emphasized that "[Mechanical analysis] has the advantage of not changing the nature of the materials in any way or causing an alteration of them; it yields them in the way that they exist in the plants." [74]

Fourcroy considered the extraction of plant materials with solvents like water, oils, alcohol, and ether to be an especially good technique of plant analysis, arguing that "these bodies have the property of dissolving certain plant substances without acting on the majority of the others." [75] Conversely, he rejected distillation as a "false analysis," emphasizing that it "produces materials which are formed by the effect of the fire, and do not exist in the plant." [76]

Why was this kind of argumentative move necessary? A look at the article "Histoire Naturelle" in Diderot's *Encyclopédie* gives a first answer to this question. Chemistry, we can read there, "commences at the point were natural history ends," and further: [77]

> The naturalist researches all natural products in their own context; he cautiously removes the veil that covers them, he observes them with an attentive eye without daring to lay an all-too bold hand; if he must touch them, then always with the fear of deforming them; if he is forced to penetrate the interior of a body, then he cuts it only with regret, he only destroys the unity in order to better know the links and in order to acquire an even more complete idea of both their inner structure and outer form. *The chemist, in contrast, studies the operations of nature only with the procedures of art; he decomposes all natural products, he dissolves them, breaks them, submits them to fire.*

In the eyes of naturalists, chemistry was the intervening science *par excellence* that deformed natural objects. Similar objections were made by physiologists of the late eighteenth and early nineteenth centuries. Timothy Lenoir observed that, for example, Karl Ernst von Baer pointed to difficulties in utilizing chemistry for studying the composition of organic structures: "The part, which the chemist removes from the plant and animal for the purpose of his investigation," von Baer asserted, "is already torn from the vital context, and from the moment of this separation, processes begin to set in which no longer belong strictly to the process of life. Although these changes may only be slight in connection with plants, they are significant with respect to animals." [78] In light of this view of chemistry, it is quite understandable that Fourcroy advanced special arguments for his claim that the chemical analysis of plant chemistry is only a natural continuation and completion of the mechanical dissections of the naturalists. In doing so, his rhetoric picked up the well-known arguments of the naturalists, who themselves did not manage totally without intervening activities and who already had fought similar defensive battles long ago. Whereas anatomic dissection used mechanical means, separating the large and visible organs from one another, the chemical operations were thought to continue this work by more refined means and thereby isolate the smallest material components.

The Testing of "Chemical Properties"

Following the extraction of plant and animal materials, chemists determined their immediately observable and their so-called chemical properties. Only after this additional step, which served to identify the substances, was the analysis completed. Besides color, odor, taste, consistency, and so on, the chemical properties of the extracted substances were considered especially relevant criteria for substance identification. The term *chemical properties* referred to the ways in which chemical substances behaved in contact with other, particularly reactive substances called *reagents*. Included among the chemical properties of a substance were, for example, its solubility in solvents, like water or alcohol, and its ability to release a gas or a precipitate after the addition of a reagent. Establishing chemical properties thus required additional experiments, which differed in some ways from the experimental study of chemical reactions but nevertheless shared some features with it. Hence, during the transformation phase of plant and animal chemistry in the 1830s, the testing of chemical properties repeatedly became a starting point for a previously unintended study of chemical reactions. In the following, I briefly discuss the likeness and disparity of both types of experimentation.

For testing the chemical properties of a plant or animal substance, the substance would, if necessary, first be dissolved in a solvent and then mixed with a reagent. In doing so, either a change in color, a precipitation, or other observable effects took place merely by mixing; or the mixture had to be shaken, warmed, distilled, or manipulated in some other way. For these experimental tests, early nineteenth-century chemists used the same type of furnaces as sources of heat and the same type of reaction vessels and instruments for storing, mixing, and combining substances as those used in the study of chemical reactions. Technological disparity between the tests of chemical properties and the study of chemical reactions laid mainly in the time required and the number of instruments used. Whereas the testing of chemical properties usually could be completed quickly and with one single reaction vessel, the experimental study of chemical reactions sometimes required more complex apparatus made up of multiple reaction vessels and could stretch out across days, especially in organic chemistry.

In conceptual terms there were important differences between the testing of chemical properties and the study of chemical reactions, which also had practical consequences. Talk of chemical reactions implied assumptions about the invisible actions of the reacting substances. First was the assumption of a symmetric interaction of the components of the transformed substances, in which these separate like stable building blocks and form the reaction products through regrouping. Second, this recombination was supposed to be directed by the mutual chemical relationships, or "affinities," of the recombining building blocks. In contrast, the concept of chemical properties explained the observed experimental phenomena not as being the result of symmetric interaction between two substances but rather as a sign of a "property" of one of the substances. The reagent, in this view, was not seen as a reaction partner but as an instrument that revealed the properties of the substance to be tested. This is clearly expressed, for example, in the term *indicator* for reagents that identify acids; the acid indicator "indicates" that a substance is acidic by taking on a red color. Neither the mechanism through which the indicator is chemically transformed nor the reaction products formed were of interest here.

The different concluding phases of the experiments corresponded to these conceptual differences. Whereas during the testing of chemical properties the produced effects were merely observed, the study of chemical reactions required the isolation and subsequent identification of the newly produced substances by chemical analysis. For example, to test whether a plant substance was acidic, the observation that a litmus solution became red was sufficient. To learn which chemical reaction took place between a plant substance and the litmus, however, the color produced and all other by-products of the reaction needed to be isolated and analyzed. This would be followed by more or less extensive work on paper, in which, based on knowledge of the composition of the reaction products and of the original substances, the recombination of the components of the original substance in the reaction products was reconstructed. In the experimental culture of organic chemistry, the final stage of the experimental study of chemical reactions also included quantitative analyses of the substances, transformation of the data of quantitative analysis into chemical formulas, and modeling the regroupings with formula models, such as linear formula equations. All in all, the testing of chemical properties in plant and animal chemistry was concluded with inscription of an effect observable to the naked eye. The study of chemical reactions of organic compounds, in contrast, ended in an epistemically complex system, which included further experiments and constructing models by using chemical formulas.

After chemists had improved quantitative analysis and employed chemical formulas from the late 1820s onward, the technological likeness between the testing of chemical properties and the study of chemical reactions facilitated spontaneous shifts from the former to the latter. If it seemed probable that the reaction products created during a test of chemical properties could be isolated without great technical difficulties, the curiosity of the chemist was awakened; isolation, purification, and analysis were attempted; and interpretive models of the reaction were built. Hence, chemists' routine tests of chemical properties of extracted organic substances were seamlessly continued with the originally unintended study of a chemical reaction.[79]

The Limits of Experimentation in Plant and Animal Chemistry

Experimental studies of chemical reactions did not belong, except for a marginal area, to the area of research objects of plant and animal chemistry. Of course, the fact that a scientific object and a corresponding experimental practice were not present at a particular time cannot be proven directly. It can, however, be indirectly supported by analysis of the claims of chemists about the possibilities and limits of experiments in plant and animal chemistry.

Chemists could observe daily in their laboratories that plant and animal substances underwent spontaneous transformations. As a rule, these transformations yielded series of different products and thus were considered decompositions. In the eyes of most chemists, this fact underpinned the assumption of a profound and characteristic disparity between organic and inorganic substances. Thomas Thomson, for example, asserted:[80]

> The most striking distinction between the substances belonging to the mineral kingdom and those which make a part of animals and vegetables, is the following: Mineral bodies show little or no tendency to change their nature; and when left to themselves, undergo no *spontaneous decomposition; whereas animal and plant substances are continually altering; and when left to themselves in favourable circumstances, always run through a regular set of decomposition.*

Although spontaneous decompositions belonged to their daily laboratory experience, chemists did not explore in any detail such processes. The plethora of decomposition products, which were difficult to untangle from one another, supported the view of the futility of any serious attempt at reconstruction and modeling of these processes. Rather than spurring experimental investigations, spontaneous decompositions of organic substances entrenched the belief that organic and inorganic substances were utterly different kinds of matter. In Fourcroy's words:[81]

> Without any doubt, one must be ready to find great differences between organic materials and mineral or fossil material. In general, the first, which are much more complicated in their composition, are also much more subject to change through the action of these bodies [the reagents], and the changes which appear are thus far more diverse and far more difficult to grasp and explain than those which had been mentioned in the history of all the other [mineral] substances.

The fact that it seemed impossible to investigate experimentally the spontaneous transformations of organic substances did not keep chemists from presenting theoretical explanations of such processes. For example, Fourcroy explained the spontaneous decomposition of organic substances as a disturbance of their equilibrium of attractions, which caused multiple "immediate degrees of decomposition" that eventually led to inorganic products of decay, such as carbonic acid (carbon dioxide) and water.[82] According to Berzelius, organic substances required the permanent influence of a specific form of electrochemical affinity that existed only within the organs of living beings and was absent in the extracted plant and animal substances; hence, the decay of the latter into inorganic compounds.[83]

If the spontaneous chemical transformations of plant and animal substances in the chemical laboratory were thus at best the subject of theories but too complex for experimental research, this was all the more true for physiological processes. The imitation of physiological processes was considered outside the realm of the chemical laboratory. Moreover, chemists also believed that the chemical synthesis of organic substances in the chemical laboratory was out of the question.[84]

"The texture of live vegetals, and their vegetating organs," Fourcroy asserted, "can alone form the matters extracted from them, and no instrument of art can imitate the compositions which are formed in the organized machinery of plants."[85] A decade later, Louis Jacques Thenard, rejecting an objection by Rousseau, pointed out that there are indeed substances that in principle could only be decomposed but not recomposed:[86]

> While taking a course in chemistry with Rouelle, Jean-Jacques Rousseau said that he did not believe in the analysis of flour, so long as he did not see that the chemists can remake it. This philosopher would undoubtedly speak differently today. One would show him that there are bodies which one can decompose but not recompose; one would explain the causes for this, and he would not fail to

understand this. The causes lie principally in the condition in which the elements are found.

On the American continent, there was no remarkable difference in the image of plant and animal chemistry as a science that is unable to imitate nature in the laboratory and to achieve the synthesis of organic substances. For example, John W. Webster observed:[87]

> They [the products of the plant kingdom] are all susceptible of decomposition by heat alone; but we cannot always as in bodies of the mineral kingdom, proceed from the knowledge of their components to the actual formation of the substances themselves. It is not probable, indeed, that we shall ever attain the power of imitating nature in these operations. For in the function of living plants, a directing principle appears to be concerned, peculiar to animated bodies, and superior to, and differing from, the cause which has been termed chemical affinity.

Spontaneous decompositions of organic substances were excluded as experimental objects of study because they were too complex; the synthesis of organic substances was considered utopian because the "directing principles" were not accessible in the laboratory.

This limitation of experimentation in early nineteenth-century organic chemistry becomes even more obvious from the comparison with inorganic chemistry of the time, where not only chemical analyses but also syntheses were part of daily experimental practice. Focusing on inorganic chemistry, Fourcroy went so far as to assert that chemistry was the science of synthesis: "Synthesis, considered to be the second general chemical means for familiarizing us with the most intimate and reciprocal actions of natural bodies, takes place far more frequently than analysis; and if one considered the use of these two means for the definition of the science [chemistry], *it would rather merit the name science of synthesis than science of analysis*."[88]

In his definition of "chemistry as the science of synthesis," Fourcroy excluded plant and animal chemistry. The term *synthesis* is not to be found in any part of his textbook dealing with plant and animal chemistry.

THE EXPERIMENTAL CULTURE OF ORGANIC CHEMISTRY AFTER 1840

The form of organic chemistry that took shape during the 1830s was a scientific culture that produced a plethora of artificial substances that did not exist outside the laboratory. At the center of its research was the experimental study of the chemical reactions and of the constitution of organic substances. To represent these invisible objects, chemists applied Berzelian formulas to construct models of constitution and reaction. Furthermore, the new carbon chemistry created entirely new classification systems based on formula models of the composition and constitution of organic compounds. In the following, in the same order as in the preceding chapter, I first analyze the entire area of research objects in carbon chemistry; second, its concept and referents of organic substance or carbon compound; third, its classification of organic substances; and, finally, its type of experiments. In doing so, I can put things much more briefly than in the first part because many of the aspects of the experimental culture of organic chemistry are discussed in detail in Chapters 4 through 7.

The Area of Research Objects

> In fact, we here touch on a fundamental characteristic that distinguishes the experimental sciences from the observational sciences. Chemistry creates its object. This creative faculty, comparable to art, distinguishes it essentially from the natural and historical sciences (*sciences naturelles et historiques*).
>
> —BERTHELOT 1860, 811

"Chemistry creates its object"; it is not an observational science such as natural history but resembles an art. The difference between the natural historical agenda of plant and animal chemistry and the experimental culture of organic chemistry after 1840 cannot be expressed more pointedly than in Marcellin Berthelot's motto to this section. With these words, Berthelot summed up his reflections about the new form of organic chemistry, also called *carbon chemistry* by many chemists.

Compared with plant and animal chemistry, carbon chemistry after 1840 displays a dramatically changed spectrum of research objects. Physiology, anatomy, and everything that could be labeled as "organization" as well as the natural history of pharmaceutically relevant materials of animal and plant origin were removed from its research agenda. All questions that exceeded the chemistry of pure carbon substances were delegated to physiology, physiological chemistry, medical chemistry, pharmacy, and so on. Simultaneously, the extracted plant and animal substances no longer were seen as deficient materials that lacked the peculiar affinity of living organs. Rather, they were treated as ordinary chemical laboratory substances that could be controlled in experiments and even experimentally combined with other substances. In 1843, for example, Frederic Daniell, professor of chemistry at King's College in London, proclaimed:[89]

> The organization or apparatus by which the different processes of organic chemistry are conducted in living plants or animals, is the peculiar province of Anatomy and Physiology. . . . When withdrawn from the immediate influence of this innovating power of life, *the products are found to be strictly chemical, and retain their composition or enter into new combinations, according to the fundamental laws of chemical affinity*: and it is to this point of them that we shall at present confine our attention.

Attempts to exclude physiology and anatomy from organic chemistry had already begun in the 1830s, especially in France. Jean Dumas was among the first to push decisively toward limiting organic chemistry to the study of pure organic compounds. The fifth volume of his chemistry textbook (published in 1835), which includes organic chemistry, begins thus: "The main goal of this book is a familiarization with the organic substances from a purely chemical standpoint."[90] A chemistry of substances, Dumas continued, implies the exclusion of the study of physiological processes and "organized substances" or "complex products."[91] "The history of those substances," he added, "which are only organs or part of organs, like wood fibers (*le ligneux*), fibers (*la fibrine*), corn starch (*l'amidon*) and all the other complex products, which interest the chemist only as a starting point for his operations, must be left to physiology."[92] Dumas described the scope of a to-be-reformed organic chemistry as follows: "I thus limit organic chemistry to the study of those particular compounds which exist in the organic kingdom or are products of reactions undertaken with substances from that provenance."[93]

With these words, Dumas also extended the scope of research in a new and completely different direction. Organic chemistry should study not only the components of plant and animal bodies but also the chemical reactions of these substances in the chemical laboratory. The experimental investigation of chemical reactions of organic substances now appeared as a promising area of research. A kind of organic substance that differs from the traditional plant and animal substance thus came into view: the organic artifact that did not exist in nature but was created in experiments as a reaction product of naturally occurring organic substances. Whereas the area of research of carbon chemistry on the one hand was diminished by excluding what can be broadly described as natural historical objects of research, it was, on the other hand, considerably enlarged by a new field of experimental inquiry, which went hand in hand with the creation of a plethora of newly created organic artifacts. When Berthelot wrote decades later that chemistry creates its object, he also referred to the artificially created organic compounds that did not exist in nature outside the laboratory. In the experimental culture of carbon chemistry that began to thrive after 1830, chemical artifacts became at least as important as organic substances found in nature. Thus, the question of the definition of organic substances, and their identification and classification, had to be posed in a new way.

Carbon Compounds

> Since all organic materials, without exception, contain carbon, one can say that this is the chemistry of carbon. It includes only organic materials in their purely chemical relations, without accounting for the role that they play in a living organization.
>
> —GERHARDT 1844–45, 1:1

Around the middle of the nineteenth century, the number of organic compounds had increased from fewer than a hundred around 1800 to several thousand.[94] Chemists' refined experimental techniques enabled them to increase the number of pure plant and animal substances extracted from organic tissues and juices. Furthermore, chemists now had access to coal tar, which was a waste product of the gas industry that proved to be an unexpectedly rich reservoir of organic compounds. The great majority of the newly introduced substances, however, were not extracted plant and animal substances or coal tar products; rather, they were artificially transformed organic compounds, which emerged as reaction products during the experimental study of organic chemical reactions.

As a consequence, in the new experimental culture of carbon chemistry, the material referents of "organic substances" fundamentally changed compared with plant and animal chemistry. Moreover, the dual natural historical and chemical meaning of the early notion of *organic substance* gave way to an unambiguous chemical definition as *carbon compound*. Organic compounds in the sense of carbon compounds were laboratory objects, just like inorganic substances, whose specific composition, constitution, and reactions were the subject of experiments. Knowledge about the composition and constitution of organic compounds acquired through their quantitative analysis and the study of their reactions became a necessary precondition for their identification and classification. Berzelian formulas became the new labels of carbon compounds, and they clearly identified and distinguished them from one another.

Jean Dumas was one of the first chemists to distance himself from the traditional definition of organic substances in the natural historical mode. "The most general characteristic regarding organic materials," he wrote, "is everything that concerns their analysis, the determination of their atomic weights and the search for their rational formulas which represent their nature and their properties."[95] The three identification criteria mentioned by Dumas required three types of experiments: quantitative analysis, density measurement (for determination of "atomic weight"), and the study of series of chemical reactions (for knowledge about the constitution of an organic substance). Furthermore, it required the transformation of analytical data into Berzelian raw formulas or "empirical formulas" and subsequent transformation of the latter into "rational formulas," that is, formula models of constitution.[96] Whereas early nineteenth-century chemists always emphasized that their chemical operations left the nature of plant and animal substances untouched, Dumas included the chemical artifacts in his definition of organic substances. Although many of his colleagues, such as Justus Liebig and Berzelius, excluded derivatives of plant and animal substances containing chlorine or bromine from the class of "organic" substances, Dumas stated, "As long as an organic material is not transformed into carbon, carbon oxide, carbonic acid, carbonated hydrogen, nitrogen, ammonia, nitrogen oxide, or water, it must remain in the array of organic matter."[97] From the new experimental perspective, the former distinction between organic substances extracted from the organs of plants or animals and artificially transformed laboratory products no longer made any sense. Dumas went even a step further, questioning the boundaries between organic and inorganic chemistry. "In my opinion," he proclaimed, "there are no organic materials."[98]

Most chemists of the generation after Dumas shared his view of organic substances. The French chemist Charles Gerhardt described organic substances found in nature and those produced in the laboratory as "members of the same chain, held together by the same laws."[99] In 1860, the Tübingen chemistry professor Julius Eugen Schlossberger wrote in his textbook on organic chemistry that the old division between organic and inorganic chemistry was outdated and that the natural origin of an organic substance was no longer relevant:[100]

> For a long time, chemists considered the origin of organic substances as their most distinct characteristic. Besides the elements, mineral chemistry included the compounds which are found in lifeless nature, or which can be produced without any help of the life process and without the participation of its characteristic products. *In contrast, chemists supposed that organic substances can be produced only by the living cell.* . . . Although today, the majority of organic bodies still can not be produced from purely mineral bodies, *the criterion of the origin of a substance has lost almost all value* in the question under consideration.

After the 1840s, the number of chemists, such as Gerhardt (see the epigraph to this section), calling for a revision of name to *carbon compounds* or *carbon chemistry* increased. The German chemist Friedrich August Kekulé defined organic chemistry as "the chemistry of carbon compounds."[101] Carl Schorlemmer, lecturer in organic chemistry at Owens College in Manchester, gave his textbook of organic chemistry the title *A Manual of the Chemistry of the Carbon Compounds* (1874), and the Russian chemist Aleksandr Butlerov, professor of chemistry at the University of Kazan, wrote around the same time that it is "advisable to consider the concepts organic compound and carbon compound to be identical. All bodies containing carbon must then belong to the scope of organic chemistry, or better: the latter should be called 'chemistry of carbon compounds.'"[102] These attempts to rename organic chemistry indicate deeper structural changes in the culture of organic chemistry. The older customary terms *organic compound* and *organic chemistry*, however, outlived the attempts at terminologic reform. The term *organic* thus took on a new meaning and referent in the second half of the nineteenth century.

The Classification of Carbon Compounds

> We ask ourselves impatiently whether it will be possible, in a few years, to navigate through the labyrinth of organic chemistry.
>
> —AUGUSTE LAURENT 1854, IX

Beginning in the late 1820s, the experimental study of the chemical reactions of organic substances proved an unforeseen "motor" for the production of new organic compounds. The first reaction products of substances obtained directly from plant and animal substances or coal tar were studied in further reactions, through which new reaction products were created, whose reactions in turn were studied and also produced new reaction products, and so on. With each of these steps, the number of organic compounds produced increased exponentially. Already in the 1840s, products of the chemical laboratory outnumbered the traditional plant and animal substances. The rapid growth in the number of organic compounds presented chemists with entirely new problems of classification. What was the connection between the experimentally produced substances, and how should they be ordered in substance classes? Should artificial laboratory substances be classified separately or grouped together with the natural plant and animal substances? If they were classified together, according to what unified criteria of classification should this take place? It was obvious that natural origin no longer would work as a criterion of classification under these new auspices. Furthermore, the earlier taxonomic problems persisted in the 1820s. With respect to the extracted plant and animal substances, it was still an open question as to whether they were genera of substances, species, or varieties of species. Friedrich Wöhler's comment that organic chemistry was a jungle[103] or Auguste Laurent's metaphor of a labyrinth,[104] as well as comments about the rapidly increasing complexity of organic chemistry—which appear over and over again in the chemical journals of the period—are expressions of the enormous taxonomic problems confronting chemists over several decades.

In 1827, the French chemists Jean Dumas and Pierre Boullay suggested an entirely new mode of classifying organic species, in which artificially produced organic compounds and natural plant substances were placed in a unified substance class.[105] As a criterion of classification, they selected the composition and the shared "binary constitution" of these compounds, which they had investigated by experiments and represented with formula models. Five years later, Liebig and Wöhler as well as Berzelius followed their example, building new classes of organic substances based on experiments and the construction of formula models of their binary constitution. The French chemists August Laurent and Charles Gerhardt continued these early attempts at a new mode of classification by developing the first more comprehensive classification systems in the 1840s.[106] Carbon chemists considered their new classifications to be "natural" as opposed to the earlier "artificial" classifications. Looking back, Aleksandr Butlerov observed:[107]

> As long as our familiarity with organic bodies was merely *superficial* and limited itself primarily to *external properties*, while *transformations and mutual relations remained almost entirely unknown*, the classifications of organic compounds, considered from the standpoint of chemistry, *could not be natural*. At that time, bodies were grouped according to their origin, color, consistency, etc., divided into volatile oils, resins, dyestuffs, etc. A closer, but not yet complete familiarity with the chemical properties of a part of the organic compounds led to the distinction between acids, alkalis, and indifferent bodies, and allowed the first to be divided into volatile and non-volatile, the second into bodies without oxygen and containing oxygen, the third into bodies containing nitrogen and free of nitrogen, etc. *The emergence of theoretical views made it possible*—at least in terms of the substances to which these viewpoints were applied—*to move towards more scientific systems*. Each theory was reflected in classification.

Butlerov emphasized that the experimental study of chemical reactions linked to their theoretical interpretations was key to both the naturalness and the scientificity of classification. A few years earlier, Kekulé had pointed out that inanimate beings and substances derived from living organisms follow the same "laws" and therefore should be classified into continuous "series" of compounds:[108]

> We have come to the conclusion that chemical compounds of the plant and animal kingdom contain the same elements as those in inanimate nature; we are convinced that their elements follow *the same laws*, and that there is no difference—neither in the substance, nor in the forces, and just as little in the number or kind of groupings of atoms—between organic and inorganic compounds. *We see a continuous series of chemical compounds*, whose individual members show such a great similarity (if one only compares the closest neighboring compounds), that it goes without saying that we cannot divide them at any point.

The "series" that Kekulé referred to were series of chemical formulas, gathered in tables that were modeled (in principle) after Dumas and Boullay's 1827 table. Compared with Dumas and Boullay's classification, Kekulé's classification paid more attention to the problem of the relationship between the invisible constitution of an organic compound represented by a formula model and its observable and measurable properties. Substances with a shared basic formula and with largely agreeing chemical properties were gathered in "homologous series," those with similar formulas but differing properties into series of "derivatives."

The observable properties of organic substances and their natural origins were the two main criteria of classification in plant and animal chemistry. By contrast, the new classification systems of carbon chemistry were based on knowledge about the composition and constitution of organic compounds, investigated by way of experiments and construction of formula models on paper. Moreover, the classifications of the experimental culture of carbon chemistry included both "naturally" found substances as well as experimentally produced artificial organic compounds. Thus, the chemical order of things in the two cultures of plant and animal chemistry and carbon chemistry differed fundamentally from each other.

The Type of Experiments

> The field of investigation which has very recently been opened to experiment in such organic metamorphoses is apparently boundless; and it has been taken possession of by labourers of the highest talent and activity; the fruits of whose industry are so abundant as to cover the ground with somewhat of that confusion which might be expected from a number of independent workmen actively engaged in the same track.
>
> —FREDERIC DANIELL, 1843, 642

At the center of the experimental practice of carbon chemistry in the second half of the nineteenth century were the study of the chemical reactions and the

Figure 2-1 A nineteenth-century laboratory bench (Schödler 1875, 38)

quantitative analyses of organic compounds. Although the procurement of organic substances from plant and animal bodies was not abandoned, it was pushed more to the background. The centuries-old pharmaceutical–chemical tradition of obtaining organic substances through the distillation and extraction of plant and animal raw materials lost its dominant role. This does not mean, however, that possible pharmaceutical or other commercial applications of organic substances ceased to play a role. Chemists continued to look for possible practical applications, but they now began to include the many new artificial organic products. Chloral and chloroform, for example, two organic reaction products containing chlorine, which Liebig first produced and identified in 1832, were used little more than a decade later as medicines and by the turn of the century were manufactured in large quantities.[109]

Seen over the long term, and considering multiple sites and researchers, the study of each chemical reaction of a substance resulted in chains of experiments, both linear and branching out, and parallel to that in an exponentially

Figure 2-2 Justus Liebig's laboratory in Giessen in 1842 (Schödler 1875, 40)

growing number of synthesized derivatives. Substitution of the hydrogen contained in organic compounds by chlorine, bromine, fluorine, or even by metals created entirely new groups of organic compounds. In other experiments, chemists oxidized organic compounds, increasing step by step their relative oxygen content, thus producing series of oxides. The new term *metamorphoses*, which chemists coined to cover these various reactions, neatly expressed the enormous array of possible transformations organic substances could undergo. Furthermore, after 1830, chemists also began to synthesize organic compounds from more simple inorganic compounds or from chemical elements.

In particular, experiments studying the so-called substitutions of organic compounds increased after 1834.[110] Many noted French and German chemists of the 1830s carried out substitution experiments, thereby producing new chemical artifacts. Leopold Gmelin, professor of chemistry in Heidelberg, wrote in 1848 in the fourth edition of his chemistry textbook that "art can produce a large number of new compounds from existing organic compounds. Many of these artificially created compounds can also be found in nature. . . . *The number of those which are not found in nature, which only exist in artificial form, is disproportionately larger.*"[111]

The artificial substances to which Gmelin was referring were substances manufactured in substitution experiments that contained elements such as chlorine and bromine. Frederic Daniell described in his textbook of chemistry a scenario of exuberant activity (see the epigraph to this section). Not all individual chemists, however, appreciated the new development that directed chemists' attention to chemical processes existing only in the laboratory rather than in "nature" outside. Justus Liebig, for example, who actually played an important role in the development of substitution experiments, abandoned the new scientific practice. The research on substitutions, he wrote in 1841, is "dreary and unsatisfying" and useless for science.[112] Although Liebig's attempt to resurrect a new form of organic chemistry oriented toward "nature" and physiological processes received some attention, he could not set a new course for carbon chemistry.

THE TWO FORMS OF ORGANIC CHEMISTRY COMPARED

The differences between pre-1830 plant and animal chemistry and post-1840 carbon chemistry affected the entire scope of scientific objects, the concept of "organic substance" and its material referents in the laboratory, the classification of organic substances, the type of experiments, the tools and modes of representation, as well as chemists' beliefs and self-image. I call the structural network of these epistemic and material elements, together with the related social institutions, a "scientific culture." In the following, I argue more explicitly for my proposition that pre-1830 plant and animal chemistry and post-1840 carbon chemistry should be considered two different scientific cultures. The first may be described as a chemical culture that overlaps with natural history and commercial practice, the second as a productive experimental culture. This distinction is based on my previous comparison, and its scope is confined to a structural comparison. All questions concerning aspects of the process of transition are excluded in the following; they are discussed in Chapter 8. Before discussing the question of why pre-1830 plant and animal chemistry and post-1840 carbon chemistry were two different scientific cultures, I examine how carbon chemists themselves reflected on the difference between the two enterprises.

The New Self-Image

> The artificial things are the realized images of abstract laws.
>
> —BERTHELOT 1860, 812

In the new experimental culture of organic chemistry after 1840, the number of chemists increased who held that the differences between organic and inorganic chemistry were inessential and instead assumed unified chemical "laws."

After 1860, this belief in the unity of chemistry and the universal application of its laws was so strong that in the rhetoric of carbon chemists, the study of unified chemical laws became a central concern. When carbon chemists addressed the issue of "natural laws," they often did so to criticize the early differences between the natural and the artificial. Moreover, carbon chemists used their criticism of the older plant and animal chemistry as a vehicle to enhance their reputation as architects of a "new world" that embraced both chemical artifacts and natural organic substances. August Kekulé, one of the most prominent carbon chemists in the 1860s, criticized Berzelius's earlier distinction between chemical laws in the living and nonliving nature with the following words:[113]

> Since the observation of plant or animal matter caused much greater difficulties than the study of minerals, it was concluded 'that in living nature the elements follow entirely different laws than in the non-living' (Berzelius's 1827 textbook); organic chemistry was defined as 'the chemistry of plant and animal substances, or of the bodies that are formed under the influence of life force.' This point of view could not keep up with scientific progress.

Charles Gerhardt wrote around midcentury that the products of substitution are "compounds which by their character and composition seem to be far removed from natural products, but which are of extraordinary scientific interest for the chemist. They in fact serve to help him discover the laws of transformation of compounds which living nature itself devises."[114] Gerhardt's French colleague Marcellin Berthelot formulated the role of the study of laws in the new scientific culture of organic chemistry in perhaps the most dramatic terms: "In a word, the sciences considered here pursue the study of natural laws, by creating an ensemble of artificial phenomena. . . . These artificial things are the realized images of abstract laws."[115]

Similar comments were made by many European chemists in the second half of the nineteenth century. For example, the Russian chemist Butlerov wrote in his textbook that "it is impossible to draw a natural clear distinction between all inorganic and all organic substances," and, he continued, the most recent syntheses of naturally existing organic compounds had proved that in the production of these substances, "the same forces are at work as those which form inorganic compounds."[116] In the same vein, Carl Schorlemmer stated, "We have now cause for the conviction that the same chemical laws rule animate and inanimate nature."[117] Finally, a look at the textbook of Josiah P. Cooke, professor of chemistry and mineralogy at Harvard University, shows that the image of carbon chemistry was largely the same among North American chemists:[118]

> It was formerly supposed that the great complexity of those substances [the organic substances] was sustained by what was called the vital principle; but, although the cause which determines the growth of organized beings is still a perfect mystery, we know that *the materials of which they consist are subject to the same laws as mineral matter* . . . although the analysis of these compounds was easily effected, the synthesis was thought impossible. But within a few years we have succeeded in preparing artificially a very large number of what were formerly supposed to be exclusively organic products; and not only this, but the processes we have discovered are of such a general application that we now feel we have the same command over the synthesis of organic, as of mineral substances.

The carbon chemists' critique and their view of the new form of organic chemistry often culminated in the claim that a scientific organic chemistry now existed for the first time. For example, in the introduction to his three-volume textbook on organic chemistry, August Kekulé wrote:[119]

> First we will gain the conviction, that *the standpoint for the question* [about the difference between organic and inorganic matter] is now entirely different than it was earlier. If earlier, *when chemistry was really part of the descriptive natural sciences*, a distinction, indeed an opposition, between the bodies of the mineral and those of the animal and plant kingdoms was correctly made, that is, [they were] correctly divided into organic and inorganic, or at least distinguished [from one another]; *now, as chemistry has begun to take its place as an independent science*, where it is increasingly conscious of its *true calling—to ascertain the metamorphoses of a substance and the laws which are thereby in force*, a division of the bodies merely based on their natural origin is truly nonsensical. . . . *That which earlier was indeed the single or central task of chemistry has now moved more and more to the background.*

By referring to a difference in "standpoints," Kekulé pointed to a kind of transformation that was not limited to a few aspects but rather affected the entire scientific culture. The rupture between the new and the old form of organic chemistry was, in Kekulé's eyes, so striking that he set the new organic chemistry apart as an "independent science" while he ordered plant and animal chemistry to the "descriptive natural sciences." By the term *descriptive natural sciences*, he meant the natural historical orientation of the earlier plant and animal chemistry, which studied naturally occurring organic substances by extracting and describing them. In the new "independent science," the collection and description of naturally occurring substance species moved to the background while in the study of chemical "metamorphoses" or reactions, the general laws became central research objects.

Berthelot used a similar distinction between experimental sciences and observational sciences to summarize the differences between plant and animal chemistry and the new form of organic chemistry:[120]

> In fact, we here touch on a fundamental *characteristic which distinguishes the experimental sciences from the observational sciences. Chemistry creates its object.* This creative faculty, comparable to art, distinguishes it essentially from the natural and historical sciences (*sciences naturelles et historiques*). The latter have an object which exists in advance and independent of the will and activity of the scientist. . . .

Like his German colleague Kekulé, Berthelot also thought chemistry as a whole was changed by the transformation of plant and animal chemistry and the development of carbon chemistry. The key point about Berthelot's comment lies in the fact that not only did he understand observation and experiment as two different methods of gaining knowledge about the same objects, but instead he emphasized the difference between the kind of scientific objects at stake. From the comparative perspective of their sheer material existence, it may be said that in the observational sciences, the objects exist largely independent of the manipulations of the scientist, whereas in the experimental sciences, scientists not only epistemically but also materially produce their objects. According to Berthelot, with the development of carbon chemistry, the last remaining natural historical realm of chemistry had disappeared; chemistry as a whole had been transformed into an experimental science that produced a world of artifacts and, as the most creative of all sciences, was an experimental science par excellence.

Two Scientific Cultures

In their own reflections, organic chemists after midcentury highlighted discontinuities with the earlier plant and animal chemistry. Their judgments confirm my thesis that plant and animal chemistry before the late 1820s and carbon chemistry after 1840 were so structurally different that they should be considered two distinct "scientific cultures." I apply the term *scientific culture* to describe the relatively coherent relations between different social, material, symbolic, and epistemic elements and practices on the communal level of a science.[121] The contours of "scientific cultures" are not identical with disciplinary boundaries. University-based chemistry of the eighteenth century, for example, was practiced largely within medicine. Although it was not an institutionalized, autonomous discipline, there were so many outstanding, distinctive features of eighteenth-century European academic chemistry that it makes sense to speak of a coherent scientific culture or "life form" in the Wittgensteinian sense. To be named are, in particular, the interior of the chemical laboratory with its furnaces, vessels, tubes, distillation apparatus, and so on; the experimental gestures linked to typical chemical procedures, such as the mixing and stirring of substances or the regulation of furnaces; the kind of scientific objects, in particular the pure chemical substances that existed exclusively within the space of a chemical laboratory; the chemical terminology; the shared conceptual framework comprising concepts such as chemical compound, affinity, chemical analysis, and synthesis; the more specific chemical theories built within this conceptual framework, such as the theory of salts and the theory of phlogiston; scholarly journals publishing articles under the label "chemistry"; and last but not least the self-image of European chemists who considered themselves a coherent group, called themselves chemists, visited each other, and exchanged pure substances and other things. When one narrows the perspective and focuses on structural patterns within the scientific culture of European chemistry in the second half of the eighteenth century, two distinct areas stand out clearly: plant and animal chemistry or organic chemistry on the one hand and mineral chemistry or inorganic chemistry on the other. Again, it is not a clear-cut disciplinary boundary that marks off the two areas; rather, the differences consist in entire networks of activities, including objects of research, concepts, classifications, types of experiments, and so on.

When we further compare pre-1830 plant and animal chemistry and post-1840 carbon chemistry, additional disparities attract attention. The plant and animal chemistry of the eighteenth century and the early nineteenth century was a scientific culture that shared many features with the natural historical enterprises of the time. Its research objects overlapped significantly with those of natural history, its goals and methods of investigation were informed by natural history, and its classification of organic substances was natural historical. The plant and animal materials were natural historical objects to the extent that they were considered the smallest constituents of plants and animals. They were identified and classified in terms of natural history, that is, on the basis of their properties and their origins from botanical or zoological species. Their isolation and collection, with the careful avoidance of any transformation, and the observation and complete description of their properties were practices in accordance with natural history. Even the chemical operations by which the substances were extracted from plant and animal tissues could be considered a continuation of natural historical practices, namely, as anatomy by chemical means. Furthermore, the anatomy of plants and animals, the plant and animal physiology, and the study of pharmaceutically relevant components of plants and animals, like herbs, roots, or seeds, also belonged to plant and animal chemistry's scope of research. On the other hand, chemists also set specifically chemical accents in their research on organic matter and processes. The material culture of their laboratories differed from naturalists' sites; their explanations for physiological processes and for life in general made use of chemical categories, such as "chemical reaction," and depicted organs and living bodies as chemical vessels and chemical workshops. Furthermore, chemists were strongly interested in the pharmaceutical and commercial application of extracted plant and animal substances. Terms like *chemical natural history* and *pluricentered culture* of plant and animal chemistry are perhaps suited for expressing this pluricentered nature of the early form of organic chemistry. In contrast, the later form of organic chemistry, which was largely framed by experimentation and the experimental production of chemical artifacts, may be called an *experimental culture.*

In carbon chemistry, which developed from the late 1820s onward, the various activities and scientific objects were closely linked to experimentation. Experiments were not epistemically fundamental in a hierarchical sense; rather, experimentation constituted a space of reference that also involved epistemic and semiotic actions. As the case of Berzelian formulas makes particularly clear, in an experimental culture, scientists mobilize symbolic and other collectively available resources both for the physical production and for representation of their scientific objects. Carbon chemistry brought together two new types of experiments: quantitative analysis and the experimental study of chemical reactions and constitution. Furthermore, it aligned the identification and classification of organic substances with experimentation. The identification of organic compounds comprised testing their chemical properties with reagents, studying their composition through quantitative analysis, and investigating their constitution by studying their chemical reactions. Based on the analytical results, chemists constructed Berzelian chemical formulas to identify clearly and to demarcate different organic species; they then used chemical formulas as paper tools for constructing models of the constitution of chemical substances and of their reactions. By means of these formula models, they then constructed new classifications of organic compounds. Although the new mode of classification required a great deal of work on paper,[122] it was based in a double sense on experimentation: First, its smallest taxonomic units, the chemical compounds or species, were by and large produced through experimentation; second, the experimental study of chemical reactions and of the constitution of organic compounds was a precondition for building interpretive formula models, which chemists further transformed to establish relationships between the species of a genus and between higher taxonomic groups.

Chemists' interpretive formula models were also the starting points for their chemical theories of constitution, such as the theory of radicals in the 1830s and the type theory in the 1840s. Although the construction of these theories required additional work on paper, this kind of work remained closely linked to the experimental practice, at least until the 1860s. It never took on the extensive scope of work on paper, which characterized, for example, the mathematical deductions in nineteenth-century heat theory. On the whole, the theoretical elements within the experimental culture of synthetic carbon chemistry did not become as autonomous as in the physics of the period. This changed slowly with the emergence of stereochemistry in the mid-1870s and development of the French school of theoretical organic chemistry, which linked organic and physical chemistry, from the 1880s onward.[123] Thus, it is not surprising that Wilhelm Ostwald, Svante Arrhenius, and other "physical chemists," who in the 1880s advocated creation of a new theoretical foundation for chemistry, saw the organic chemists as mere empiricists lacking theory.[124] In sum, research within the form of organic chemistry that took shape in the 1830s and continued to thrive in the 1850s and 1860s was framed largely by experimentation, hence my characterization as an "experimental culture."

A few general remarks about "scientific cultures" might be helpful at this point. The connotations of the term *culture* are particularly useful for underlining the historical contingency and local diversity of the kind of scientific networks at stake here. There is no one privileged, universal pattern of a scientific culture; nor are the relations between its individual elements and practices absolutely fixed and temporarily stable. Scientific cultures are coherent in some way or another; yet nothing like complete fits of all individual elements or even logical consistency is presupposed here. Tensions and frictions may result, for example, from the different historical origins of the various items and from their individual relations to bordering scientific cultures and the broader social and cultural context. As a consequence, despite the relative stability and longevity of a scientific culture, stability is never complete, and shifts of elements and reorganizations of practices may occur all the time. For example, in the late eighteenth century, a few chemists concerned with plant and animal chemistry began to elaborate the quantitative analysis of organic substances. Quantitative analysis of organic substances, which was further developed from the 1810s onward, had historical roots independent from plant and animal chemistry (that is, stoichiometry) and was an activity that would have been dispensable from the perspective of an overall natural historical type of inquiry. Furthermore, already in the late eighteenth century, there were also a few experimental investigations of chemical reactions in organic chemistry, such as the reactions of alcohol and ethers,[125] which exceeded the natural historical horizon. All these practices contributed to epistemic shifts and ruptures, which gathered momentum in the 1830s, when chemists also extended Berzelian formulas from inorganic to organic chemistry and used them as tools for modeling. Moreover, the decline and disintegration of plant and animal chemistry and the emergence of the experimental culture of carbon chemistry exemplify that it would be wrong to assume that the heterogeneous elements of a scientific culture are completely determined by the overall system and hence can exist only within the boundaries of a particular system or culture. Experimental practices and research objects of plant and animal chemistry, such as extractions and extracted substances, shifted to the periphery of the new experimental culture or were transferred into other scientific subcultures without being completely erased. The focus of this chapter was on the comparison of patterns before and after the transition period between the late 1820s and the early 1840s, and so it was necessary to neglect any detailed discussion of such shifts, tensions, and ruptures. The resulting picture of two more or less static cultures, however, should not be confounded with the claim that scientific cultures actually are static systems deprived of any dynamic processes. This will become clearer in Chapters 4 through 8, where new conjunctions, shifts, and reorganizations in the transition period are scrutinized.

Experimental Cultures and Classification

Concluding, I wish to discuss an issue that has been repeatedly addressed by historians and philosophers of science: How are experimentation and classification related to each other? Carbon chemists after 1840 continued to occupy themselves intensively with problems of the classification of organic substances, and in this context they explicitly adopted terms from "naturalists," such as Georges Cuvier, Etienne Geoffroy Saint-Hilaire, and Augustin P. de Candolle. This raises the question of whether my distinction between a culture of plant and animal chemistry strongly shaped by its natural historical agenda and a culture of carbon chemistry circumscribed in a space of experimentation does not overemphasize the differences between the two forms of organic chemistry at the expense of their shared features. Are there not, from the perspective of other sciences of the period, such as physics, significant areas of agreement between both forms of organic chemistry? Several historians of chemistry argue that organic chemistry until the last two decades of the nineteenth century was natural historical in character as a result of its prevalent concern with classification. David Knight, for example, wrote that "natural history provided a paradigm for chemistry" in the nineteenth century.[126] Besides the adoption of terms from natural historians, the main argument seems to be that every type of classification is intrinsically natural historical.

In a revised version, Knight wrote: "But Rutherford's heirs have themselves been forced into classifying fundamental particles; and every science has to get its categories right, and thus involves classification. . . . the urge to classify is as basic as that to explain, part of that desire to make sense of things which make us human."[127]

The classification of objects, Knight argues in this later version, is a component of all sciences; indeed, classification, he suggests, is part of everyday life. Further arguments that classification in and of itself is not necessarily a natural historical activity are provided by the current philosophical debate about "scientific kinds."[128] For example, based on Thomas Kuhn's later works on scientific taxonomies, Jed Buchwald writes about classification in nineteenth-century physics: "Scientific practice (at least) is characterized by the separation of whatever scientists working in a particular area investigate into special groups, or *scientific kinds*. Physicists might examine kinds of light, or kinds of electric conductors, or, in general, any set of kinds that are thought to constitute a related group."[129]

Gaston Bachelard is one of the few philosophers who have dealt with the relationship between experimentation and classification in experimental sciences. In the introduction to his *Le Pluralisme Cohérent de la Chimie Moderne*, he points out:[130]

> The thought of chemists seems in fact to oscillate between pluralism on the one hand and reduction of plurality on the other. . . . The chemist multiplies in a certain way the "distinguo." It is through this work of experimental distinction that he creates or finds new substances. . . . Behind each pluralism one can recognize a system of coherence. This system is of course always more or less hypothetical. The hypotheses, assuming that they are fruitful, have after all two functions: the coordination of knowledge and impetus for new experiences.

Bachelard emphasizes the following characteristics of substance chemistry: the plurality of specific things and their permanent multiplication through experiments, the simultaneous reduction of plurality through the creation of a classificatory system, and the fact that these systems contain hypotheses that generate new experiments.

In the subsequent discussion, Bachelard makes clear that the experimentally manufactured particulars of classical substance chemistry were not only identified and classified by observing physical properties and chemical behavior with reagents but rather primarily through the study of chemical reactions.[131] The experimental study of chemical reactions allowed chemists to draw conclusions about the "composition" and "constitution" of the substances. Composition and constitution, in turn, were the basic criteria of classification in eighteenth-century and nineteenth-century inorganic chemistry and in cabonic chemistry after the 1830s. As Bachelard points out, the experimental study of the chemical reactions of a substance, however, went hand in hand with an exponential multiplication of things. Every study of a chemical reaction of a particular substance produced new substances: the reaction products. Even more, to ascertain the constitution of a specific compound, the study of one single reaction did not suffice. Instead, series of reactions needed to be studied whereby in the end, like in a puzzle, the interpretation of all individual experiments was required to produce a coherent image of the building blocks recombining in chemical reactions. Bachelard describes the network, both experimental and conceptual, in which the multiplication of things and their classification was (is) embedded, as follows:[132]

> All of *modern chemistry thus rests on a notion of composition ("composition"). The plan of combination ("combinaison") is further systematically expanded.* It is no longer at issue to study a body by way of its reactions with some other bodies, which are chosen as reagents (*réactifs*) because of the rapidity of their indications. Increasingly, *a great number of reactions need to be coordinated, compounds (composés) multiplied, and all possibilities for grouping need to be studied.* One can say that the bodies presented in chemical phenomena interest us primarily *as building blocks (pièces d'une construction).*

In the experimental cultures of substance chemistry, classifications were characterized by the experimental study of series of chemical reactions, interpreted and represented within a conceptual system comprising the chemical categories of composition and combination.[133] Hence, the "laboratory style" and the "taxonomic style" are not necessarily mutually exclusive styles of reasoning.[134] Furthermore, if Bachelard speaks of a systematic expansion of experiments and their coordination for classificatory purposes, he does not refer to local activities by individuals but to the coordinated work in transregional scientific communities. Series of experiments, which culminated in statements about the constitution of a specific organic compound and their classification, required communication, the exchange of substances and experimental devices, as well as corresponding institutional and other social conditions.

If we consider the type of scientific objects and the order of these things (or their "kind structure") as the decisive criterion for comparing scientific cultures, as Thomas Kuhn did,[135] it becomes obvious that there was a rupture between organic chemistry before the late 1820s and after 1840. The natural historical mode of identifying and classifying organic materials, based on their observable properties and their natural origin, did not spur chemists to distinguish between pure chemical compounds, mixtures, and complexes of organized substances. As a consequence, the objects identified and ordered in plant and animal chemistry were different from those in the later experimental culture of carbon chemistry. The former culture dealt with extracted materials of any chemical state (mixtures, complexes of organized substances, and pure compounds), the latter, as a rule, exclusively with pure chemical compounds. Furthermore, the natural historical mode of identifying and classifying organic materials referred exclusively to natural organic substances, produced by the organs of plants. The ordering of these things according to their properties and natural origin resulted in two taxonomic trees: the one representing the order of plant substances, the other that of animal substances—whose branches represented groups of phenomenologically similar materials stemming from the same plant or animal species or genus. In contrast, in the experimental culture of carbon chemistry, classification was based on knowledge of the composition and constitution of substances, gained by experimental investigations of their chemical reactions. "Natural origin" was no taxonomic criterion. Hence, naturally occurring and artificially produced pure compounds were grouped together according to similarities in composition and constitution. Only one taxonomic tree was created in this way whose branches represented groups of substances that could be phenomenologically different but had similar chemical constitutions.

CHAPTER THREE

Experiments on the Periphery of Plant Chemistry

> Alcohol and its numerous derivatives, ethers, acetic acid, etc. play a similar role in the new chemistry to that of the process of combustion in the chemistry of the previous century; that is, theoretical chemistry has developed through the explanation of the composition of these bodies.
>
> —LIEBIG 1861, 673

In plant and animal chemistry, experimental practice focused on the extraction of organic substances from plants and animals and the identification and description of these materials. Compared with the overwhelming dominance of this kind of experimentation, investigations of the chemical reactions of organic substances played almost no role at all. As a rule, chemists viewed both spontaneous and reagent-initiated transformations of organic substances as chaotic decompositions, which, after an experimentally uncontrollable cascade of organic intermediate products, yielded simple inorganic compounds. These decompositions were considered too complex for any construction of an interpretive building-block model representing the recombinations of the components of the initial substances into the reaction products.

A few exceptional experiments, however, were performed by French, German, and other European chemists, and these experiments did investigate organic chemical reactions, such as experiments on fermentation, the experimental manufacture of acids from extracted plant substances, and in particular those that studied the reactions of alcohol with various kinds of acids, yielding the so-called ethers.[1] In this chapter, I trace the development of the experiments with alcohol and ethers between 1797 and 1820. I also reconstruct the relationship of these experiments to commercial pharmacy and the ways chemists varied them and created extended series of experiments.[2] A historian of science studying the transformation of organic chemistry between the late 1820s and 1840 in some detail can hardly overlook the fact that experimental investigations of the reactions of alcohol, ethers, and other derivatives of alcohol played a prominent role. It seems as though chemists were obsessed with these experiments in the 1830s, which raises questions of why alcohol and ethers became so important in this period instead of any other possible group of substances and how these experiments developed. I argue in more detail in Chapter 8 that the function of the experiments with alcohol and ethers in the 1830s was the result of a contingent convergence of various factors. Only after these different factors had been associated did the experiments with alcohol and ethers move to the center of the experimental practice of organic chemistry. Then they began to play the dual role of model objects intended by the historical actors and an unintended paradigm for the emerging experimental culture of organic chemistry.

A particular focus of this chapter is on the material obstacles that confronted chemists in their rare attempts to study and represent chemical reactions of organic substances, such as alcohol and ethers, before they began to apply chemical formulas in the late 1820s. Thus, the following reconstruction of experiments is also part of my larger argument about the generative function of Berzelian chemical formulas. Within the traditional conceptual framework, studying a chemical reaction meant reconstructing or interpreting the recombinations of the constituents of the transformed substances into the reaction products.[3] The traditional strategy, which developed in inorganic chemistry from the early eighteenth century, was to isolate, purify, and analyze *all* reaction products and to compare their composition with that of the initial substances. In organic chemistry, there were three main obstacles to this strategy. First, it was often difficult to isolate all the reaction products, either because the produced amount of some of them was too small or because a suitable technique of isolation and purification was not available. The other two obstacles emerged after chemists had recognized that organic substances differ mainly in the quantitative aspects of their composition, and, as a consequence, that the construction of a building-block model of a chemical reaction requires information about the proportions of the elements constituting the organic compounds and about the recombining masses of the entire compounds. Before the invention of Liebig's "Kaliapparat"[4] and the stabilization of its application, the quantitative analysis of organic substances posed many difficulties and yielded uncertain results. An even more severe problem was measurement of the masses of the transformed and produced substances that could not be solved on the technical level of laboratory instruments.

In the following, I analyze and reconstruct how chemists attempted to remove these obstacles by varying or "accommodating"[5] experimental manipulations (if they did not give up, as in most cases) and by reducing the complexity of the measurement of the masses of reacting substances. To ascertain the number of the reaction products and to isolate all of them, chemists pursued the following main strategies. They varied the proportions of the initial substances to obtain a greater amount of a specific reaction product; they varied the temperature at the beginning and during the process; they applied carefully dried ingredients (this was a particular problem in the case of alcohol, which usually was mixed with some water); they introduced new solvents for specific reaction products or new reagents to form solid, crystalline products that were easy to isolate. My analysis, however, will show that chemists were often either unsuccessful in their endeavors to isolate *all* the reaction products, or their success was dubious. As to simplification of their measurement of the masses of the reacting substances, chemists tried to find a way out by focusing on measurement of the most important initial substance (such as alcohol) and reaction product (such as an ether). Yet the cost of this simplification was the introduction of new speculative hypotheses concerning the conservation of recombining constituents.

The following depiction of the construction of interpretive models of organic chemical reactions also reveals another aspect: the limits of the traditional mode of representing a building-block model of a reaction by means of verbal language. Here chemical formulas also offered an advantage by presenting the recombinations of the building blocks at a glance, at least for the experienced eye. Thus, we can distinguish two sites where Berzelian formulas eventually became extremely productive paper tools. First, they were used to obtain and stabilize the physical traces and inscriptions of the investigated invisible chemical reaction. On this level, chemical formulas took on the traditional functions of laboratory tools. Second, they were indispensable tools in the subsequent work of interpreting or modeling on paper, beginning with inscriptions stemming from quantitative analysis.

ETHER PRODUCTION AND COMMERCIAL PHARMACY

As so often was the case in the history of chemistry in the eighteenth and nineteenth centuries, experimentation investigating the reactions of alcohol and ethers did not begin in the context of scientific research but instead emerged from the world of commercial pharmacy. Made from alcohol and sulfuric acid, "ordinary ether" had a long history, going back to the sixteenth century and possibly even earlier.[6] In the early nineteenth century, ordinary ether belonged to the standard inventory of the pharmacist. Pharmacists also supplied other products deriving from the production of ordinary ether, such as "sweet oil of wine." Furthermore, pharmacists were highly interested in extending their stock of different kinds of ether produced with acids other than sulfuric acid and in simplifying and reducing the costs of their production. Ethers were thus first and foremost produced in the laboratories of pharmacists, and from here they gained entry into the laboratories of academic chemists.

In their influential paper on the formation of ordinary ether published in 1797, Antoine F. de Fourcroy and Louis Nicolas Vauquelin observed that "The preparation of ether is a complicated *pharmaceutical operation*, the results of which are as well known as its theory remains obscure."[7] Fourcroy and Vauquelin pointed to the fact that pharmacists' empirical knowledge preceded the theoretic inquiries of academic chemists and often stimulated further scientific research. Although the products of the ether manufacture were known to pharmacists and chemists, the "theory" of its formation, that is, the reconstruction and explanation of the invisible reactions underlying the phenomenological level of the operations, was still an open field of inquiry. Hence, academic chemists repeated, varied, and extended the pharmacists' production processes to achieve their explanatory goals and to improve the theory of ether formation.

It would be wrong, however, to assume that academic chemists pursued only theoretical subjects. The close entanglement of scientific research and commercial pharmaceutical practice entailed the fact that academic chemists, too, never lost sight of possible implications of their research for commercial practice. Alternatively, pharmacists took the results of scientific research and tried to make them fruitful for commercial manufacturing. For example, in their study on the formation of ordinary ether (also called *sulfuric ether*), Fourcroy and Vauquelin had divided the production of ether into three phases, claiming that the multiple by-products of ether formation were located primarily in the second and third phases (see later discussion). Based on this study, the Parisian pharmacist Pierre F. G. Boullay suggested in 1807 the practical use of this theory to avoid the unwanted by-products and thus to render the manufacturing of ether more economical:[8]

> The use of sulfuric ether is today quite extensive, and its consumption is considerable; it has become a true product of art produced on a grand scale. The operation, although it is quite simplified, still deserves attention, and it *seems it could be improved, especially in regard to economics* and in terms of the purity of the product. . . . According to the wise research and theory of Messrs. Fourcroy and Vauquelin, the attraction of sulfuric acid for water, with the help of heat, determines the transformation of alcohol into ether. This reaction of the principles of alcohol, which takes place under the influence of sulfuric acid, precedes the carbonization of the mixture, the formation of sweet wine oil, the release of

> sulfurous acid and other phenomena towards the end of the operation. . . . *Therefore, it would be advantageous to prevent or at least delay the appearance of these products* which announce the complete decomposition of alcohol.

Boullay then suggested that the alcohol be periodically replenished during the distillation "to preserve the proportions of ether formation" and extend the first phase of pure ether formation (ibid., 243). His suggestion was in turn taken up by the chemist Louis Thenard, who in his textbook suggested replacing the used-up alcohol from time to time to block the formation of by-products and increase the produced amount of ether. He added that "this has been confirmed by experience, and is exactly what is practiced by many pharmacists in the laboratories."[9]

Prevention of the formation of by-products in the manufacture of ether continued to interest both academic chemists and pharmacists in the following decade. When in 1827 Jean Dumas and Polydore Boullay—the former an academic chemist, the latter a pharmacist—introduced a new mode of representing the reaction procuring ordinary ether by applying Berzelian formulas, they also speculated about the possible pharmaceutical applications of their research and argued against the attempts of their colleagues to prevent the formation of economically unwanted by-products by adding magnesium peroxide or chromic acid:[10]

> One sees that there is no advantage in adding magnesium peroxide to the usual mixture for the production of ether, as has often been suggested. . . . the sweet wine oil would accompany the ether throughout the entire distillation process; . . . Therefore, until means are found to use fluorboric acid at a low price or to replace it, the current technique will remain preferable.

A further example of the close link between scientific-experimental and commercial ether production concerns the variations in ether formation. Ordinary ether was made with sulfuric acid, but pharmacists also tried to produce ether with other mineral acids as well as with acetic acid. When, for example, the chemist Louis Thenard in 1807 began a systematic study of such variations, he repeatedly and explicitly referred to the experience and concepts of pharmacists; his comments were both positive and negative.[11] In connection with his experiments on the production of ether with nitric acid, he wrote that the pure ether of nitric acid was a gas, not the ether-like fluid sold by pharmacists.[12] On the other hand, Thenard introduced the experience of pharmacists in the production of ether with acetic acid as an argument against the conviction of his academic colleagues that it was impossible to produce an ether with this acid: "But it is obvious that Schéele, Schultze, etc. are mistaken, since ether is made daily in the laboratories of pharmacies by the method used by Lauraguais and de Pelletier [with acetic acid]."[13]

We can conclude that academic chemists often based their experiments and interpretative models on the experience of pharmacists and that both pharmacists and academic chemists retained an interest in the commercial applicability of scientific research. This exchange was not fully symmetric. By and large, pharmaceutical practice was more important to academic chemistry than the reverse. Most of the laboratory instruments and many of the substances used in academic chemical laboratories were rooted in pharmacies. Moreover, the intertwining of scientific and commercial pharmaceutical practice stimulated chemists' interest in the investigation of the reactions of alcohol and ethers. It thus explains to some extent why chemists actually pursued this kind of inquiry, despite the fact that it did not belong to any of the three systematic areas of mineral, plant, or animal chemistry.[14]

INTERPRETIVE MODELS OF THE FORMATION OF ORDINARY ETHER AROUND 1800

In the first three decades of the nineteenth century, French chemists and a few German chemists made several attempts to reconstruct and explain the reactions underlying the production of ordinary ether. Although alternative suggestions were made, Antoine F. Fourcroy and Nicolas L. Vauquelin's theory became widely accepted after 1800, both inside and outside France.[15] This theory fitted the formation of ether into the order of reactions formerly developed in inorganic chemistry. In keeping with the traditional reasoning referring to inorganic reactions, their theory focused on the causal role of affinities rather than on quantitative aspects of the reaction. Fourcroy and Vauquelin's traditional verbal representation of the recombinations taking place in the reaction sheds some light on the historical and epistemological limits of this mode of representation in organic chemistry.

Structuring Organic Reactions

Ordinary ether was produced by distilling a mixture of alcohol and sulfuric acid. The results of this operation depended heavily on the proportions by weight of these two initial substances and the temperature at the beginning of and during the distillation process. Other than the main product, ordinary ether, many by-products occurred in the experiment, depending mainly on weight relations and temperature. Fourcroy and Vauquelin tried to explain not only the reaction that produced the ether, but also the formation of these various by-products. Because sulfurous acid was one of the by-products, many French chemists believed that ether resulted from the oxidation of alcohol by sulfuric acid, yielding sulfurous acid as the simultaneous reduction product. Carbonic acid and water were viewed as further oxidation products of alcohol, whereas the other by-products were considered decomposition products of alcohol. Fourcroy and Vauquelin argued against this common belief, suggesting an alternative theory that was based on a series of experiments, which they described in detail.

In these experiments, they systematically varied the conditions of the production of ether to clarify some questions concerning the number and identity of the reaction products and their occurrence during the distillation process. As a result, they found altogether eight reaction products: ordinary ether, sulfurous acid, "carbonic acid," water, "sweet oil of wine," acetic acid, "oil building gas" (or "olefiant gas"), and carbon, which they identified by their physical and chemical properties. In one of their experiments, however, which they performed using an unusual proportion of alcohol and sulfuric acid, they did not find any sulfurous acid. This result—together with their further observation that in the usual ether manufacturing process, sulfurous acid occurred only after the ether—became one of their main empirical arguments for rejecting the prevailing explanation of ether formation as an oxidation of alcohol by sulfuric acid and the simultaneous reduction of the acid into sulfurous acid. Fourcroy and Vauquelin stated that the entire chemical operation could be divided into three phases. According to this, during the first phase, a little ether was produced together with water and carbon without heating the distillation vessel; in the next phase, the temperature was increased and the entire amount of ether was distilled but without the simultaneous creation of sulfurous acid; only in the third phase was this acid observed (smelled), as well as sweet oil of wine, acetic acid, and carbonic acid; in addition, water was produced in all phases.[16] Based on their division of the distillation into three different phases, Fourcroy and Vauquelin went on to present their own version of the reaction.

Within the collectively shared conceptual framework of the time, any explanation of a chemical reaction had to reconstruct the recombinations of the components of the initial substances—in this case, alcohol and sulfuric acid—into the products of the reactions, eight in the case of ether production. This was not a simple task, given the fact that chemists represented these recombinations in verbal language. Moreover, to solve this puzzle, knowledge about the composition of the total of ten substances was necessary. Certain and reliable knowledge, however, did not exist, in particular not with respect to the composition of the organic substances. Acetic acid and alcohol had been formerly analyzed by Lavoisier, who had found that both of these two substances were made up of different proportions of carbon, hydrogen, and oxygen.[17] Fourcroy assumed, however, that Lavoisier's quantitative analysis of alcohol was not accurate;[18] in fact, even the qualitative composition of alcohol was uncertain.[19] Ether had also been analyzed, but Fourcroy and Vauquelin remarked only in passing that the comparative analysis of alcohol and ether (both substances are made up of carbon, hydrogen, and oxygen in different proportions by weight) confirmed their own reconstruction of the reaction without giving any details about the experiment and its quantitative results.[20] In the case of the so-called olefiant gas, Fourcroy and Vauquelin relied on the results of experiments performed a few years earlier by a group of Dutch chemists who had claimed that this substance consisted only of carbon and hydrogen.[21] Analogous experimental results did not exist for sweet oil of wine; its composition was completely unknown.

Nevertheless, Fourcroy and Vauquelin set out to reconstruct the reaction underlying the formation of ether based on their division of the operation into three phases and the available knowledge about the composition of the substances; they hoped their reconstruction of the reactions would also yield more detailed information about the composition of the substances. A third pillar of their reconstruction was the collectively shared knowledge about affinities (embodied in eighteenth-century affinity tables), in particular that sulfuric acid had a strong affinity to water, and therefore would attract the water contained in many chemical compounds. The three presuppositions led to the simple hypothesis that the chemical function of sulfuric acid was to withdraw water (or rather oxygen and hydrogen in the proportions which form water) from alcohol. Reconstruction of the reaction, however, became more complicated because of the simultaneous production of carbon. Thus, ether was not viewed simply as alcohol minus water. Based on the estimation that the mass of the precipitated carbon was greater than that of the hydrogen withdrawn from alcohol, Fourcroy and Vauquelin concluded that ether contained comparatively more hydrogen and oxygen than alcohol; it was "alcohol plus hydrogen and oxygen."[22]

As a result, Fourcroy and Vauquelin reconstructed and explained the formation of ether as a "displacement reaction"—a type of reaction belonging to the order of reactions established in inorganic chemistry—in which hydrogen and oxygen were withdrawn by sulfuric acid and the rest of the alcohol together with "unsaturated" carbon were released (or displaced by sulfuric acid). By ordering the formation of ether and water into the first two phases of the distillation process, Fourcroy and Vauquelin then ascribed the production of all other by-products to the last phase. In the last phase, in which the temperature was increased considerably, the substances remaining in the distilling flask would undergo reactions characteristic of organic substances, namely, stepwise decompositions.[23] In contrast to the formation of ether, which followed the ordered pattern of an inorganic displacement reaction, the later decomposition was comparatively chaotic, yielding a cascade of reaction products: sulfurous acid, water, sweet wine oil, acetic acid, olefiant gas, and carbon.

Despite the confusing number of reaction products, Fourcroy and Vauquelin reconstructed almost all the recombinations taking place in the last phase of the operation. According to this reconstruction, the creation of sulfurous acid was due to the decomposition of sulfuric acid at high temperatures, and the oxygen resulting from this decomposition recombined with hydrogen (from alcohol or ether) to form water.[24] The simultaneous production of sweet wine oil was explained as a synthesis from undistilled ether and carbon (from the previous ether formation). Hence, sweet wine oil was viewed as ether plus carbon. This was supported by the fact that less carbon was produced in the last period of the operation. Finally, the two chemists explained the creation of olefiant gas as the result of the decomposition of sweet wine oil into oxygen and the hydrocarbon caused by the high temperature.

Style of Reasoning: Arguing with Affinities

To a remarkable extent, the argumentation that accompanied Fourcroy and Vauquelin's representation of ether formation reactions focused on "affinities" rather than on certain knowledge about the composition of the substances. Invoking "affinities" was a style of reasoning established and practiced in inorganic chemistry, where the eighteenth-century affinity tables comprised and ordered affinities between pairs of different substances. Affinity tables, however, did not cover many of the recombinations at stake in the formation of ordinary ether. In particular, they did not cover recombinations between qualitatively similar or equal elements, or groups of them, which differed only in quantitative aspects. Furthermore, in their 1797 article, Fourcroy and Vauquelin did not pay much attention to exact quantitative considerations. Instead, they were satisfied with approximations about the quantities of recombining substances and the resulting quantitative composition of the reaction products.

Another striking feature of Fourcroy and Vauquelin's representation of the reactions for today's readers is the apparent limit of verbal language as a suitable mode of representation. If many reaction products occurred, as was usual in the case of organic substances, reconstruction of the recombinations of the components of the initial substances into the reaction products by using verbal language required many pages and often left the reader confused. Hence, it is not surprising that several of their contemporaries misunderstood Fourcroy and Vauquelin's theory of ether,[25] although it was also supported by many chemists inside and outside France[26] and became the most prominent theory in the first half of the nineteenth century. Dabit, a French pharmacist from Nantes, was one of the few who openly criticized the theory. In his attempt to rescue the older oxidation theory, Dabit focused on arguments concerning the affinities between the substances involved in the reaction.[27] Furthermore, he assumed that another reduction product of sulfuric acid was created in an intermediate state of oxidation between sulfuric and sulfurous acid[28] and that this explained why sulfurous acid was not observed at the beginning of the operation (when ether was produced). Critics like Dabit, however, did not attract much interest in the chemical community, in part because a pharmacist from Nantes was not very influential.

QUANTITATIVE APPROACHES IN THE STUDY OF ORGANIC REACTIONS

In his textbook, Fourcroy mentioned that it was possible to produce ethers using acids other than sulfuric acid, but he also asserted that these ethers were not different species but rather variations of one and the same ether.[29] Ten years later, a former student of Fourcroy and Vauquelin, Louis Jacques Thenard, took up this question and repeated experiments with alcohol and various acids, such as nitric acid, muriatic acid, oxidized muriatic acid (later chlorine), and acetic acid.[30] He concluded that different species of ether actually did exist. He also extended the series of experiments by using new vegetable acids as well as nonacid plant and animal substances.[31] Around the same time, Parisian pharmacist Pierre Boullay published the results of similar experiments, which by and large confirmed Thenard's claim that different species of ether could be produced by using different acids.[32] Both Thenard and Boullay also suggested new explanations of the reactions underlying the formation of the different ethers. These suggestions implied new analogies with inorganic reactions, in particular with the synthesis of salts. Thenard based his claims about the different species of ether and his reconstructions of the reaction on the careful quantitative analyses of the substances as well as attempts to measure the quantities of the reacting masses. Again, he was not the only one to undertake this endeavor; similar experiments were carried out by the Geneva chemist Nicolas Theodore de Saussure.[33] The emphasis of argumentation now began to shift to quantitative aspects of the composition of substances and the reaction, at the expense of arguments based on affinities.[34]

Quantitative Analysis

In their 1807 quantitative analysis of alcohol and ethers, Thenard and de Saussure used both the traditional technique of dry distillation and the combustion method introduced by Lavoisier and Berthollet; the latter technique was to become the successful route in the following decades.[35] They burned a previously weighed sample of organic compound by adding a determined volume of oxygen, thus transforming the compound into water and "carbonic acid" (carbon dioxide). After this, they measured the volume of the oxygen consumed and the carbonic acid produced; the proportion of the water produced was determined indirectly. From these data, the weight percentages of carbon, hydrogen, and oxygen contained in the organic compound could be calculated based on knowledge about the density of the two gases (oxygen and carbonic acid) and the quantitative composition of carbonic acid and water. If nitrogen was present in the compound, it was determined in an independent analysis with the same kind of procedure.[36] The combustion was performed in a special tube (called a *eudiometer*) over mercury using an electric spark to initiate the process. Such a technique was not only dangerous (because of possible explosions), but it also often did not yield results that satisfied chemists' expectations. It was improved stepwise, in particular by Gay-Lussac and Thenard's mutual work of 1810, by Berzelius's technique developed in the second decade of the nineteenth century, and by Liebig's introduction of the "Kaliapparat" in 1830, which was seen as a breakthrough by the chemical community in terms of both its accuracy and time demand.

In 1810, Gay-Lussac and Thenard substituted a new oxidizing agent—potassium chlorate—for oxygen.[37] They oxidized the sample with potassium chlorate in a vertical tube that was strongly heated at the bottom and collected the carbonic acid and excess oxygen over mercury. This method yielded satisfying analytical results for nineteen different organic substances, but it was time consuming and nearly as dangerous as the application of free oxygen. Jacob Berzelius transformed the Frenchmen's apparatus into a safer horizontal tube and slowed the combustion by diluting the sample and the oxidizing potassium chlorate with common salt. He also replaced the volumetric estimations by directly weighing carbonic acid and water after these two combustion products had been captured in the apparatus; water was bound by calcium

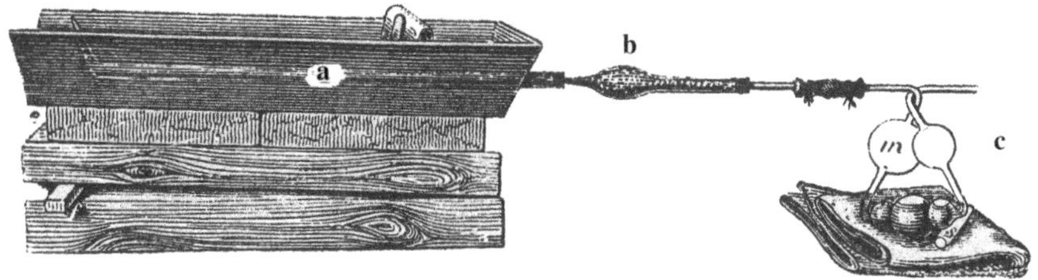

Figure 3-1 The sample mixed with copper oxide was oxidized in the glass tube a. Water was collected in the calcium chloride tube b. Carbonic acid (carbon dioxide) was collected in c, i.e., the "Kaliapparat" (a glass tube consisting of five bulbs, and containing a potassium hydroxide solution; after Liebig 1853).

chloride, and the gaseous carbonic acid was absorbed in a potassium hydroxide solution.[38] Berzelius's technique underwent modifications by Gay-Lussac, who suggested in 1815 the use of copper oxide as oxidizer, a substance that was far more stable physically. In the next fifteen years, the Gay-Lussac/Berzelius apparatus and procedure (with copper oxide as oxidizer) was used by French, German, and British chemists, such as Michel Chevreul, Jean Dumas, Johann Wolfgang Döbereiner, Andrew Ure, and William Prout. Finally, in 1830, Justus Liebig introduced a new version of the apparatus that became known as the "Kaliapparat."[39] As Alan Rocke pointed out, Liebig's apparatus allowed an increase in sample size, which meant higher accuracy of analytical results. "The increase in precision and ease of measurement was dramatic."[40] Large samples, however, yielded large amounts of carbonic acid gas, which Liebig managed to collect by their absorption in an array of five glass bulbs containing potassium hydroxide (Fig. 3.1).[41] The measurement and calculation principles remained the same as in Berzelius's method. The two products of oxidation—carbonic acid and water—were determined gravimetrically, and, if required, a second volumetric determination of nitrogen was performed. From these data, the weight percentages of carbon, hydrogen, and oxygen (determined indirectly) were calculated based on knowledge about the quantitative composition of carbonic acid and water.

Thenard's and Boullay's Variations of the Experiments

In his experiments on the production of ether from alcohol and nitric acid, Thenard wanted primarily to explore the question of whether different ethers were produced by using acids other than sulfuric acid.[42] He first produced an ether by mixing equal parts of alcohol and nitric acid and distilling the mixture. As a result, he obtained a series of different reaction products. In the distillate receiver, he found a liquid distillate, which pharmacists viewed as an ether, and in the distilling flask, he discerned a liquid substance, which he identified as a mixture of water, an unknown substance that carbonized easily, and untransformed alcohol and nitric acid. Furthermore, he observed the development of a gas, which he collected in another distillate receiver to be analyzed in additional experiments.

After his analysis of the liquid substance, which smelled like ether, Thenard was convinced that this liquid was not a pure ether but rather a mixture of different substances containing some ether. His attention was now drawn mainly to the gas, which had a similar smell and was a good candidate for an ether, although gaseous ethers were unknown up to that time. Because analysis of the gas did not yield constant proportions of the components, he concluded that it was also a mixture, namely, a mixture of the gaseous ether of nitric acid, nitrogen, gaseous nitrous acid, nitrogen oxide, and carbonic acid. In the next step, Thenard succeeded in isolating and analyzing the gaseous ether component. To ensure that the nitrogen found in his analysis was in fact a component of the ether (and thus that the ether was a new species, different from ordinary ether) rather than an admixture of foreign nitrogen, he performed the quantitative analysis of the ether. The result was affirmative; it showed that the ether was made up of 16.41% nitrogen, 39.27% carbon, 34.73% oxygen, and 9.59% hydrogen.[43] The repeated quantitative analysis of different samples, Thenard wrote, yielded identical results, "and consequently it was certain that the ether which I employed was homogeneous."[44]

Thenard's new claim, that the ether of nitric acid was a species different from the ordinary ether produced with sulfuric acid, contradicted not only his teachers' opinion but also the interpretations of a group of Dutch chemists who had already performed similar experiments. The Dutch chemists believed that the gas obtained from alcohol and nitric acid was not a new species of ether but a mixture of ordinary ether and gaseous nitrous acid. In his article, Thenard mentioned this objection, only to reject it immediately on the basis of his quantitative analysis. Contrasting the high "degree of exactitude" of his own experiments to the experiments of his opponents, he criticized their results by unmasking their "inexact" methods. The new relevance that arguments referring to the exactitude of chemical measurements had gained in the ten years after the publication of Fourcroy and Vauquelin's article also can be seen from a commentary by Guyton de Morveau, Vauquelin, and Berthollet—all members of the Société d'Arcueil, which saw quantification and precision measurements as two of their main goals.[45] In this commentary, the three scientists supported Thenard's view without reservation, praising his "patience and talent," his "ingenious and exact methods," and his "very delicate analyses."

Thenard's Representation of the Formation of the Ether of Nitric Acid

If, according to Thenard's results, the ether of nitric acid, or "nitric ether," contained nitrogen, and hence differed in composition from ordinary ether, it followed that the interaction of nitric acid with alcohol and the recombinations of the constituents of these two substances differed from the interaction of alcohol and sulfuric acid. Following the model of Fourcroy and Vauquelin, Thenard aimed to represent the formation of all eight reaction products: the ether, water, nitrous acid (both liquid and gaseous), nitrogen, nitrogen oxide, carbonic acid, acetic acid, and the substance that carbonized easily. As a result, he wrote:[46]

> It is evident that the oxygen of nitric acid combines with *a large part* of the hydrogen and *a very small quantity* of the carbon of alcohol. This results in: 1. *a large amount* of water, *a large amount* of nitrogen oxide, *a little* carbonic acid, *a little* nitrous acid and a little nitrous gas; 2. the separation of *a small quantity* of nitrogen and the formation of *a large amount* of nitric ether through the combination of *a sufficiently large quantity* of the two principles of nitric acid with the dehydrogenized and slightly decarbonized alcohol; 3. furthermore, the hydrogen and the carbon of alcohol also combine with the oxygen of nitric acid *in such proportions* that a little acetic acid is formed, and *a small quantity* of the material which carbonizes easily.

According to Thenard, the formation of nitric ether and the other reaction products was the result of a complicated, mutual decomposition of alcohol and nitric acid and a subsequent recomposition of some of their parts. First, a part of the oxygen of nitric acid combined with a part of the hydrogen of alcohol to form water; the other products of this first step were nitrogen oxide and nitrous acid (both nitric acid minus oxygen) and carbonic acid, which was a product of the synthesis of carbon (a decomposition product from alcohol) and oxygen (from nitric acid). In the subsequent step, nitric ether was formed by the combination of the remaining part of alcohol (consisting of carbon, hydrogen, and oxygen) and nitrogen and oxygen (stemming from nitric acid). Simultaneously, a part of the nitrogen of nitric acid was set free. The formation of acetic acid resulted from the combination of carbon, hydrogen, and oxygen, the latter two elements being decomposition products of alcohol and nitric acid.

Finally, the unidentified substance that carbonized easily was interpreted as a residue stemming from alcohol.

Although Thenard's verbal representation of the recombinations taking place in the formation of nitric ether was brief compared with Fourcroy and Vauquelin's representation of the formation of ordinary alcohol, the limits of verbal language for representing the rather complex recombinations of organic substances were obvious. Furthermore, given the fact that Thenard presented himself as an exact quantitative experimenter, his semiquantitative representation of the formation of nitric ether is astounding. It is even more surprising considering that he also reported about delicate measurements of several substances involved in the reaction. Why did he not use the results of these measurements for a more exact quantitative reconstruction of the reaction? The following analysis of the problems that confronted Thenard gives a possible answer to this question.

Obstacles to Measuring the Masses of Reacting Substances

In his experimental report, Thenard gave exact information about the masses of the initial substances used in the experiments and some of the reaction products. He used 300 g of alcohol and 300 g of nitric acid, and he measured the volume of the gas containing ether ("approximately" 12 liters). He also measured the fluid in the distillate receiver, which smelled like ether (75 grams) as well as the mixture remaining in the distilling flask (384 grams). In his eventual representation of the reaction, however, he did not use the results of these measurements. It very well may be that Thenard performed the measurements only to enhance his reputation as an exact scientist, but another possibility suggests itself: He stopped halfway because of insurmountable technical problems. The latter is especially likely because the measurements are in fact incomplete for any quantitative reconstruction of the reaction.

To represent the reactions in an exact quantitative way (within the conceptual building-block scheme of the time), it would have been necessary first to perform quantitative analyses of the two initial substances and of all eight reaction products and then to measure the masses of the initial substances that actually had been transformed in the reaction as well as the masses of all reaction products. From the combination of both kinds of information, conclusions could be drawn as to the masses of the building blocks of the initial substances recombining into the reaction products. Given the fact that one of the reaction products could not be identified qualitatively (the substance carbonizing easily), it is clear that Thenard had to put together a kind of puzzle with some pieces missing. This would not have prevented a solution of the puzzle if a sufficient number of pieces had been known; however, additional problems occurred. For example, referring to his attempts to isolate the different substances contained in the gas, besides nitric ether, and to measure their masses, Thenard wrote that the gas "contains several principles which I could not isolate completely."[47] The task of isolating different organic substances from a mixture (in particular from a liquid mixture) in a quantitative way, that is, without any loss of the mass of a substance, was technically not feasible. Given the fact that the experiment yielded two liquid mixtures—one in the distilling flask, the other in the distillate receiver—it becomes obvious that Thenard was confronted with problem after problem, until he gave up. We will see in one of the following sections that his colleague de Saussure undertook a similar endeavor, which was more successful. Yet the costs of his success and the fact that his strategy could not be generalized illuminate the deeper, fundamental problem of measuring the masses of reacting substances, which chemists eventually solved by using chemical formulas.

A Surprise

Thenard's second experimental report dealt with the formation of the ether of muriatic acid, or "muriatic ether," which he also viewed as a specific species different from ordinary ether. In his report, he described the following surprising event:[48]

> When, on February 18, I gave a lecture on the ether of muriatic acid at the Institute, all members of the Institute—Messrs. Berthollet, Chaptal, Deyeux, Fourcroy, Guyton, Vauquelin, Gay-Lussac, etc. considered the results to be very new; they were struck by the consequences which can be drawn. Mr. Proust, who is currently staying at Paris and for whom, in accordance with his wish, I repeated my experiments previously done at the Institute . . . completely shared the surprise, as well as the opinion of the French chemists.

What was so surprising about Thenard's new experiments, and what were their consequences? In contrast to the previously known ether production, the experiment with alcohol and muriatic acid yielded only two reaction products: an ether and water. The combustion analysis of the ether again yielded muriatic acid. Yet the means normally used for testing whether muriatic acid was contained in a compound (such as the addition of a solution of silver nitrate procuring the precipitation of white silver chloride) did not work. To decide whether muriatic acid was actually a component of the ether or, alternatively, an impurity, Thenard performed a series of additional experiments. The first one was to exclude the possibility that muriatic acid was synthesized during the combustion from oxygen and the "radical" of the acid contained in the ether (the collectively shared presupposition was that muriatic acid was an oxygen acid). To that purpose, Thenard measured and compared the volume of the muriatic acid consumed in the formation of ether with the volume of the acid set free in the combustion analysis. From these data, he calculated that 900 grams muriatic acid consumed in the synthesis approximately equaled 896 grams of muriatic acid produced in the analysis. Hence, Thenard concluded that "all of the elements of muriatic acid do exist in the ether and are combined with the elements of alcohol in the same way as the elements of water, carbonic acid, and ammonia, etc. exist in plant and animal materials."[49]

In the conclusion of his report, Thenard alluded to some broader consequences:[50]

> If it is possible to prove that muriatic acid preexists as a whole in the gaseous ether, we have created a compound whose existence could not be predicted by theory; perhaps we will find similar [compounds] in nature; and if we do find muriatic acid or other acids in our studies, [in places] where nothing causes us to suppose them, we will be cautious to conclude that the bodies we retrieve them from do not contain them.

What was not predicted by theory and difficult to imagine for Thenard and his contemporaries was the possibility that an aggressive mineral acid, such as muriatic acid, would combine with an organic substance rather than destroy it. Moreover, chemists now took it for granted that the plant and animal substances found in nature were made up of carbon, hydrogen, oxygen, and sometimes nitrogen. "Radicals" of mineral acids, such as chlorine, never had been found in these substances.

Analogies Between Organic Chemical Reactions and the Synthesis of Inorganic Salts

Shortly after this, Thenard published a second article on the formation of muriatic ether.[51] In the introduction, he mentioned that for three months he had been working daily for several hours studying the reaction. A particular problem to be solved was the origin of the second reaction product, water, in connection with the question of whether alcohol was preserved in the reaction or not. Either water was synthesized in the reaction—with the consequence that alcohol was decomposed to some extent—or it came from water mixed but not combined with alcohol (with the consequence that alcohol was preserved). Again, Thenard was looking for a quantitative method to decide between the alternatives. A month before, de Saussure had published his results of the quantitative analysis of alcohol.[52] Thenard took up these results; in addition, he performed a series of quantitative analyses of muriatic ether, aiming at comparing the quantitative composition of alcohol with that of the muriatic ether produced from it. If muriatic ether contained the elements carbon, hydrogen, and oxygen in the same proportions as alcohol, it was probable that alcohol combined with muriatic acid without losing water from its composition. As was usual at this time, the quantitative analysis of the ether turned out to be technically difficult; the results of the analysis and the subsequent comparison of the quantitative composition of alcohol and the ether did not allow unambiguous conclusions. Thenard summarized, "Nevertheless, considering everything, I am inclined to believe that alcohol [as a whole] or its elements are part of the composition of muriatic ether."[53]

Given that muriatic ether was the product of a simple synthesis of alcohol and muriatic acid, Thenard was seeking to answer another question by experimentation: Was it possible that alcohol and muriatic acid were preserved as integral building blocks within the ether rather than decomposing into their constituting elements? If the former actually was the case, the formation of muriatic ether would have been fully analogous to the synthesis of inorganic salts from an acid and a base, and muriatic ether would have a binary constitution analogous to the salts, the most characteristic and even paradigmatic inorganic compounds. This was a bold hypothesis that Thenard hesitated to state plainly, for organic substances were commonly held to be utterly different—in composition, constitution, and reactions—from inorganic compounds. His total of sixteen experiments, in which he aimed to separate either alcohol or muriatic acid from muriatic ether by mixing in any of the well-known reagents, again did not yield satisfying, unambiguous results. Contemplating the consequences if alcohol and muriatic acid actually did preserve their integrity while recombining into the ether, Thenard wrote:[54]

> If the muriatic ether were a combination ("*combinaison*") of muriatic acid and alcohol, as some people do not fear to claim, it seems that these two bodies had to unite like the acids and the alkalis; and consequently had to neutralize themselves as soon as they were in contact, for they had to be viewed as having a much stronger mutual affinity than muriatic acid and potash.

Thenard remained cautious about the analogy between the binary constitution of inorganic salts and that of muriatic ether. He alluded to some "other people" who did not "fear" to utter such a conclusion, however. By this he meant, the pharmacist Pierre Boullay, who shortly before had given a lecture at the

Paris Academy of Sciences on the very same issue—muriatic ether and its constitution—as well as on the ether of acetic acid.

In the introduction to this report, Boullay referred to Thenard's earlier experiments with muriatic ether, provocatively remarking that his own opinion considerably deviated from that of the "talented chemist."[55] Boullay asserted that he had decomposed muriatic ether as well as acetic ether into the respective acid and alcohol—these were the experiments that Thenard repeated without getting unambiguous results. He concluded from these experiments that the ethers of muriatic acid and acetic acid were "true compounds in the manner of salts, in which alcohol functions as a base."[56] Moreover, Boullay suggested dividing the known ethers into two groups: one containing the ether produced with sulfuric acid and phosphoric acid, the other containing ethers produced with volatile acids, such as the muriatic acid and acetic acid. In the first group, the acid caused the decomposition of alcohol without entering the composition of the ether; the second group comprised products of the synthesis of the alcohol and the acid.

In a third article, immediately following the second and published in the same volume of the journal, Thenard accepted Boullay's classification. Meanwhile, Thenard had performed additional experiments to decompose the ethers of nitric, muriatic, and acetic acid. The experiments with acetic ether confirmed Boullay's claim that this ether could be decomposed into alcohol and the acid. Thenard now generalized that the ethers of all three acids were products of the synthesis of the alcohol and the acid. He left the question open, however, as to whether these two initial substances preserved their integrity as components of the ethers or disintegrated into their elements.[57] Three years later, Boullay extended the series of ether production with various acids, describing two new ethers produced with arsenic acid and "fluoric acid." He ordered the former into the group of ethers created by decomposing the alcohol and the latter into the group of synthesis products.[58]

The Extension of the Series of Experiments—New Products of Synthesis

In the fall of 1807, Thenard reported to the Paris Academy of Sciences a series of further experiments to produce ethers. Here he had systematically used a new set of acids to produce ethers: the vegetable acids, such as acetic acid, oxalic acid, and acid of lemon.[59] These acids produced ethers if a mineral acid was added to the distilling mixture.[60] In a further series of experiments, Thenard succeeded in decomposing these ethers by distilling them with a solution of potash. As decomposition products, he obtained alcohol and the vegetable acid. From the results of both types of experiments—the synthesis of the ethers and their subsequent analysis in which the initial substances of the synthesis were recovered—he concluded: "There we are, these [ethers] are new combinations ("*combinaisons*") of a vegetable acid with an alcohol."[61]

What appeared to be an astonishing exceptional case a half year earlier—the fact that an organic substance such as alcohol combines with an acid—was now viewed as a new principle that probably could be extended to all kinds of plant and animal substances: "After this principle has been recognized [that vegetable acids combine with alcohol], there is no longer any reason that it could not be applied to animal acids. . . . perhaps, it allows us to combine all plant and animal substances, if not with all acids, then at least with the strong and concentrated ones."[62] Again, Thenard set out to study this possibility by repeating several experiments that had been performed previously by himself or by other European chemists. The first experiment was done with an oil produced from a mixture of alcohol and "oxidized muriatic acid" (later chlorine).[63] A year earlier, Thenard already had attempted to produce an ether by means of this acid. The oil created in this way was apparently very different, however, particularly in smell, from any ether.[64] Now Thenard studied whether this oil, although not a kind of ether, was analogous in composition to ethers and consisted of an as yet unknown plant substance and muriatic acid. At the same time, referring to his earlier experiments, he now clearly stated that alcohol "produced the same effect as part of this kind of compound as a true base that can be salinized."[65]

Although Thenard's experiment to decompose the oil with an alkali into muriatic acid and the plant substance did not yield an unambiguous result—even with strong alkalis only very small quantities of muriatic acid and of oxidized muriatic acid were produced, and produced very slowly—and although he failed to identify the plant component, he now stated that the oil contained muriatic acid as a component and that the muriatic acid "was intimately combined with the other substance."[66] He continued, it is "certain" that the unknown plant substance "neutralizes the acid in the manner of alkalis," thus expressing his growing conviction that there were analogies between inorganic and organic compounds.[67]

The next substance studied by Thenard was so-called artificial camphor, produced from oil of turpentine and muriatic acid. In contrast to the vast majority of chemists, who believed that "artificial camphor" was a decomposition product of oil of turpentine, Thenard now hypothesized that this substance was also a product of the synthesis of the oil of turpentine and the muriatic acid. His experiment, not described in detail, supported this view,[68] as did his further experiments with "essence of lemon," "essence of lavender," and "tannin," as well as with five animal substances, such as casein, albumin, and urea. Thenard concluded: "Hence, there are five plant materials and five animal materials which can combine intimately with the acids. Three among them . . . neutralize the acids as well as the strongest alkalis. The other seven form compounds with the acids, which are themselves acids, similar to the metal salts and several salts of earths."[69]

In the end, Thenard was fully convinced that there was a strong analogy between the inorganic salts and a large group of organic compounds. In analogy to the creation of salts, he explained the creation of these organic compounds as a synthesis, and their constitution as a binary one, made up of two immediate constituents, the acid and the organic component. Thus, he contradicted the prevailing view of chemists that organic and inorganic substances were utterly different, the organic substances being exclusively subject to decompositions in the laboratory made up directly from elements rather than from larger building blocks.[70]

SIMPLIFIED QUANTITATIVE EXPERIMENTS

In 1807, the same year Thenard began to perform exact quantitative experiments with various kinds of ethers, de Saussure also attempted to ground Fourcroy and Vauquelin's theory of ordinary ether in quantitative experimentation. De Saussure reduced the experimental complexity of the ordinary ether formation by focusing exclusively on the two most interesting substances, alcohol and ordinary ether, into which alcohol was transformed. He suggested that the comparison of their composition and the measurement of their reacting masses should yield more precise information about the reaction.[71]

The first step was the quantitative analysis of both alcohol and ordinary ether, which de Saussure performed for each of the two substances in different, independent ways, to obtain reliable results. A table containing the results of the different analyses facilitated comparison of the different data. From this comparison, de Saussure obtained the result that 100 parts by weight of alcohol consisted of 43.5 parts (by weight) carbon, 38 parts oxygen, 15 parts hydrogen, and 3.5 parts nitrogen. In comparison, ordinary ether contained 59 parts (by weight) carbon, 19 parts oxygen, 22 parts hydrogen, and "a little" nitrogen. The comparison of the composition of the two substances showed "that, given the same weight, the ether contained much more carbon and hydrogen than alcohol, but less oxygen."[72]

This result contradicted the assumption of Fourcroy and Vauquelin that alcohol contained relatively more carbon than ether—a hypothesis they made based on the observation that carbon precipitated during the ether production. "Undoubtedly one must ask," de Saussure wrote, "how it is possible that the ether contains more carbon than alcohol, although a part of this element [carbon] precipitates while the alcohol is transformed into the ether."[73] Because hydrogen and oxygen (forming water) were set free simultaneously, the only possible solution was that the proportions of these two elements were greater than the proportion of the precipitated carbon. To test this hypothesis, de Saussure set out to measure the mass of the alcohol transformed as well as the mass of the produced ether. Only if both the quantitative composition of the two substances and their reacting masses were known could conclusions be drawn about the quantities of the building blocks of alcohol recombining into ether.

A similar approach was taken in the same year by the French chemist Louis Jacques Thenard to reconstruct the reaction yielding nitric ether (see preceding). Thus, de Saussure confronted the same technical problem as Thenard: the complete quantitative isolation of the substances at issue from a liquid mixture. Thenard had failed to solve this problem, but de Saussure considerably simplified the task, focusing on only two substances rather than on all reaction products and initial substances. He invented an ingenious but also time-consuming procedure to measure both the alcohol consumed in the experiment and the ether produced. In the usual procedure, only a part of the alcohol was transformed into ether; the unconsumed part remained in part in the distilling flask, in part in the distillate receiver. De Saussure managed to transform the entire mass of alcohol, which he had determined beforehand, by a stepwise repetition of the distillation. First, the ether produced was separated by distilling ("rectifying") the solution contained in the distillate receiver. Then the solution remaining in the distillation apparatus—containing untransformed alcohol and sulfuric acid, in addition to the by-products of the reaction—was distilled again; the whole procedure was repeated three times until the alcohol was consumed. In each step of the procedure, de Saussure had to avoid carefully any portion of the substances being lost. How successful he actually was is hard to say on the basis of his published paper, which does not contain any information concerning these quantitative aspects of the experiment.

De Saussure finally presented the following results: 80 parts (by weight) of alcohol had been transformed into 38.75 parts of ether. He rounded up to 200 parts of alcohol and 100 parts of ether, concluding that 200 parts of alcohol had been transformed into 100 parts of ether and 100 parts of all other reaction products. Based on the results of the quantitative analysis of alcohol and ether, he then showed that 200 parts of alcohol contained 87 parts of carbon, 76 parts of oxygen, 30 parts of hydrogen, and 7 parts of nitrogen; the same operation for 100 parts (by weight) of ether yielded 59 parts of carbon, 19 parts of oxygen, and 22 parts of hydrogen. Balancing this result with the rest of the 100 parts of all other reaction products, de Saussure showed that these contained a total of 28 parts of carbon, 57 parts of oxygen, 8 parts of hydrogen, and 7 parts of nitrogen.[74]

From this result, he drew further conclusions about the reaction by which ordinary ether was created. First, the balance proved both that the relative content of carbon in the ether was in fact higher than in alcohol and that simultaneously carbon was set free from alcohol because the rest of the reaction products contained a total of 28 parts of carbon. Second, de Saussure stated that the weight ratio of oxygen and hydrogen in this remainder was 57 : 8, or approximately 7 : 1, that is, the same weight ratio of oxygen and hydrogen contained in water. He concluded, "One must concede that 100 parts of ether equal approximately 200 parts of alcohol minus 28 parts of carbon and 65 parts of water."[75] In his conclusion, de Saussure had deleted the nitrogen previously found in the analysis of alcohol to adjust the experimental results to his presupposition that Fourcroy and Vauquelin's theory of ether—which stated that ordinary ether was created by the withdrawal of water from alcohol, with a simultaneous separation of carbon from alcohol—was basically right. There was, however, a fly in the ointment, as de Saussure admitted frankly in his final remarks. Despite his enormous efforts, his quantitative results were only approximations. Seven years later, he published another article dealing with the same issue.[76] He then gave up the experimental way of measuring the masses of reacting substances, seeking the solution in a new mode of representing the composition of alcohol and ether. I discuss this in the next chapter, after completing the story of the expansion of the series of experiments performed with alcohol before 1814. These experiments contributed to de Saussure's new mode of representing the formation of ordinary ether.

THE MANUFACTURE OF AN ARTIFICIAL OIL IN THE CHEMICAL LABORATORY

In 1794, a group of Dutch chemists had created an oil from a gas, which in turn they had obtained in an experiment with alcohol and sulfuric acid to produce ordinary ether. It was well known by chemists of the period that at the end of the distillation of alcohol and sulfuric acid, a gas was produced because it often caused explosions, which destroyed the distillation apparatus. "This gas exhibits remarkable and curious properties," the Dutch chemists remarked.[77] To obtain larger quantities of it, they varied the usual procedure of manufacturing ordinary ether by simulating the conditions at the end of the distillation process; instead of equal parts by weight of alcohol and sulfuric acid, they used four parts by weight of sulfuric acid and one part of alcohol. In the subsequent testing of the chemical properties of the gas with several reagents, they observed that the gas was combustible but did not react with any of the reagents used—with the exception of one: the oxidized muriatic acid (later chlorine). It was this reagent, not often used in tests of chemical properties, that exhibited the unexpected phenomenon: "The only reagent that has produced an all the more curious effect, for the phenomenon is new and hitherto unknown, is oxidized muriatic acid . . . Equal parts of the inflammable gas and of the oxidized muriatic acid gas were mixed; we saw the mixture began to diminish its volume . . . drops of a thick oil, with the color of pearls, appeared at the surface."[78] Because of this effect, the gas was named *olefiant gas*. Based on analysis of its composition, the Dutch chemists also called it *gaz hydrogène carboné huileux* (*oily carbonated hydrogen gas*).

In his report to the Paris Academy three years later, Fourcroy emphasized the far-reaching consequences of the experiment:[79]

> What is so very important about this study for the general theory of science is *the new light it sheds on the formation of oil*; . . . on the nature and the composition of the vegetable oily bodies; . . . hence, *one is on the way to achieving the artificial composition of oil. It seems that for fabricating it by art, no plant material is necessary*, but only a carbonated hydrogen gas weighing 0.909; the latter is to be mixed with oxidized muriatic acid gas, and one can hope to achieve the production of this olefiant gas with mineral materials which are very carbonized.

Chemical experiments illuminating natural processes occurring in plants, chemistry on its way to the artificial production of natural organic materials from mineral material: This all sounded like bold fantasies in 1797. "Oils" were typical objects of the natural historical agenda of plant and animal chemistry, meaning natural materials extracted from plants or animals that share a set of characteristic observable properties. Thus, in the late eighteenth and early nineteenth centuries, the fact that an oil could be produced from a nonoil in the chemical laboratory was an unexpected event that challenged the existing chemical order. Moreover, chemists viewed olefiant gas as an inorganic or mineral decomposition product of alcohol because it was made up of only carbon and hydrogen. When the Dutch chemists set out to analyze olefiant gas, they had expected that it also contained oxygen, as did all the organic substances that had been analyzed before then. Yet, in their repeated analyses, they concluded that this was not the case; the gas consisted only of carbon, ranging from 74 to 80%, and of hydrogen, ranging from 20 to 28%.[80] Hence, like marsh gas, identified in 1776 by Alessandro Volta, it was viewed as an inorganic or mineral compound, which implied that a vegetable oil could be produced from a mineral substance. Fourcroy did not dare mention this in his textbook, nor did any other chemical textbook author at the time.

The experiments of the Dutch were repeated and varied by other chemists, in particular by the French—as was usual when someone claimed to have produced a new substance. For example, Vauquelin and Hecht investigated the question of whether the oil could be produced from oxidized muriatic acid and ordinary ether instead of the olefiant gas; they believed the oil they had obtained was identical to the oil produced from the gas.[81] In 1811, de Saussure repeated the quantitative analysis of the olefiant gas.[82] The results of this experiment would become highly important in 1814, when de Saussure again constructed a model of the reaction of alcohol with sulfuric acid, yielding ordinary ether. The latter study, as well as Gay-Lussac's work on the same topic in 1815, created a closer link between the series of experiments investigating ether formations and those studying olefiant gas and its reactions. In turn, the results of de Saussure's and Gay-Lussac's work propelled further research on both objects.

In 1816, the French chemists Pierre Jean Robiquet and Jean Jacques Colin picked up where these research results left off, again examining the quantitative composition of the oil produced with "chlorine" and the olefiant gas. They stated that the oil was the product of the synthesis of these two substances and that it consisted of equal volumes of both components.[83] In the same year, similar experiments were performed by Claude Louis Berthollet. His specific goal was to examine whether the oils produced with chlorine and different organic substances were identical or different species.[84] This question was further investigated by European chemists in the 1820s, particularly by the French. In 1832, Justus Liebig focused his attention on an experiment on the interaction of alcohol and chlorine yielding an oil. In his experiments, Liebig created and identified two new substances—chloral and chloroform—which themselves became new objects of research, along with the investigation of their reactions.

When Dumas took up these experiments in 1834, they became the unintended site of a new kind (and concept) of chemical reactions: substitutions.[85]

NEW ATTEMPTS TO BALANCE THE MASSES OF REACTING SUBSTANCES

In 1814, de Saussure again took up his quantitative experimental investigation of the formation of ordinary ether. This time, he renounced any attempt to measure the masses of the alcohol transformed and the ether produced in the experiment; instead, he determined a new way to represent the quantitative composition of alcohol and ether and to balance the masses of both substances with respect to the transformation of alcohol into ether. A year later, a similar approach was taken by Gay-Lussac, who also integrated his law of gaseous volumes. The conclusions both chemists drew about the quantitative aspects of the ether production reaction implied, however, an uncertain hypothesis. This contributed to the precarious status of their quantitative reasoning, which was put into question in 1819 by new experiments and subsequently was set aside by Gay-Lussac in 1820.

Chains of Representation: The Composition of Organic Compounds

In 1814, de Saussure suggested a new mode of representing the composition of alcohol and ordinary ether, embedded in a series of repeated and modified experiments, and drew new conclusions about the reaction underlying the formation of the ether. He repeated his quantitative analysis of the olefiant gas, which he had performed three years earlier, and he transformed the result, representing the composition in weight percentages, into the weight ratio 1:5.68 of hydrogen and carbon.[86] He also repeated his earlier quantitative analyses of alcohol and ether; the weight percentage of the elements contained in alcohol was 51.98% carbon, 34.32% oxygen, and 13.70% hydrogen.[87] De Saussure then transformed this first representation of the composition of alcohol by dividing the 13.70 parts of hydrogen into 4.55 plus 9.15 and combining 4.55 parts with the 34.32 parts of oxygen to form 38.87 parts of water (Fig. 3.2). When he combined the rest of 9.15 parts of hydrogen with the 51.98 parts of carbon, he obtained a striking result: There was approximately the same weight ratio of hydrogen and carbon as in olefiant gas, 1:5.68. Based on this, in a third transformation, de Saussure added 9.15 parts of hydrogen and 51.98 parts of carbon to form 61.13 parts olefiant gas. He concluded that "It can be easily seen that 100 parts of alcohol by Richter [i.e., dehydrated alcohol] are represented by the elements of 61.13 parts of olefiant gas and of 38.87 parts of water."[88] In his "chain of inscriptions"[89]—which were in this case different representations of the composition of chemical compounds—de Saussure had linked the results of two formerly rather independent series of experiments and created a relationship between the composition of alcohol and that of olefiant gas and water. He did not claim that alcohol had a binary constitution containing olefiant gas and water as integral building blocks—as Gay-Lussac did shortly afterward, as well as Jean Dumas and Polydore Boullay in 1827—but said more cautiously that alcohol can be represented as consisting of "the elements" of these two parts.

The quantitative analysis of ordinary ether and the analogous transformation of the mode of representing its composition led to another astounding result: 100 parts ordinary ether could be represented by 80.05 parts olefiant gas and 19.95 parts water.[90] Again, de Saussure transformed the representation of the composition of ordinary ether and alcohol, a step that was crucial for the subsequent new representation of the reaction yielding ether. He now calculated the weight ratio not by referring it to a constant weight of the compound, but to a constant weight of one of its components. The composition of alcohol was represented by 100 parts of olefiant gas and 63.58 parts of water, that of ether by 100 parts of olefiant gas and 25 parts of water. He concluded:[91]

> Since alcohol is represented by olefiant gas, combined with a quantity of water which is approximately half of the quantity of this gas—whereas ether is composed of olefiant gas united with water having a quarter of the weight [of the olefiant gas]—and olefiant gas does not contain any water, the action of sulfuric acid on alcohol to form ether or olefiant gas becomes obvious: in both cases it is restricted to subtracting the essential water from alcohol.

In a series of elegant transformations of the representation of the composition of alcohol and ether, de Saussure was able to confirm his 1807 statement on the formation of ether without any additional experiment.

In his new representation of the reaction of alcohol with sulfuric acid, de Saussure applied the method of balancing the masses of the initial substances and of the reaction products without experimentally determining the reacting masses. It is important to note that the way he did this, that is, without assuming theoretical combining weights (be they called *equivalents*, *portions*, *atomic weights*, or whatever), which had been introduced shortly before by Dalton and Berzelius,[92] required another kind of hypothesis. He had to presuppose that the olefiant gas was preserved in the reaction, qualitatively and quantita-

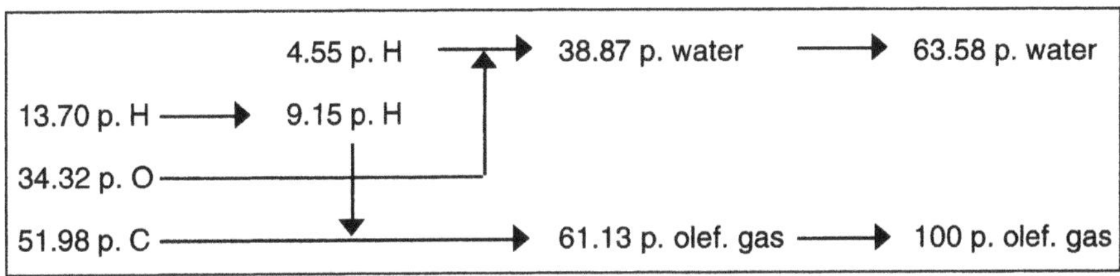

Figure 3-2 Simplified reconstruction of de Saussure's chain of representations of the composition of alcohol

tively. Thus, the idea was entrenched that olefiant gas was in reality an integral "immediate constituent" of alcohol and ordinary ether and that both substances had a binary constitution. This contradicted the earlier result—confirmed by the observation of a simultaneous precipitation of carbon—that in the formation of ether not only water but also carbon was set free.[93] De Saussure remained silent on this point—his trust in the new results of quantitative analysis was clearly strong enough.[94]

The Integration of Gay-Lussac's Law of Gaseous Volumes

In 1815, the French chemist Joseph Louis Gay-Lussac added another transformation in the chain of representations of the composition of alcohol and ether begun by de Saussure: He applied his law of gaseous volumes to the two compounds. More than ten years earlier, Gay-Lussac had formulated this law based on a series of experiments studying the quantitative relations of two reacting gaseous substances in terms of their volumes.[95] According to the law, the ratio of the volumes of two combining gaseous substances was always in small and integral numbers. If the density of a gas was measured, the law also gave information about the relative combining masses (by multiplying the density by the volume), or, alternatively, if the masses and densities were known, the volumes could be calculated. The latter possibility was realized by Gay-Lussac in the following way:[96]

> Mr. Saussure . . . has found that alcohol is composed in [parts by] weight of
> olefiant gas 100,
> water . . . . .63.58.
> If these weights are reduced into volumes, dividing the first by 0.978 (the density of olefiant gas), and the second by 0.625 (the density of water vapor), one finds for the composition of alcohol [represented] in volumes:
> olefiant gas 102.5,
> water . . . . .101.7.

> It is evident, because of the approximate identity of these two figures, that alcohol can be viewed as consisting of equal volumes of olefiant gas and water.

In his new representation, Gay-Lussac obtained a result that corresponded to an astonishing degree with the law of gaseous volumes. To add further empiric evidence, he compared the measured density of alcohol vapor, which was 1.613, with the calculated sum of the density of olefiant gas and water vapor, which was 1.603. Again, the approximate equivalence of the two figures was impressive. Gay-Lussac concluded that in the hypothetical synthesis of alcohol from equal volumes of olefiant gas and water, a condensation of two volumes into one took place.[97]

In the analogous transformation of the representation of the composition of ordinary ether, 100 parts by weight of olefiant gas and 25 parts by weight of water yielded 102.49 parts by volume (or "volumes") of the former and 40 parts by volumes of the latter. Although this result did not fit the requirements of the law of gaseous volumes as well as in the case of alcohol, in Gay-Lussac's eyes, it did not contradict this law. Hence, he brought the data into the round figures 100 and 50. He further confirmed this result by comparing the density of ether vapor with that of the sum of its two hypothetical components, concluding that "the vapor of sulfuric ether is made up of two volumes of olefiant gas and one volume of water vapor, condensed into one volume."[98]

The Assumption of a Binary Constitution of Alcohol and Ether

Based on these two new representations of the composition of alcohol and ordinary ether by the volume ratio of their two constituents, Gay-Lussac confirmed de Saussure's quantitative model of the formation of the ether, that "it can be seen, that in order to transform alcohol into ether, only half of the water contained in the former needs to be withdrawn."[99] Again, this model included the hypothesis that the component "olefiant gas" was preserved, qualitatively and quantitatively, in the reaction. It further reinforced the idea that olefiant gas and water in reality were integer building blocks of alcohol and ether and that both compounds had a binary constitution. Both assumptions were underdetermined by empirical evidence. The experimental observation that carbon was produced from alcohol simultaneously with ether in fact threatened both hypotheses.

In 1807 and 1808, Louis Thenard and Pierre Boullay had already assumed that alcohol as well as other plant and animal substances yielded binary compounds if they reacted with acids and that these binary compounds were fully analogous to salts, the then paradigmatic inorganic compounds. The compounds studied by these two chemists, however, were artificial products of the chemical laboratory. Boullay emphasized that they were "veritable products of art"[100] without speculating about their existence outside the laboratory; Thenard cautiously raised the question of whether analogous organic substances existed in nature.[101] This was now different because alcohol was considered to be a plant substance occurring also in nature outside the laboratory. The assumption that a truly "organic" substance had a binary constitution analogous to inorganic compounds and moreover that olefiant gas, viewed as an inorganic decomposition product of alcohol, was its immediate constituent, further blurred the border between the inorganic and the organic substances. Gay-Lussac even went a step further: He speculated about the consequences for the constitution of sugar and the reactions underlying its fermentation.

Jacob Berzelius, who was an unquestioned authority in the first two decades of the nineteenth century, responded critically.[102] In particular, he refuted the hypothesis that organic substances were "binary inorganic compositions," that is, made up of two inorganic compounds, such as water and olefiant gas. The vast majority of chemists outside France shared Berzelius's view. It was only after 1833, when Berzelius himself, as well as other leading chemists, revised their view about the constitution of organic compounds, that organic substances were commonly held to have a binary constitution, analogous to inorganic compounds.[103]

THE OVERTHROW OF THE ACCEPTED INTERPRETIVE MODEL OF THE FORMATION OF ORDINARY ETHER

In 1820, the *Annales de Chimie* anonymously published an article claiming that Fourcroy and Vauquelin's theory of the formation of ordinary ether, which had previously been confirmed by the quantitative experiments of de Saussure and Gay-Lussac, was wrong: "Today, Fourcroy and Vauquelin's theory of the formation can no longer be accepted. Sulfuric acid actually gives its oxygen to alcohol."[104]

It is very probable that Gay-Lussac, who edited the journal together with Arago, was himself the author of the pamphlet.[105] His revision was caused by the discovery of a new reaction product, called *sulfovinic acid*, which occurred at the beginning of the distillation process, simultaneously with the ether.

In his historical introduction, Gay-Lussac reminded his readers that the newly discovered product actually had a longer history and that already in

1799, a relatively unknown pharmacist named Dabit had attacked Fourcroy and Vauquelin's theory, asserting that sulfuric acid actually did relinquish oxygen to alcohol. Dabit had assumed that sulfuric acid was reduced to an acid in an intermediary state of oxidation between sulfuric and sulfurous acid that did not possess the characteristic smell of sulfurous acid.[106] His suggestion was largely ignored and subsequently forgotten. In 1819, two Germans, the pharmacist Friedrich W. A. Sertürner and the chemist Heinrich August Vogel, claimed they had found a new acid containing sulfur, which largely confirmed Dabit's assertion. They had repeated and further modified the ether experiment, in particular by varying the proportions of alcohol and sulfuric acid. Instead of using one part of alcohol and one part of sulfuric acid, Sertürner performed the experiment with one part of the former and two of the latter.[107] He believed that the acid he had found at the beginning of the experiment was a compound consisting of alcohol and sulfuric acid. In contrast, Vogel, who had repeated Sertürner's experiment, claimed that it was an intermediate acid between sulfuric and sulfurous acid combined with an unknown organic oil.

Vogel's interpretation was supported by Gay-Lussac, who repeated the experiment and made further modifications. He used the same proportion of alcohol and sulfuric acid but a different temperature; after a moderate increase in temperature, he began to distill the mixture only after sulfurous acid had appeared. He then isolated the new acid product from the residue contained in the distilling flask by adding a baryta solution. Additional experiments with the barium salt of the acid obtained in this way yielded olefiant gas and acetic ether, or, in another experiment, sulfurous acid and barium sulfate. Gay-Lussac concluded from these decomposition experiments that the acid, which he called *sulfovinic acid*, consisted of an organic component, very similar to sweet oil of wine, and the intermediate "hyposulfuric acid." Referring to the entire reaction, he wrote:[108]

> The result of the formation of ether seems to be ether, hyposulfuric acid, and a new oily plant material which is very analogous to sweet wine oil. In fact, a large quantity of the hyposulfuric acid is formed relative to the ether, and the sweet wine oil manifests itself only at the same time as sulfurous acid; this means, that it is very probable that the [latter] two bodies result from the decomposition of the sulfovinic acid. In order to be transformed into ether, alcohol has only to release hydrogen and oxygen in the proportions necessary to form water; but since sulfuric acid actually relinquishes oxygen to it, carbon would precipitate, and it is in sweet wine oil where we find it.

Despite his earlier, very accurate quantitative experiments supporting the theory of Fourcroy and Vauquelin, Gay-Lussac completely dismissed this theory in 1820 because a new acid had been found that seemingly contradicted it. According to his new view, alcohol was oxidized by sulfuric acid, yielding ether, hyposulfuric acid, and probably sweet oil of wine, and the latter two substances combined into sulfovinic acid. Gay-Lussac did not support his new theory with any quantitative experiment. Neither sweet wine oil nor sulfovinic acid was submitted to a quantitative analysis. This sheds some light on the precarious status of arguments derived from quantitative experiments in general but also on the uncertainties of the more specific arguments, previously derived from the comparison of the composition of alcohol and ether and the balancing of the masses. Gay-Lussac might have been well aware that a comparison of the quantitative composition of alcohol and ether, both represented in ratios of combining volumes, implied the speculative assumption that the proportion of olefiant gas was preserved in the reaction. Now a new qualitative aspect of the reaction sufficed to upset the delicate quantitative reasoning. It was only seven years later, when the chemist Jean Dumas and the pharmacist Polydore Boullay, the son of Pierre Boullay, again took up the topic, that quantitative reasoning was given a new and much stabler form.

CHAPTER FOUR

Paper Tools for the Construction of Interpretive Models of Chemical Reactions

> Organic chemistry now drives you completely crazy. It appears to me like a jungle in the tropics, full of curious things, an immense thicket, without entrance and exit, which one does not dare to enter.
>
> —WÖHLER 1835, in Wallach 1901, 1:604

Beginning in the late 1820s, Berzelian formulas took on various functions as paper tools for representing chemical reactions of organic substances, constructing models of their constitution, and classifying them. In this chapter and the following three chapters, I scrutinize these functions. Only these detailed analyses provide the historical evidence for the central hypothesis of this book: Berzelian formulas did not merely serve to illustrate already existent assumptions, but they were primarily productive tools for work on paper, which was intimately linked to experimental practice. Despite the degree of self-evidence these formula models and the experiments linked to them have for today's readers, from the current perspective, they were a bundle of truths and errors. In the following, I do not attempt to untie this knot for the sake of a historical reconstruction of the performances of the historical actors. I also want to point out beforehand that the emphasis of my reconstruction will not be so much on the particularities of individuals but on the collectively available resources and the newly emerging collective style of work. I wish to show that chemists in fact did apply Berzelian formulas as productive paper tools and how they performed this application.

Continuing from the depiction of the development of the experiments investigating the reactions of alcohol and ethers in the preceding chapter and of the obstacles chemists encountered while attempting to construct interpretive models of these reactions, this chapter demonstrates how Jean Dumas and Polydore Boullay for the first time used Berzelian formulas and removed these obstacles. At issue here is a form of constructing models so closely intertwined with experimentation that it is not an exaggeration to claim that the experiments themselves—in the production of traces of the research object, the processing of these traces, and the representation of the end result—included paper tools in the form of Berzelian formulas. Berzelian formulas, in their turn, presupposed quantitative chemical analyses of the substances at stake. They were products of the transformation of the analytical results into small integral numbers of "portions" (or "atoms") of elements in that the elemental composition of the substances expressed in weight percentages was divided by the theoretical combining weight (also "atomic weight") of each element. Thus, a complete historical reconstruction of the material preconditions of the study of chemical reactions of organic substances also would have to scrutinize the improvements in the methods of quantitative analysis during the first decades of the nineteenth century. Detailed historical work on this subject, which unquestionably deserves a whole monograph of its own, has begun only recently.[1] Therefore, here I can merely point out that the construction of quantitatively exact interpretive models of organic chemical reactions required two different kinds of tools and technologies, apart from the experimental investigation of the reaction: refined and stabilized quantitative analysis and Berzelian formulas.

THE HISTORICAL PROBLEM

As a consequence of the discovery of sulfovinic acid, Gay-Lussac in 1820 had questioned Fourcroy and Vauquelin's theory, according to which ordinary ether was formed from alcohol and sulfuric acid by the mere removal of water from alcohol. In doing so, however, he did not suggest a new quantitative model of the reaction. In 1827, Jean Dumas and Polydore Boullay returned to the problem of explaining the ether experiment. In the introduction to their experimental report, they provided a historical overview of the experiments that had been undertaken up to that point by their French and German colleagues, ensuring in this way the continuity of the collective enterprise. They pointed out that Fourcroy and Vauquelin's "theory of ether" had long gone almost unchallenged and was corroborated by de Saussure's and Gay-Lussac's comparisons of the quantitative composition of alcohol and ether. Subsequently, however, the theory had been rejected on the basis of the discovery of sulfovinic acid. Now the task was to find a new interpretive model of the reactions. As the two authors noted in their introduction, the multiplicity of reaction products complicated the problem: "The strongly varying transformations which alcohol undergoes due to the effect of different amounts of concentrated sulfuric acid represent one of the strangest phenomena of organic chemistry."[2]

The problem was in particular to explain which interactions and regroupings took place in the simultaneous formation of ordinary ether and sulfovinic acid and the subsequent formation of sweet wine oil.[3] Up to that point, the exact quantitative composition of sulfovinic acid was unknown; thus, the two chemists in their introduction pointed to Gay-Lussac's earlier supposition that it could possibly be a compound of hyposulfuric acid and an oily substance as yet not clearly identified. At the same time, even in the introductory passages, they left no doubt about the direction of their enterprise. Fourcroy and Vauquelin's explanation of ether production, they wrote, should be modified but by no means thrown overboard entirely:[4]

> Since the composition of alcohol and ether is well known, it is evident that Fourcroy and Vauquelin could have ignored the existence of hyposulfuric acid, and that they could have made a mistake concerning the production of sweet wine oil without it having to be the case that the cause to which they attribute the formation of ether ceases having to be true. The following experiments will give us the means to prove this most rigorously.

The two chemists were able to save the core of Fourcroy and Vauquelin's ether theory by showing that ordinary ether and sulfovinic acid were the reaction products of two simultaneously occurring but fully distinct reactions between alcohol and sulfuric acid and that sweet wine oil was a decay product of sulfovinic acid. In doing so, they ordered the cascade of reaction products by distinguishing between independent parallel reactions and successive reactions. To elaborate and demonstrate this, they applied Berzelian formulas and constructed a formula model of the formation of sulfovinic acid. This formula model was embedded in additional experiments, in particular the quantitative analyses of the substances. In the case of alcohol and ether, they repeated previous analyses; in the case of sweet wine oil and sulfovinic acid, they performed the first analyses. These analyses fulfilled a minimum requirement for the experimental embedding of their work on paper with formulas and the construction of formula models.

MODELING SEPARATE REACTION PATHWAYS

Dumas and Boullay's repetitions of the quantitative analysis of alcohol and ether, which they claimed to have carried out with "much more rigorous precision"[5] than ever before, confirmed the previous results of de Saussure and Gay-Lussac. The transformation of the composition in weight percentages into "theoretical volumes" led to the conclusion, in agreement with Gay-Lussac's presentation from 1815, that alcohol had a binary constitution from one "volume" of olefiant gas or "bicarbonated hydrogen," as the gas was now more often called, and one "volume" of water vapor. The analogous transformation of the analytical results of ordinary ether resulted in a model of its binary constitution from one volume of bicarbonated hydrogen and a half-volume of water vapor. The first quantitative analysis of sweet wine oil and the transformation of its composition in weight percentages into theoretical volumes resulted in a composition of four volumes of carbon vapor and three volumes of hydrogen. Furthermore, the result of the quantitative analysis of two samples of sulfovinic acid, which also was undertaken for the first time (in its combination with "baryta," that is, barium oxide, and barium sulfate determined separately by calcination) was 53.30% (54.00%) barium sulfate, 14.65% (14.85%) sulfuric acid, 11.32% (10.33%) carbon, 1.46% (1.39%) hydrogen, and 19.31% (20.00%) water.[6] This analytical result showed that the weight relation of carbon to hydrogen corresponded to that of sweet wine oil.

Based on these analytical results, Dumas and Boullay transformed the composition in weight percentages into a Berzelian formula as follows:[7]

> Baryta of sulfovinic acid is thus represented by one atom of hyposulfate, two atoms of sweet wine oil and five atoms of water
>
> $\dot{\mathrm{Ba}}\overset{\therefore}{\mathrm{S}}^2 + 2\,\mathrm{H}^3\mathrm{C}^4 + 5\,\mathrm{Aq}.$

Dumas and Boullay's formula for sulfovinic acid was a model of its binary constitution, which showed that this compound was made up of two complex "immediate constituents," the barium salt of hyposulfuric acid ($\mathrm{BaOS^2O^5}$) and sweet wine oil ($\mathrm{H^3C^4}$). It also showed the addition of water of crystallization in this binary compound, represented by the signs + and Aq.

Equipped with the formula for sulfovinic acid, Dumas and Boullay went on to construct a model of its formation reaction. This model enabled them to impose a chemical order on the confusing transformation processes taking place in the simultaneous production of sulfovinic acid and ordinary ether. It made possible the following division of the phenomenologically uniform productions of ordinary ether and sulfovinic acid into two independent parallel reactions:[8]

> The theory of ether formation is made much more simple by the following fact: the acid and the alcohol *divide into two parts, one of which forms sweet wine oil and hyposulfuric acid* as well as a certain amount of water in the following proportions:
>
> $$\left.\begin{array}{l}\text{2 at. sulfuric acid } 2\,\dddot{\mathrm{S}}\\ \\ \text{4 vols. of alcohol vapor}\end{array}\right\} = \left\{\begin{array}{l}\overset{\therefore}{\mathrm{S}}^2 + 2\ \mathrm{C}^4\mathrm{H}^3 \text{ or 1 at. sulfovinic acid}\\ \\ \text{2 vols. of water formed,}\end{array}\right.$$
>
> 4 vols. of water released.
>
> The other portions of acid and alcohol provide through their reaction dilute acid and ether.

According to this schema of balancing the masses of the initial substances of the reaction (alcohol and sulfuric acid) and the reaction products (sulfovinic acid and water),[9] Dumas and Boullay explained the simultaneous formation of sulfovinic acid and ordinary ether by arguing that the masses of alcohol and sulfuric acid divided into two parts, of which one part formed ordinary ether, and the other part sweet wine oil and hyposulfuric acid, which combined into sulfovinic acid. Hence, the simultaneous formation reactions of ordinary ether and hyposulfuric acid were entirely independent of one another. Agreeing with Gay-Lussac, the two chemists further proposed that in a reaction following these two parallel reactions, sweet wine oil and sulfuric acid were formed as decay products of sulfovinic acid[10] (see my reconstruction in Fig. 4.1). In particular, the distinction between two simultaneous but independent reactions in the first period of the experiment offered a solution to the problem, first raised in 1820, of explaining the simultaneous formation of sulfovinic acid alongside ether. This claim and its plausibility depended heavily on the balancing of the masses involved in the interpretive model of the formation reaction of sulfovinic acid. The balance of the number of element symbols C, H, O, and S representing the theoretical combining weights ("atom weights") of the elements on the left and right sides of the equation shows that the masses of the initial substances and the reaction products are equal. It followed from this that the formation reaction of sulfovinic acid was completely reconstructed, a result that was decisive in the given context. Thus, Dumas and Boullay could eliminate the possibility that additional, still-unknown reaction products existed. Moreover, they could exclude the possibility that the formation reaction of sulfovinic acid interfered in any way with the simultaneous production of ether.

At the end of their article,[11] Dumas and Boullay represented the reaction by way of the following equation, which instead of alcohol only shows its immediate constituent bicarbonated hydrogen H^2C^2:

$$2\,\overset{\cdot\cdot\cdot}{S} + 4\,H^2C^2 = \overset{\cdot\cdot}{\overset{\cdot\cdot\cdot}{S}}{}^2 + 2\,H^3C^4 + H\,\dot{H}$$

If the use of dots for oxygen is converted into the letter O, this is equivalent to the following:

$$2\,SO^3 + 4\,H^2C^2 = S^2O^5 + 2\,H^3C^4 + H^2O$$

This new form of representation not only fulfilled the task of balancing reacting masses; it was also an interpretive formula model of the reaction, showing how the components of the initial substances recombine to form the reaction products. The steps necessary to construct this model from the Berzelian raw formulas of the substances were, first, comparison of the formulas of the two

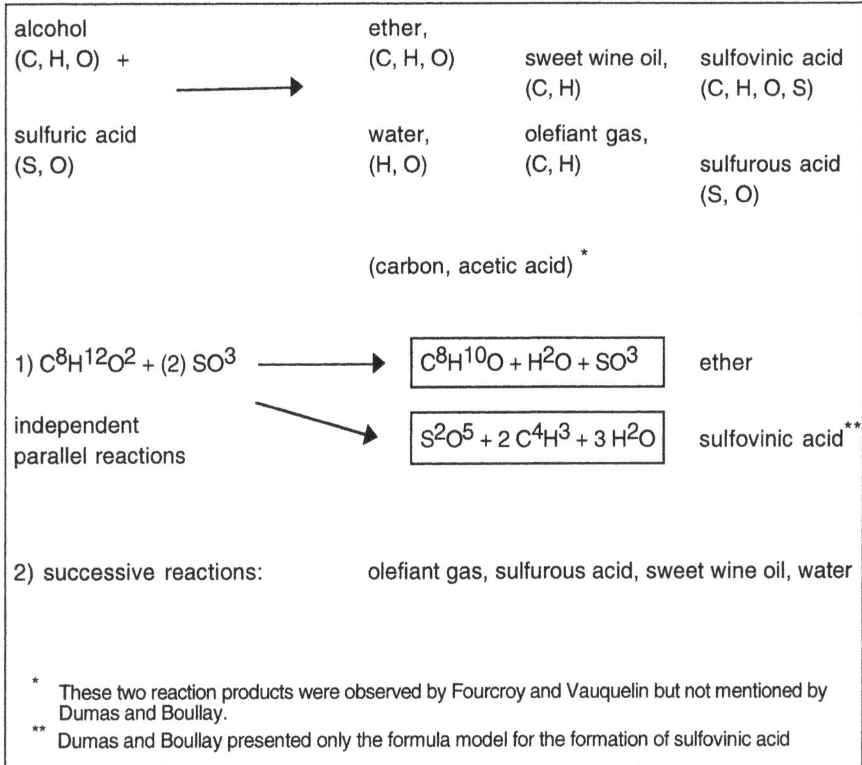

Figure 4-1 Dumas and Boullay's structuring of a cascade of reaction products into parallel and successive reactions (my reconstruction)

initial substances sulfuric acid (SO^3) and alcohol or bicarbonated hydrogen (H^2C^2) and the reaction product sulfovinic acid ($S^2O^5 + 2\,H^3C^4$); second, a multiplication of formulas in order to balance the masses; and, third, their regrouping as reconstructed in Fig. 4.2.

Expressed in words, Dumas and Boullay's formula model shows that during the formation of sulfovinic acid from alcohol and sulfuric acid, oxygen (one "atom" or "volume") is removed from sulfuric acid, forming hyposulfuric acid; at the same time, hydrogen (two "atoms" or "volumes") is removed from the bicarbonated hydrogen of the alcohol, leaving sweet wine oil; furthermore, water is synthesized from the removed hydrogen and oxygen; finally, all three substances combine to form sulfovinic acid. All these regroupings, which are awkward to describe by verbal language, were readily comprehensible in the formula model of the reaction. To a certain extent, the formula model was a visual building-block model, for there was a one-to-one correspondence be-

1)	$SO^3 + H^2C^2 = S^2O^5 + 2\,H^3C^4$	comparison of all formulas
2)	$2\,SO^3 + 4\,H^2C^2 = S^2O^5 + O + 2\,H^3C^4 + 2H$	multiplication
3)	$2\,SO^3 + 4\,H^2C^2 = (S^2O^5 + 2\,H^3C^4) + H_2O$	regrouping

Figure 4-2 Dumas and Boullay's steps for constructing the model of the formation reaction of sulfovinic acid (my reconstruction)

tween a symbol (letter) and the recombining chemical building block to which it referred (Fig. 4.3).[12]

In their comments in the text, Dumas and Boullay mentioned that, based on the results of the analysis of sulfovinic acid, a second, alternative constitution of this acid could not be excluded as a possibility, which would require a different explanation of its formation. At around the same time as Dumas and Boullay, Henry Hennell, a chemist in the London Apothecaries' Hall, was carrying out various experiments with sulfovinic acid. He concluded, however, that sulfovinic acid was a simple synthesis product made up of olefiant gas (from alcohol) and sulfuric acid, in which the acid was half saturated by the olefiant gas.[13] The assumption that sulfuric acid could combine in a simple synthesis with olefiant gas did not seem implausible to Dumas and Boullay because of an analogy to an experiment reported by Michael Faraday. Faraday had performed experiments with naphthalene and sulfuric acid; in his interpretation of these experiments, he claimed that naphthalene could combine directly with sulfuric acid.[14] Dumas and Boullay considered Hennell's alternative reaction model less likely, however, and at first cast it aside. Only a year later, they returned to it in the context of their proposed classification.[15]

Dumas and Boullay's 1827 article had solved a vexing problem in the study of the chemical reactions underlying the production of ordinary ether. Jacob Berzelius, who reported on their article in his annual report of 1829, recognized the quality of their work and remarked in particular about their analyses of alcohol, sulfovinic acid, and sweet wine oil that the two French chemists had "now finally made this subject clear."[16] This does not mean that all differences in opinion had been set aside, nor does it mean that there were no attempts at minor revisions of the reaction model accompanying the repetition and variation of the experiments in the years after 1827. Dumas and Boullay

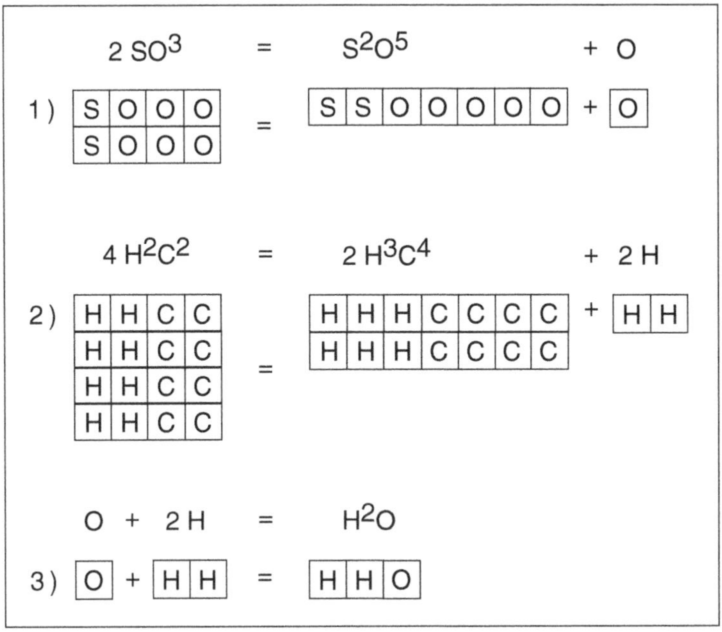

$2\,SO^3 = S^2O^5 + O$

1) | S | O | O | O | S | O | O | O | = | S | S | O | O | O | O | O | + | O |

$4\,H^2C^2 = 2\,H^3C^4 + 2\,H$

2) | H | H | C | C | H | H | C | C | H | H | C | C | H | H | C | C | = | H | H | H | C | C | C | C | H | H | H | C | C | C | C | + | H | H |

$O + 2\,H = H^2O$

3) | O | + | H | H | = | H | H | O |

Figure 4-3 The implicit building-block model of the regrouping in the formation of sulfovinic acid (drawn after Berthelot 1864)

themselves undertook such a revision a year later.[17] In 1833, Théophile J. Pelouze suggested a further possibility when he proposed the hypothesis that sulfuric acid first combined with alcohol—according to him, this compound was sulfovinic acid—and then decomposed during heating into ether, water, and sulfuric acid. This was, however, not a fundamental revision of Dumas and Boullay's model, as Johann Chr. Poggendorff remarked in a commentary on Pelouze's article: "The explanation always returns to the old hypothesis on the dehydration of alcohol by sulfuric acid which Fourcroy and Vauquelin derived from the composition of alcohol and ether."[18] A year later, Eilhard Mitscherlich explained the working of sulfuric acid as a "special kind of decomposition" in which the sulfuric acid removes the water from alcohol by "contact" (*Kontakt*), without itself undergoing a change.[19] In the same year, however, Liebig again declared unmistakably, "Ether is formed when alcohol is dehydrated by a powerful chemical affinity; about this there was never any doubt."[20] In the two decades after 1827, it was largely unquestioned that ordinary ether was a product of the analysis of alcohol formed via dehydration by sulfuric acid. It was only in 1850 that Alexander Williamson interpreted the formation of sulfuric acid ether as a "synthesis," but this time as a synthesis on the submicroscopic level of two molecules of alcohol with the removal of one molecule of water.[21]

THE PERFORMATIVE FUNCTION OF BERZELIAN FORMULAS

What role did Berzelian formulas play in the interpretation and representation of the formation reaction of sulfovinic acid and ordinary ether? Were the preceding formula models of the formation reaction of sulfovinic acid merely illustrations that served to represent results in a comprehensible way after they had been obtained by different means—as is still today assumed almost unanimously in the historiography of chemistry—or were they an essential research tool for constructing the models? I claim the latter for the following reasons. Dumas and Boullay's proposition that two different reactions underlie the simultaneous experimental production of ordinary ether and sulfovinic acid implied an important innovation. The two chemists asserted and proved that the two simultaneous reactions did not interfere with one another but rather were entirely independent. That two chemical reactions might take place simultaneously already had been stated occasionally before, for example, by Fourcroy and Vauquelin using verbal language. The more specific claim, however, which is especially relevant here, is that two simultaneous reactions of the same initial substances were entirely independent of one another. This claim rested on a specific precondition: the balancing of the masses of the initial substances and reaction products for at least one of the reactions at stake. In the collectively shared intellectual framework, only a balancing schema could prove that a reaction was completely represented and thus independent of a simultaneous reaction of the initial substances.

My argument is that, as a rule, balancing schemata of organic reactions could not be constructed based on the measurement of the masses of the reacting substances but required the acceptance and application of theoretical combining weights such as those represented by Berzelian formulas. In the great majority of organic chemical reactions, it was technically extremely difficult, if not entirely impossible, to isolate in quantitatively complete form the masses of all reaction products and the remaining masses of the initial substances. In Chapter 3 of this book, in which I analyzed early attempts at such isolation procedures and subsequent measurements, I discussed this problem. At issue here is a historical problem of obtaining the experimental traces necessary to represent a chemical reaction of organic substances in an exact quantitative fashion.[22] As a rule, there was no fundamentally different way to construct exact balancing schemata in organic chemistry other than using chemical formulas (Berzelian formulas or alternative symbolic systems, if the latter had been historically available and collectively accepted). This presupposes that chemists intended the construction of exact models of organic reactions rather than quantitative approximations.

Berzelian formulas provided the necessary information about the masses of the reacting initial substances and the reaction products in theoretical form. They also gave exact information about the quantitative composition of the individual substances. Because any overlapping in quantitative composition was ruled out—in contrast to the results of quantitative analysis—Berzelian formulas made possible the clear identification of the substances involved in the reaction.[23] Furthermore, Berzelian formulas were especially suited to the work of modeling on paper a chemical reaction because they were very easy to manipulate. The only syntactic rule that had to be followed was the additivity of the symbols. The alternative interpretations of a reaction could be worked through by sketching out on paper the possible recombinations of symbols in the formulas, that is, by removing (subtraction) and inserting (addition) individual symbols in formulas and by multiplying or dividing formulas to balance the masses of the denoted substances.[24] Dumas and Boullay's formula model of the formation of sulfovinic acid also exemplifies that additional hypothetical information could be obtained from the final formula model. The model shows that water was produced as a second synthesis product along with sulfovinic acid. This additional information was provided only by the model. In principle, it was experimentally verifiable, but in this case such experimental verification was not possible because water was formed simultaneously in two other ways in the experiment (as a mere decomposition product during formation of sulfovinic acid and in the parallel reaction of ether formation).

Dumas and Boullay's article is among the first publications in which the chemical formulas introduced by Berzelius were used to complete the experimental data and to model chemical reactions. In fact, the European chemical community accepted Berzelian formulas rapidly in the 1830s and 1840s, after further paradigmatic cases of the uses of chemical formulas for the construction of reaction models had been published. In the preceding analysis of the performative function of Berzelian formulas, I was not concerned with Dumas and Boullay's various ways of tinkering with chemical formulas that might have preceded the publication of their paper. This detail will be left to future analyses of the two chemists' manuscripts. I was interested in the more general sub-

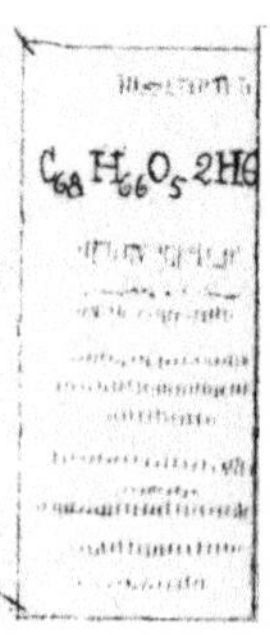

Figure 4-4 A mid-nineteenth-century chemical laboratory containing tables with Berzelian chemical formulas (from Morfit 1850, 24)

ject of the practical function of Berzelian formulas in chemists' experimental investigations of organic chemical reactions. This epistemological focus might become clearer through comparison with the role played by mathematical paper tools in modern physics. It would hardly occur to us to claim that, as a rule, the mathematical formulas with which physicists represented (and represent) their invisible research objects in their published papers are added after the fact—thus serving merely as a means of illustration and rhetoric in a strategy of gaining acceptance—whereas their "investigative pathway"[25] as reconstructed from the manuscripts was fashioned from entirely different resources. In the emerging experimental culture of synthetic organic chemistry, chemical formulas had a similar epistemic status as mathematical formulas for the genesis and shaping of the research objects of modern physics. They were essential paper tools both for securing experimental traces and for the subsequent work of constructing interpretive models of the chemical reaction.

The increasing use of Berzelian formulas after 1827 had a strong impact on chemists' style of reasoning, which I want to mention briefly. In the eighteenth century and early nineteenth century, chemists argued for their interpretations of chemical reactions by invoking assumptions about affinities between pairs of chemical substances. In contrast, in their model of the formation reaction of sulfovinic acid and their distinction of parallel reactions and successive reactions, Dumas and Boullay relied entirely on the evidence of their balancing of the reacting masses; arguments based on affinities played no role here. Whereas little more than a decade earlier, de Saussure, for example, occasionally advanced ideas about affinities in addition to his balancing of reacting masses, the word *affinity* or terms that referred to the related causes of chemical reactions do not appear in the work of Dumas and Boullay. This phenomenon stabilized in the 1830s as part of a collective style of reasoning. In light of the persuasiveness of the balancing schemata of chemical reactions, in particular the mathematical form of equations, which were the balancing form of first choice, arguments concerning affinities were obviously seen as superfluous. *Affinity* disappeared almost completely from texts and for a certain period was lost as a possible research object. This changed again with the introduction of the valence concept in the 1850s.

CHAPTER FIVE

Paper Tools for the Classification of Organic Substances

> A world may be unmanageably heterogeneous or unbearably monotonous according to how events are sorted into kinds.
>
> —GOODMAN 1978, 9

> Order is, at one and the same time, that which is given in things as their inner law, the hidden network that determines the way they confront one another, and also that which has no existence except in the grid created by a glance, an examination, a language.
>
> —FOUCAULT 1997, XX

> Classification is usually treated as an outcome of an ordering process as if the organisation of thoughts comes first, and a more or less fixed classification follows as an outcome. But the ordering process is itself embedded in prior and subsequent social action.
>
> —DOUGLAS AND HALL 1992, 2

In the 1820s, the problem of identifying and classifying organic substances, which various European chemists already had addressed around 1800, became more urgent. There are various reasons for this. The number of plant and animal materials mentioned in Fourcroy's *Systême* (approximately thirty) had grown enormously as a result of the refinement of extraction techniques and the production of chemically transformed organic substances. After the experiments with opium performed by the German pharmacist Friedrich Wilhelm Sertürner in 1805, German and French chemists isolated many new alkaline plant substances, such as morphine, narcotine, strychnine, quinine, cinchonine, caffeine, and codeine.[1] Michel Eugène Chevreul's experiments with fats of various plant and animal origins had enlarged the group of vegetable acids to include margaric acid, oleic acid, stearic acid, butyric acid, capric acid, caproic acid, and delphinic acid (today known as *valeric acid*).[2] Chevreul was also one of the chemists who isolated new dyestuffs and pigments, such as quercitrin and morine, from the wood of various trees. To this partial list of new "natural" organic substances isolated during the first three decades of the nineteenth century, we need to add coal tar products and the many new chemically transformed organic substances, such as "ethers," which did not exist in nature outside the chemical laboratory.

In eighteenth-century plant and animal chemistry, organic substances were identified and classified on the basis of their natural origins and observable properties. These criteria did not allow a clear distinction between species of organic substances, as chemists had already observed in the early nineteenth century, because overlapping in the sets of properties of various substances was unavoidable. Furthermore, chemists had also raised the question of whether the existing classification referred to genera, species, or varieties of species of substances.[3] By the end of the 1820s, problems of this kind had increased. For example, a clear demarcation among the dozens of known vegetable acids required additional criteria besides their observable properties and natural origins because several vegetable acids could be found in multiple species of plants. In the case of the rapidly growing class of substances manufactured in the chemical laboratory, such as the derivatives of alcohol, the overlap of properties was often even greater. In this case, it was obvious that the traditional classification by "natural origin" was not applicable. Moreover, many chemists were uncomfortable having the identification and classification of organic substances rest on entirely different criteria than those found in inorganic chemistry, where substances were classified on the basis of experimentally obtained knowledge about their chemical composition and constitution.

Contrary to all expectations, the first attempts at the classification of organic substances based on knowledge of their quantitative composition, which accompanied the elaboration of methods of quantitative analysis after 1810, brought no satisfactory results. On the contrary, they made the problem of identification and classification worse, as Berzelius made unequivocally clear in his monograph about the theory of chemical proportions or "portions."[4] His first precise quantitative analyses had shown that the analytical results, expressed in weight percentages of elements, overlapped in the cases of various organic substances. Not only was a distinction between different *species* by quantitative analysis a problem, but in extreme cases, the composition of two different substance *genera* differed so marginally from one another that, according to the standards of exact quantitative analyses for inorganic compounds, they would have to be considered identical substances.

In 1828, Jean Dumas and Polydore Boullay proposed for the first time a new, alternative classification of organic species, celebrated later as a breakthrough by the community of chemists. This classification was based on the composition and "constitution" of organic compounds, that is, the invisible grouping of their constituting elements into "immediate constituents."[5] Based on the work of de Saussure and Gay-Lussac on the binary constitution of alcohol and ordinary ether and on that of Chevreul on the binary constitution of fats,[6] as well as on their own experiments with ethers and other alcohol derivatives (analyzed in Chapter 4), Dumas and Boullay made the daring suggestion that all the organic substances they had grouped together were not directly composed of the elements carbon, hydrogen, and oxygen but shared the same binary constitution. The common denominator of the new class of binary compounds, which was given the name *compounds of bicarbonated hydrogen*, was the immediate constituent bicarbonated hydrogen. The differences between the species were explained by the differences in the second constituent. This new mode of classification assimilated the classification of organic substances to the classification of inorganic ones.[7] Dumas and Boullay increased the plausibility of the implied analogy between the constitution of organic compounds and inorganic ones by constructing the more specific analogy between the new class and a special class of inorganic compounds, the salts of ammonia.

In retrospect, Dumas and Boullay's work was the first extensive classification of organic species based on principles of classification that required the experimental study of the chemical reactions of substances as a precondition. This intertwining of experimenting and classifying became characteristic of the experimental culture of carbon chemistry that began to emerge in the late 1820s. Justus Liebig, who was often critical of Dumas's work, recognized the importance of the accomplishments of his French colleagues: "This work was the first in the field of the new organic chemistry, where it was attempted to link with a shared tie one series of bodies with another; in this light, it marked the beginning of the current era, and is important for the history of chemistry."[8] Decades later, Liebig stood by this judgment. In the *Handwörterbuch der Chemie*, which he edited together with Friedrich Wöhler and Johann Poggendorff, he wrote, "The first attempt to explain the constitution of organic compounds by assuming well-defined groups of atoms as components of these compounds was made by Dumas and Boullay."[9] Similarly, August Kekulé, one of the most prominent exponents of synthetic carbon chemistry in the 1860s, remarked that Dumas and Boullay's classification was "the first grouping of a large number of derivatives" in organic chemistry.[10]

Dumas and Boullay presented their new class of organic substances in the form of a table containing the formula models of the constitution of the classified substances. They included this table at the end of an article in which they reported primarily the results of new and repeated experiments. Thus, even a fleeting glance reveals that the new classification system was closely tied to experimental practice and required work on paper with chemical formulas. In the following, I analyze the structure of the table, the experiments, the construction of formula models of binary constitution based on these experiments, and the additional work on paper involved in classification. My analyses reveal that Berzelian formulas were essential paper tools in the enterprise of classification. Dumas and Boullay's experiments, which were limited to quantitative analyses (with one exception), concluded with formula models of the binary constitution of the organic compounds, which they constructed by manipulating formulas on paper. These formula models were then the immediate points of departure for the classification of the substances. The analogies of constitution on which the classification was based, both the analogy of the classified organic compounds with one another and the analogy to the ammonium salts, rested exclusively on the mutual fitting together of the formula models. To achieve that fit, the formulas had to be manipulated again in organizing the final table.

THE STRUCTURE AND FUNCTION OF DUMAS AND BOULLAY'S TABLE

The majority of the total of thirteen organic compounds included in Dumas and Boullay's table (Table 5.1) are the so-called ethers, obtained from alcohol and an acid. Also included are alcohol, two additional alcohol derivatives, a substance known as "ethal," which Chevreul had obtained from fats, as well as two vegetable substances in the narrower natural historical sense, namely, two species of sugar. The table expresses the conviction of the two chemists that the constitution of the classified organic compounds is analogous to the binary constitution of ammonium salts (according to the common view of early nineteenth-century chemists, ammonium salts were composed of the base ammonia and an acid). Hence, every formula of an organic compound is preceded by the formula of the analogous ammonium salt (except the last three formulas), which helps the reader to grasp the postulated analogy by way of a direct comparison.

With their hypothesis that the ethers were binary compounds, analogous to the inorganic salts, Dumas and Boullay followed a French tradition that went back to the work of Louis Thenard and Pierre Boullay in the first decade of the nineteenth century.[11] New here was the claim that olefiant gas, or bicarbonated hydrogen, was the immediate constituent of all species of ether without exception. New was also the assumption that sulfovinic acid and the hypothetical oxalovinic acid, as well as ethal and the two listed species of sugar, were

NOM DU COMPOSÉ.	BASE.	ACIDE.	EAU.
Hydro-chlorate d'ammoniaque........................	Az H^3	2 H Ch	
Hydro-chlorate d'hydrogène bi-carboné (éther hydro-chlorique)...............................	2 H^2 C^2	2 H Ch	
Hydriodate d'ammoniaque.....	Az H^3	2 H I	
Hydriodate d'hydr. bi-carb. *(éther hydriodique)*............	2 H^2 C^2	2 H I	$\dot{H}$ H
Hypo-nitrite d'ammoniaque hydraté...................................	2 Az H^3	$\dddot{A}$ Az	$\dot{H}$ H
Hypo-nitrite d'hydr. bi-carb. hydraté *(éther nitrique)*......	4 H^2 C^2	$\ddot{A}$ Az	$\dot{H}$ H
Acétate d'ammoniaque hydraté...................................	2 Az H^3	H^6 C^4 O^3	$\dot{H}$ H
Acétate d'hydr. bi-carb. hydraté *(éther acétique)*.....	4 H^2 C^2	H^6 C^4 O^3	$\dot{H}$ H
Benzoate d'ammoniaque hydraté...................................	2 Az H^3	H^{12} C^{30} O^3	$\dot{H}$ H
Benzoate d'hydr. bi-carb. hydraté *(éther benzoiq.)*.....	4 H^2 C^2	H^{12} C^{30} O^3	$\dot{H}$ H
Oxalate d'ammoniaque cristallisé et desséché..........	2 Az H^3	C^4 O^3	$\dot{H}$ H
Oxalate d'hydr. bi-carb. hydraté *(éther oxalique)*.....	4 H^2 C^2	C^4 O^3	$\dot{H}$ H
Bi-sulfate d'ammoniaque........	2 Az H^3	2 $\dddot{S}$	
Bi-sulfate d'hydr. bi-carb. *(Acide sulfo-vinique)*..........	4 H^2 C^2	2 $\dddot{S}$	
Binoxalate d'ammoniaque......	2 Az H^3	2 C^4 O^3	
Binoxalate d'hydr. bi-carb. *(acide oxalo-vinique)*..........	4 H^2 C^2	2 C^4 O^3	
Bi-carbonate d'ammoniaque hydraté...................................	2 Az H^3	4 $\dot{C}$	$\dot{H}$ H
Bi-carbonate d'hydr. bi-carb. hydraté *(sucre de cannes)*..	4 H^2 C^2	4 $\dot{C}$	$\dot{H}$ H
Bi-carbonate d'hydr. bi-carb. bi-hydraté *(sucre de raisins)*.................................	4 H^2 C^2	4 $\dot{C}$	2 $\dot{H}$ H
Hydrate d'hydr. bi-carb. octo-basique *(éthal)*...........	16 H^2 C^2	....	$\dot{H}$ H
Hydrate d'hydr. bi-carb. bi-basique *(éther sulfurique)*............................	4 H^2 C^2	....	$\dot{H}$ H
Hydrate d'hydr. bi-carb. *(alcool)*.................................	4 H^2 C^2	....	2 $\dot{H}$ H
Ammoniaque liquide..............	Az H^3	....	2 $\dot{H}$ H

Table 5-1 Table of classification, after Dumas and Boullay 1828

binary compounds that contained bicarbonated hydrogen as a preexisting immediate constituent. Despite all their differences in observable properties, Dumas and Boullay considered these substances to belong to the same class of organic substances because of their common immediate constituent. Moreover, despite the overwhelming phenomenologic differences between organic compounds and inorganic salts, the former were interpreted as "salts" or hydrates of bicarbonated hydrogen, analogous to the inorganic ammonium salts.[12]

The table has four columns, the first of which lists the substance names. The new, systematic names of the organic compounds, which express their binary constitution and their analogy to the ammonium salts, are followed by the traditional names placed in parentheses. The next column lists the formula H^2C^2 for the shared basic constituent of the class, bicarbonated hydrogen (referring to one volume), and AzH^3 for ammonia (referring to two volumes). The third column includes the formulas for the acidic constituents particular to the individual species. Often, but not always, oxygen is represented by dots, as suggested by Berzelius, so that, for example, instead of SO^3, the symbol $\dddot{S}$ is used. Finally, there is a column for the formula of water, which either symbolizes water of crystallization, not considered part of the binary constitution, or, in a few cases (ethal, ordinary ether, and alcohol), the second immediate constituent of the compound (accordingly, in these three cases there is empty space in the third column for the acids, and the systematic name of the compound is *hydrate*).

Given chemists' assumption that water of crystallization did not enter the constitution of a compound, the division of the table and the separation of the two partial formulas in two columns visually represent the binary constitution of the classified substances. In addition, the table makes it possible to grasp the entire class of substances and the principle of their classification at a glance: "One can see in this table that all the compounds which it contains, except for the hydrates, correspond exactly to one another. In each there are the same quantities of acid, base and water."[13]

The following particular organic compounds are included in the table. The first two are ethers of hydrochloric acid (HCh) and hydroiodic acid (HI), whose formula models of constitution, following Thenard's 1824 models,[14] show analogous compositions, that is, two volumes each of a halogen acid and bicarbonated hydrogen. In both cases, the analogy with the salts of ammonia is also quantitatively complete (one AzH^3 stands for two volumes of ammonia). Identical volumes of the bases ammonia and bicarbonated hydrogen thus saturated the same amount of acid. The next eight formulas symbolize a second group of ethers—the ethers of nitric acid and of three vegetable acids—and the four analogous ammonium salts. Louis Thenard and Pierre Boullay already had manufactured these compounds two decades before and interpreted them as binary compounds of alcohol and the respective acid.[15] In the table, these compounds are represented for the first time as compounds of bicarbonated hydrogen, which additionally contains water, interpreted as water of crystallization. Again, comparison of the formula of each ether with the preceding formula of the ammonium salt shows a perfect analogy, even in terms of quantitative details. It should be noted that this group of ethers played a key role in the entire classification. Dumas and Boullay went to great lengths to support their new models of the constitution of this group of ethers empirically, and the major part of their article hence refers to the experiments that were carried out with these substances (see the following discussion).

Next in the table are the formula models for sulfovinic acid, which was a by-product in the manufacture of ordinary ether, and for a hypothetical substance (oxalovinic acid) that until then had not yet been isolated experimentally. Both substances are represented as water-free compounds of bicarbonated hydrogen, with sulfuric acid and oxalic acid, respectively, analogous to ammonium bisulfate and bioxalate. After this, the table lists the binary formula models for two species of sugar, cane sugar and grape sugar, which are represented as compounds of bicarbonated hydrogen with carbonic acid and have a constitution analogous to the constitution of ammonium carbonate. From the formula models, it is clear that the two kinds of sugar, by analogy to ammonium carbonate, contain water of crystallization and that they differ only in their amounts of water of crystallization. The last three formula models represent the "hydrates" alcohol, sulfuric acid ether (ordinary ether), and "ethal," and analogy is made to "liquid ammonia" dissolved in water, which, however, does not include the quantitative relations in this case.

What role did the chemical formulas play in this new mode of classification? In the text accompanying their table, Dumas and Boullay wrote, "We have summarized all facts treated in this paper in the following table, where they are expressed by means of atomic formulas in order to make the exposition concise. One notices here such an agreement between ammonia and bicarbonated hydrogen that we have good reason to hope that one might regard our opinion to be dictated by facts."[16]

According to Dumas and Boullay, formulas are especially suited to constructing an easily comprehensible table. They are also "concise" to the extent that their division into two partial formulas depicts the binary constitution of the substances and simplifies the recognition of the analogy to the ammonium salts; however, the role of chemical formulas was by no means exhausted by such illustrative functions. Subsequent analyses will show that Dumas and

Boullay used Berzelian raw formulas also as productive tools for the construction of the binary formula models and that the comparison and alteration of the latter were decisive steps in the performance of classification. Before these analyses, we explore how the new claims about the binary constitution of the classified compounds were or were not supported by experiments.

EXPERIMENTATION AND THE CONSTRUCTION OF FORMULA MODELS FOR THE "COMPOUNDS OF BICARBONATED HYDROGEN"

In the text accompanying their table, Dumas and Boullay referred to the earlier experiments on alcohol and ethers performed by European chemists, thus securing the coherence of a collective experimental enterprise. Additional immediate historical preconditions of their classification were their quantitative analyses and their new explanation of the simultaneous formation of ordinary ether and sulfovinic acid, published in 1827. The previous experiments gave some empirical support for the assumption that the ethers of sulfuric acid, arsenic acid, and phosphoric acid and the group of halogen acid ethers were composed of bicarbonated hydrogen and a second constituent. Among the decomposition products of these ethers, chemists had found bicarbonated hydrogen and, in the case of the halogen acid ethers, also the acid. In contrast, analogous experiments on the decomposition of the ethers of vegetable acids and nitric acid produced not bicarbonated hydrogen but alcohol and the respective acid. Thus, as late as 1824, Louis Thenard had stated that these ethers were composed of alcohol and an acid and had demarcated them as a particular group of ethers.[17]

In their experiments, Dumas and Boullay concentrated on this group, with the goal of showing that all ether species had the same constitution and thus could be grouped in one class. They repeated the production of these ethers and their quantitative analyses, and they performed additional experiments studying their chemical reactions, which were successful, however, in only a single case. The repetitions of the quantitative analyses of the ethers yielded results that deviated from those of Thenard. They consistently showed a lower proportion of hydrogen and oxygen. Based on these new results, it was now possible to construct formula models of constitution containing the partial formula not of alcohol, but that of ordinary ether. Thus, Dumas and Boullay had come a step closer to their goal of representing the ethers of vegetable acids and nitric acid as binary compounds of bicarbonated hydrogen. In the next step, they transformed the first formula model into a model containing the partial formula for bicarbonated hydrogen. In the following section, I analyze this procedure using the example of acetic acid ether.

Quantitative Analyses and Their Association with Work on Paper

In their analysis of acetic acid ether, Dumas and Boullay obtained as a result a composition of 54.82% carbon, 36.425% oxygen, and 8.755% hydrogen. The two chemists transformed this result into theoretical volumes by dividing the relative combining weights, expressed in weight percentages, by the density of the gaseous elements. They obtained a composition of sixteen volumes of carbon, sixteen volumes of hydrogen, and four volumes of oxygen.[18] At the end of the 1820s, the tacit revision of Gay-Lussac's law of combining volumes, extending it to larger numbers, was no longer unusual. Nevertheless, the question of why Dumas and Boullay did not prefer the alternative of the smaller ratio 4:4:1 must be raised. The answer lies in the next step of the transformation of inscriptions in which the following formula model for the binary constitution of acetic acid ether was constructed: "Acetic acid ether can thus be represented by one atom of sulfuric acid ether $H^{10}C^{8}O$ and one atom of acetic acid $H^{6}C^{8}O^{3}$."[19]

Dumas and Boullay went immediately from representing the composition of acetic acid ether in volumes to the formula model $H^{10}C^{8}O + H^{6}C^{8}O^{3}$. The formation of this formula model can be reconstructed easily based on the collectively available resources. (In the text itself, there is only one explicit example of the construction of a formula model, this being grape sugar. I will return to this.) The two chemists took the partial formula $H^{6}C^{8}O^{3}$ for acetic acid from Jacob Berzelius.[20] They arrived at the formula $H^{10}C^{8}O$ for ordinary ether (sulfuric acid ether), which they had represented in 1827 with the formula $2\ H^{2}C^{2} + \frac{1}{2}\ H^{2}O$, by way of two simple manipulations of the original formula. Both partial formulas needed only to be grouped as one formula ($2\ H^{2}C^{2} + \frac{1}{2}\ H^{2}O = H^{5}C^{4}O^{1/2}$) and then multiplied by two. The new formula for ordinary ether $H^{10}C^{8}O$ hence referred no longer to two but to four volumes of this substance. This transformation of the formula of ordinary ether had become necessary to fit it as a partial formula into a formula model of the binary constitution of acetic acid ether that both represented the analytical results and contained the accepted Berzelian formula for acetic acid (see my reconstruction in Fig. 5.1).

Dumas and Boullay's way of working could be compared with assembling a puzzle in which the individual parts are modified by joining them together.

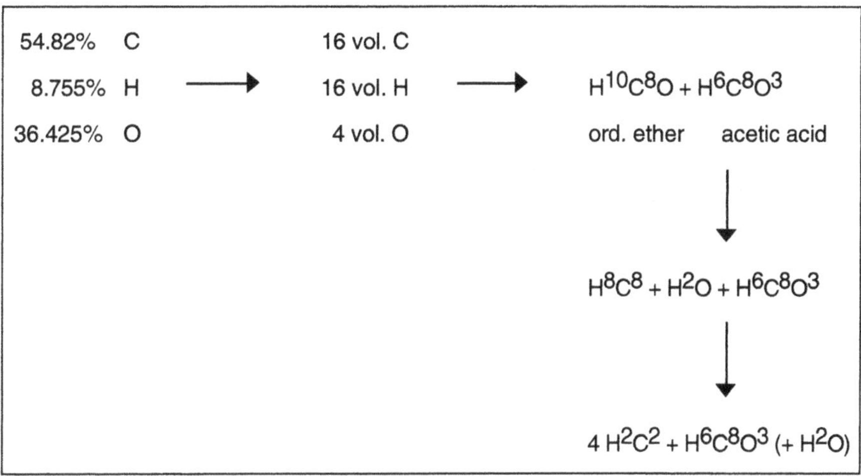

Figure 5-1 Dumas and Boullay's chain of inscriptions for constructing the formula model of binary constitution for acetic acid ether (my reconstruction)

The result of transforming the composition in weight percentages into volumes was adjusted after the fact to match the result of the following step in the transformation, in which the formula model $H^{10}C^{8}O + H^{6}C^{8}O^{3}$ was constructed. Only from the perspective of this final formula model does it make sense that Dumas and Boullay did not express the transformation of the composition according to weight percentages into volumes in the simpler volume relation 4:4:1 but 16:16:4. What in the first transformation step was presented as a necessary inductive conclusion thus proves to be "theory laden" in the broadest sense in that the formula model of constitution had an effect back on the so-called empirical formula.

The new formula model of acetic acid ether, $H^{10}C^{8}O + H^{6}C^{8}O^{3}$, required a few simple manipulations to transform it into the formula model represented in the table that contains the partial formula for bicarbonated hydrogen $H^{2}C^{2}$. The Berzelian raw formula for ordinary ether, $H^{10}C^{8}O$, had to be transformed into its binary formula model $H^{8}C^{8} + H^{2}O$ by subtracting $H^{2}O$ for water; thus, the original formula model, $H^{10}C^{8}O + H^{6}C^{8}O^{3}$, was transformed into $H^{8}C^{8} + H^{2}O + H^{6}C^{8}O^{3}$. As becomes clear from the table, Dumas and Boullay interpreted the partial formula $H^{2}O$ as a symbol for water of crystallization. They thus established the analogy with ammonium salts, which contain water of crystallization as well.[21] In the last step, fitting the constitution formula into the table, the partial formula $H^{8}C^{8}$ for bicarbonated hydrogen that refers to four volumes had to be transformed into 4 H^2C^2 (the formula for acetic acid is printed in the table as $H^6C^4O^3$, presupposing Berzelius's "atomic weight" 12 for carbon).

Dumas and Boullay's new claim that acetic acid ether and the ethers of other vegetable acid and nitric acid were binary compounds containing bicarbonated hydrogen relied mainly on work on paper. The formula models of these ethers to be found in the table were not after-the-fact illustrations previously figured out in an entirely different way, but they were the result of manipulating chemical formulas. Only in one case did Dumas and Boullay present additional experimental support for their new models, which, according to then-current argumentation standards, was necessary. They decomposed oxalic acid ether, and they obtained results that they interpreted in favor of their models.

The Experiment on the Decomposition of Oxalic Acid Ether

Thenard had presented strong experimental evidence for the hypothesis that the ethers of vegetable acids and nitric acid were composed of alcohol and a corresponding acid. He had shown that these ethers could be decomposed by bases like potash into alcohol and a salt of the acid. Analogous to the displacement reactions characteristic of inorganic salts, these decompositions could be interpreted as a displacement of the "base" alcohol by the stronger base potash and the formation of a compound of potash with the acid. The new models of the constitution of this group of ethers as compounds of bicarbonated hydrogen now required reinterpretation of Thenard's experiments. Dumas and Boullay solved this problem by no longer explaining the formation of alcohol in these reactions as the result of an analysis but by explaining it as the result of a synthesis from the organic constituent of these ethers and water. In their experimental report, they first left it open as to whether this organic constituent was ordinary ether or bicarbonated hydrogen.[22] They then presented results of a new and refined experiment, which they viewed as strong evidence that in fact bicarbonated hydrogen was the immediate constituent of oxalic acid ether.

This refined experiment was a variation of the decomposition experiment with alkalis, accompanied by quantitative investigations. Instead of potash, Dumas and Boullay passed gaseous ammonia through oxalic acid ether and observed that after a certain period, a white solid precipitated from a fluid residue. They identified the solid as a salt of oxalic acid and the fluid as alcohol.[23] The additional quantitative studies showed that in the decomposition of oxalic acid ether with ammonia, only half as much alcohol was produced as in analogous decompositions of ethers with potash and that the relative nitrogen content of the oxalic acid salt was lower than in the neutral ammonium oxalate. Based on these results, Dumas and Boullay argued that in the reaction of oxalic acid ether with ammonia, only half of the bicarbonated hydrogen contained in oxalic acid ether was displaced by ammonia. Thus, an oxalic acid salt was produced in which the oxalic acid was saturated with the same volumes of the base ammonia and the base bicarbonated hydrogen. Simultaneously, the bicarbonated hydrogen displaced by ammonia combined with water to form the alcohol found in the experiment. The fact that only half as much alcohol was produced as in the decomposition of ethers with potash supported the interpretation that bicarbonated hydrogen was an immediate constituent of oxalic acid ether and that only half of its amount contained in the ether was released. Of course, this was not an inductive conclusion but rather a kind of plausible experimental reasoning, where the model of binary constitution for oxalic acid ether, made up of bicarbonated hydrogen and oxalic acid, was the precondition. Given that oxalic acid ether was made up of bicarbonated hydrogen and oxalic acid, the experimental results could be reasonably explained in Dumas and Boullay's fashion.

According to Dumas and Boullay, their experiment also showed that bicarbonated hydrogen had properties of a base and possessed exactly the same capacity for saturation as ammonia.[24] The subsequent generalization of this result to the other ethers of the vegetable acids and nitric acid was based on a comparison of the formulas of the ammonium salts of the respective acids with the formula models of the ethers listed in the table. This comparison showed the same agreement in saturation capacities between bicarbonated hydrogen and ammonia as in the case of oxalic acid ether and ammonium oxalate. For example, comparison of the formula for acetic acid ether, $4\ H^2C^2 + H^6C^8O^3 + H^2O$, with the formula of ammonium acetate, $2\ AzH^3 + H^6C^8O^3 + H^2O$, in which two atoms of AzH^3 corresponded to four "volumes" of ammonia, showed that both "bases" had the same saturation capacity.

THE ENLARGEMENT OF SUBSTANCE CLASSES BY THE CONSTRUCTION OF FURTHER FORMULA MODELS

Based on their results for the group of the ethers of vegetable acids and nitric acid, near the end of their article, Dumas and Boullay included sulfovinic acid, the hypothetical oxalovinic acid, ethal, and two species of sugar in the group of "compounds of bicarbonated hydrogen." In all four cases, the classification was based exclusively on manipulations of Berzelian formulas and the construction of formula models of binary constitution, which contained the partial formula of bicarbonated hydrogen. In none of the cases were new experiments undertaken, nor were any new quantitative analyses done. The only experimental bases for the suggested models of constitution were the results of quantitative analyses, which had been carried out earlier by other chemists. If one ascribed to bicarbonated hydrogen the properties of a base, Dumas and Boullay remarked in the introduction to the expansion of their substance classes, it is possible "to grasp at a glance the most various compounds belonging to the same order; we consider this point significant, and its simplicity causes us to prefer it to others, which we previously indicated."[25] The idea expressed in the last part of the sentence was that the more compounds containing bicarbonated hydrogen that could be listed, the more likely became the claim that this compound was in fact a relatively stable, immediate constituent of organic compounds. Thus, not only Dumas and Boullay, but later also chemists such as Liebig and Berzelius, attempted to create classes as large as possible with the same immediate organic constituent (later also called *organic radicals*). The new classes were constructed on the basis of formula models of the binary constitution of the particular substances; the size of the group functioned as a supporting argument for the chosen formula models.

In 1827, Dumas and Boullay had declared sulfovinic acid to be a compound with a binary constitution, made up of hyposulfuric acid and sweet wine oil, which they represented with the formula model $S^2O^5 + 2\ H^3C^4$ for the anhydrous sulfovinic acid and $S^2O^5 + 2\ H^3C^4 + H^2O$ for the hydrated acid.[26] They had also considered it possible that sulfovinic acid was made up of sulfuric acid and bicarbonated hydrogen.[27] Analogous experiments undertaken by Faraday, interpreted as a simple synthesis of naphthalene with sulfuric acid, as well as experiments by Hennell with sweet wine oil, supported this alternative model of constitution.[28] Considering the results of these experiments, however, in 1827 Dumas and Boullay clearly favored the formula model $S^2O^5 + 2\ H^3C^4$. This changed in 1828, but not because of new empirical evidence for or against either of the models. Dumas and Boullay now simply claimed that sulfovinic acid could be interpreted optionally as either a compound of hyposulfuric acid and sweet wine oil or a compound of sulfuric acid and bicarbonated hydrogen. The only new argument, which they introduced in favor of the new formula model, was the analogy to a salt from the class of ammonium salts. The formula model of ammonium sulfate $2\ AzH^3 + 2\ SO^3$ was perfectly analogous to the new formula model $4\ H^2C^2 + 2\ SO^3$, even in the quantitative sense and in the agreement of their saturation capacities, because $2\ AzH^3$

stood for four volumes of ammonia.[29] In fact, it was not far from one model to the other if one takes the maneuverability of Berzelian formulas as a starting point. Dumas and Boullay merely had to reintegrate the partial formula for water into the formula for hydrated sulfovinic acid, in principle as follows:

$$S^2O^5 + 2\,H^3C^4 + H^2O = S^2O^5 + O + H^6C^8 + H^2 = S^2O^6 + H^8C^8 = 2\,SO^3 + 4\,H^2C^2.$$

$2\,SO^3 + 4\,H^2C^2$ was the new formula model for sulfovinic acid, which Dumas and Boullay placed in the fourteenth place in their table.

Based on their experiment with oxalic acid ether and ammonia, Dumas and Boullay also included the free "oxalovinic acid" in their table. The acid was an entirely hypothetical construct, identified exclusively by its formula model. The basic idea was that the ammonia contained in the oxalic acid salt produced from decomposing oxalic acid ether could be displaced by bicarbonated hydrogen H^2C^2 while retaining the quantitative relations between the acid and the basic component. The formula model representing this idea was $4\,H^2C^2 + 2\,C^4O^3$, with C^4O^3 being the formula introduced by Berzelius for oxalic acid. The fact that the formula model $4\,H^2C^2 + 2\,C^4O^3$ was analogous to the formula model of sulfovinic acid $4\,H^2C^2 + 2\,SO^3$ was viewed as an additional argument for the possible existence of oxalovinic acid.[30]

The substance termed *ethal* denoted by the formula $16\,H^2C^2 + H^2O$ in Dumas and Boullay's table had been discovered by Michel Eugène Chevreul in connection with his studies on the saponification of fats and oils; he had obtained this substance by saponification of the fat of the whale. In his 1823 treatise describing these experiments, Chevreul asserted that the fatty oils were compounds of water-free glycerin (or glycerin minus water) and acids, analogous to the ethers of the vegetable acids and nitric acid.[31] He also claimed that ethal played a role as a base in its water-free form, similar to water-free glycerin, and that it was made up of bicarbonated hydrogen and water, but in different ratios than in alcohol and ether. If 100 parts of bicarbonated hydrogen were combined in alcohol with 63.23 parts water, and in ether with 31.61 parts water, then in ethal they were combined with 7.61 parts water. This corresponded to an integral relationship of 8:4:1 in the proportion of water. Presupposing that the formula of alcohol was $4\,H^2C^2 + 2\,H^2O$, Chevreul's quantitative assumptions could be transformed into the simplest formula $16\,H^2C^2 + 1\,H^2O$ for ethal, which is the formula presented in the table.

The inclusion of ethal in the class of compounds of bicarbonated hydrogen was especially important because this substance had been obtained by the decomposition of a fat, that is, from a typical organic substance of natural origin. Thus, a bridge was built from the artificially produced ethers to animal substances. Dumas and Boullay summarized that fats, like ethers, were saltlike organic compounds.[32] For the same reason, placing the two sugar species, "ordinary cane sugar" and "grape sugar," became important for Dumas and Boullay's classification. In this case, the artificially produced ethers were linked to natural substances extracted from plants. Again, the inclusion of sugars in the group of compounds of bicarbonated hydrogen was based exclusively on the construction of formula models of constitution. The way in which Dumas and Boullay manipulated the initial formulas toward this end can be followed in the example of grape sugar and starch, where the individual transformation steps are represented and explained in the text.

The starting point was de Saussure's quantitative analyses of grape sugar and starch, which produced identical results. Dumas and Boullay only hinted at these results in the text and immediately introduced the formula they had constructed from it: $C^6H^7O^{3\frac{1}{2}}$. In three additional steps, which by and large are identical to the transformation steps reconstructed in the preceding for the construction of the formula model of acetic acid ether, they constructed the formula model $4\ H^2C^2 + 4\ CO + 2\ H^2O$ for starch and grape sugar, which is listed in the nineteenth place in the table:[33]

> Indeed, according to the analysis of Th. de Saussure, these sugars contain $C^6H^7O^{3\frac{1}{2}}$, which we represent as $H^6C^6O^3 + HO^{\frac{1}{2}}$ *viewing the half-atom of water as water of crystallization.* It remains: $H^6C^6O^3 = C^2O^2 + (H^4C^4 + H^2O)$; this means that starch can be represented by equal volumes of carbonic acid and alcohol. Seen this way, cane sugar and starch can be considered carbonates of bicarbonated hydrogen, which differ from one another only in the fact that *the former contains two times less water of crystallization than the second.*

In the first transformation step, Dumas and Boullay separated the formula $HO^{\frac{1}{2}}$, which stood for a half "atom" (or one volume) of water, from the initial raw formula for grape sugar and starch and interpreted $HO^{\frac{1}{2}}$ as a symbol for water of crystallization. According to general opinion, water of crystallization was not included in the constitution of salts and could be removed by strong heating; therefore, in the next step, the formula $HO^{\frac{1}{2}}$ was omitted. In this second step, Dumas and Boullay also transformed the remaining formula $H^6C^6O^3$ by regrouping the individual symbols and adding plus signs and parentheses to achieve the binary constitution formula $(H^4C^4 + H^2O) + C^2O^2$. $(H^4C^4 + H^2O)$ represented alcohol, and the parentheses symbolized that its two constituents—bicarbonated hydrogen and water—formed a

$C^6H^7O^{3\ 1/2} \longrightarrow H^6C^6O^3 + HO^{1/2} \longrightarrow H^6C^6O^3 \longrightarrow$

$C^2O^2 + (H^4C^4 + H^2O)$

Figure 5-2 Dumas and Boullay's construction of the formula model of binary constitution for sugar (my reconstruction)

more complex constituent of grape sugar and starch (see my reconstruction in Fig. 5.2).

As can be seen in the verbal commentary that followed, Dumas and Boullay interpreted the formula C^2O^2 as the symbol for carbonic acid, which in the end was transformed in the table to the formula 2 CO and doubled to represent four volumes of carbonic acid (4 CO). Analogously, the formula for the component alcohol, $(H^4C^4 + H^2O)$, was converted to $(2\ H^2C^2 + H^2O)$ and multiplied by a factor of 2, giving $(4\ H^2C^2 + 2\ H^2O)$. The result of these manipulations of formulas was that grape sugar (and starch), $4\ H^2C^2 + 4\ CO + 2\ H^2O$, was now represented as a binary compound made up of four volumes of bicarbonated hydrogen and four volumes of carbonic acid, analogous to the neutral substance ammonium carbonate (the partial formula $2\ H^2O$ was interpreted as water of crystallization). Comparison with the formula model $4\ H^2C^2 + 4\ CO + H^2O$ for cane sugar, which could be constructed in the same way, showed that both species of sugar differed only in the proportion of water of crystallization.[34]

MANIPULATIONS OF FORMULAS AND THEIR SIGNIFICANCE

The preceding analysis of the generative function of chemical formulas for constructing models of the binary constitution of organic compounds and their classification raises several questions. What manipulations were allowed? Which had a chemical significance, and which did not? On the one hand, there were semantic constraints for regrouping individual symbols (letters), for their exclusion or inclusion in formulas, and for the multiplication or division of formulas. It was crucial that the final formula models of constitution could be divided into two meaningful partial formulas to represent both the binary constitution of the substance and an analogy to the constitution of ammonium salts. The general concept of binary constitution as well as the more specific construction of analogies to the ammonium salts thus determined what was or was not allowed in the manipulations. In this context, Dumas and Boullay often commented on the result of a particular manipulation, namely, the separation of the partial formula H^2O for water, which they interpreted as water of crystallization.

There were also manipulations of formulas to which Dumas and Boullay clearly did not assign any chemical significance and that were done without comment. For example, they converted the formula C^2O^2 for carbonic acid without commentary to 2 CO or H^4C^4 to 2 H^2C^2. Further, they multiplied almost all formula models so that they each contained the partial formula 4 H^2C^2 to fit them together in the classification. From today's point of view, it is utterly incomprehensible how such manipulations of the formulas were possible without limitation because formulas like C^2O^2 and 2 CO today have different chemical meanings. The first signifies a single molecule comprising two atoms of carbon and two atoms of oxygen, and the second signifies two molecules, each made up of one atom carbon and one atom oxygen. Not only did Dumas and Boullay manipulate chemical formulas quite liberally in this respect, but they also saw this kind of manipulation on paper as a purely formal, algebraic procedure.

This fact, which seems odd from today's perspective, becomes immediately understandable in the historical context of the use of Berzelian formulas. Whereas the now current chemical significance of formula manipulations results from submicroscopic–atomic theories (and the corresponding meaning of chemical formulas, according to which they stand for atoms and molecules in the sense of submicroscopic entities), the significance of Dumas and Boullay's manipulations resulted from the contemporary meaning of the concept of binary constitution, which was formed in an experimental practice dealing with macroscopic substances. Thus, the two "immediate constituents" making up the binary constitution of a compound were macroscopic substances that could be isolated in experiment. In this conceptual framework, transformation of the symbol C^2O^2 into 2 CO, for example, merely meant a different way of writing the same theoretical combining weight of carbonic acid relative to a given combining weight of a second macroscopic component. In the same way, the doubling of a partial formula, for example, H^2C^2 to H^4C^4, meant doubling the theoretical combining weight of this substance relative to the combining weight of another one. In sum, Dumas and Boullay gave only realistic, chemical significance to manipulations of chemical formulas if they could fit them into the collectively shared, traditional conceptual framework. This changed in 1834 with the introduction of the concept of substitution.

RECEPTION OF THE NEW CLASSIFICATION AMONG EUROPEAN CHEMISTS

Berzelius's immediate reaction to Dumas and Boullay's 1828 classification was ambiguous. He gave unqualified approval to the new results of the quantitative analysis of the ethers of the vegetable acids and nitric acid, but he rejected the new models of constitution, asserting that the ethers of the vegetable acids and nitric acid were binary compounds made up of bicarbonated hydrogen and an oxygen acid. Instead, he saw them as binary compounds of sulfuric acid ether and the acid.[35] In particular, Dumas and Boullay's hypothesis that bicarbonated hydrogen was a base analogous to ammonia and, as a consequence, that the organic compounds of bicarbonated hydrogen were either hydrates or salts were bold "theoretical speculations" in the eyes of Berzelius.[36] Moreover, he was very passionate in opposing Dumas and Boullay's inclusion of organic compounds of natural origin in the new substance class. The opinion, already expressed by Chevreul, that fats and oils shared an analogous binary constitution with the artificially produced ethers and the further claim that ethers, fats, and sugars were in general analogous to the salts for him went much too far. "This kind of comparison is crippled," he remarked.[37]

The following comments by Berzelius on the role of chemical formulas in Dumas and Boullay's constitution theory and in their research style in general are particularly revealing:[38]

> But we can only consider them [the opinions of Dumas and Boullay] an attempt to make the mutual proportions of the elements comprehensible and easier to remember *by way of association to the formulas of other, better known compounds.* If this view moves into the philosophy of science, if one assumes that ethers, oils, and sugars are salts of acids with hydrocarbon, one would introduce to science all-too-limited views, which, even if sometimes justified, can never be true in all cases. . . . *If the properties of the compounds do not count,* or if these properties are not taken into consideration in the groupings, which one makes the subject of science, and in the analogies, which one chooses for naming; *and everything instead is to be judged exclusively according to formulas of composition, then the philosophy of science has come to an end*; the use of such a foundation for the study of organic chemistry would ultimately lead to absurdity.

These remarks of Berzelius largely confirm my previous analyses of the productive function of chemical formulas for the construction of models of constitution of organic compounds and their classification. Dumas and Boullay's research was, according to Berzelius, based too much on formulas; he even claimed that everything in their work revolved around formulas. The inventor of this sign system now distanced himself from an application he had not intended. The chemical formulas, he remarked, were designed to make existent knowledge about the composition of substances perceptible;[39] they were "symbols of composition, which make the conceptualization and remembering [of composition] easier."[40] Berzelius had never intended to use them as paper tools for playing out possible models of binary constitution or making analogies that also could not be made plausible by sensual experience.

Berzelius's indignation lasted only a short time. As the next chapter shows, after 1832, he himself began to use chemical formulas as tools for work on paper, often with much less experimental evidence for the models of constitution he constructed than had been the case earlier in the work of Dumas and Boullay. This made him generally somewhat more sympathetic to the work of his French colleagues. In 1833, when he already had presented alternative models of constitution and a new classification proposal, Berzelius wrote that the two French chemists' models of constitution and their classification provided "simple views about a series of phenomena, and seem to follow the usual method of combination of the bases in so many cases, that they certainly deserve all the attention which they have received."[41] Similarily, Justus Liebig, who also proposed alternative formula models of alcohol and its derivatives in 1834, commented that Dumas and Boullay had provided convincing experimental evidence for their models.[42] Théophile Jules Pelouze wrote quite emphatically at the same time: "It was unavoidable that such a fundamental consideration, supported by so many facts, won the conviction of the chemists, since never before had a closer analogy been shown between two bodies [ammoniac and olefiant gas]."[43] By and large, Dumas and Boullay's models of the constitution of alcohol, alcohol derivatives, and some additional organic compounds and their associated classification of these substances enjoyed a thoroughly positive reception in the first five years after their publication. A little more than a decade later, Dumas and other chemists abandoned both this classification in its concrete form and the hypothesis that organic compounds have a binary constitution analogous to inorganic salts.[44] At the same time, chemists rejected all variants of the so-called radical theory of the 1830s and the alternative classifications based on it, which contained very similar claims about the binary constitution and electrochemical duality of organic compounds and their analogy with inorganic compounds. Nevertheless, they still viewed Dumas and Boullay's 1828 classification as a breakthrough in organic chemistry.

CHAPTER SIX

Paper Tools for Modeling the Constitution of Organic Compounds

> Indeed, almost all the systems constructed in organic chemistry in the past twenty-five years have this common and singular characteristic of being based nearly exclusively on the combination of signs and formulas.
>
> —BERTHELOT 1860, CXXIV

In the previous chapter, we saw that Dumas and Boullay used Berzelian chemical formulas as paper tools for constructing an entirely new classification of several organic compounds, analogous to the classification of inorganic compounds, based on formula models of their binary constitution. Unlike the French chemists, most European chemists believed that organic and inorganic compounds were significantly different and hence also have a different type of constitution. This changed after 1833, when Jacob Berzelius accepted unequivocally that organic compounds also have a binary constitution. By the mid-1830s, the great majority of chemists were convinced of the binarity of organic compounds. This conviction was expressed by the formula models of constitution and the new proposals for classification.

In this chapter, I first analyze the historical meaning of the concept of binary constitution and then explore in more detail the extension of this concept to organic compounds and the construction of particular models of binary constitution before and after 1833, the year when the famous controversy about radicals between European chemists began. In this controversy, for example, Dumas suggested the formula model $C^8H^8 + H^4O^2$ for the binary constitution of alcohol (with the "atomic weight" of C = 6), Berzelius divided the formula in half and suggested $C^2H^6 + O$ (with C = 12), and Liebig's model was $C_4H_{10}O + H_2O$ (with C = 12). I argue that these alternative models for the binary constitution of an organic substance and for organic radicals were realizations of various possible manipulations of Berzelian formulas and that none of the formula models was superior in terms of its empirical foundation. In the last part of this chapter, I discuss questions concerning the acceptance of alternative models of constitution, and I argue that three factors contributed to this: the different role assigned to overarching chemical theories, such as Berzelius's electrochemical theory; different metaphysical beliefs tied to classification; and the need to fit a new formula model to an existing class of substances.

The role of Berzelian formulas in the construction of models of constitution and, as a consequence, in the controversy about radicals has been largely overlooked in the historiography of chemistry. No historian of chemistry can avoid mentioning chemical formulas because there is no other way to denote the radicals in question. Nevertheless, this has not been an incentive to further exploration of the question of why chemical formulas were so essential for denoting organic radicals and, further, whether chemical formulas played a generative role in the construction and justification of models of constitution. The following analyses show that Berzelian formulas (or slightly different modes of representing the theory of chemical portions) were not coincidentally the best means of denotation but had, in fact, a generative function. In each case, tinkering with Berzelian formulas and regrouping the individual symbols into a two-part formula model of constitution were decisive steps in the construction of the models.

The productive function of Berzelian formulas as tools for constructing models of constitution was reflected, for example, in Berzelius's 1833 distinction between "empirical" and "rational formulas" (see later discussion) as well as in contemporary comments. In a letter to Jean Dumas published by the latter, the Genevese physicist de la Rive asked:[1]

> Allow me to ask you *whether chemists are not being a bit facile when they group their symbols any which way.* There is in this facility of permutation something that does not totally satisfy us physicists, *and it seems to us to lend itself complacently to all combinations. Is there not some arbitrariness in the way chemists make these choices?* To attack the electrochemical theory, you put your formulas together in a certain way; immediately, to defend this theory, Mr. Berzelius puts them together in a different way: *where is the law of nature?*

Shortly after this, Charles Gerhardt wrote: "Six or seven different formulas have been proposed for alcohol; each author seeks, through numerous reactions, to support his own, which he believes to be the best, as if one could give any idea of the grouping of molecules *by putting down on paper, a bit more to the left or to the right, this or that symbol.*"[2]

The construction of models of the binary constitution of organic compounds was not an autonomous enterprise separate from experimental practice. There is, however, no simple answer to the question of whether the models of constitution were a part of experimental practice or whether they immediately preceded or followed it. On the one hand, because these models were always presented at the end of a typical experimental report and did not require exorbitant additional work, it makes sense to think of them as part of experimentation. On the other hand, after 1833, the models of constitution constructed with chemical formulas only rarely were accompanied by a corresponding experimental study of chemical reactions. The models of constitution at the center of the controversy about radicals were usually embedded in experiments only to the extent that the respective Berzelian raw formula was based on results of the quantitative analysis of the substance in question. As a rule, chemists transformed these analytical data into an integral number of "atoms" and then transformed the result of this into a Berzelian formula, which they arranged and rearranged for constructing two partial formulas representing the binary constitution of the substance.

THE SHARED CONCEPTUAL PRECONDITIONS FOR THE MODELS OF CONSTITUTION

In the late eighteenth and early nineteenth centuries, chemists shared the belief that *inorganic* compounds were always made up of two components and thus had a binary "constitution." Whereas the concept of *composition* referred to the final components of a chemical compound that could not be further decomposed, that is, the "chemical elements" in Lavoisier's sense, the notion of *constitution* expressed the assumption that the chemical elements of a compound grouped together to form more complex components, the *immediate constituents* of a compound. Whenever a chemical compound was composed of more than two different elements, the various elements would combine in such a way with one another that two immediate constituents of the respective chemical compound emerged.[3] The concept of the binary constitution of inorganic compounds was based on the chemical behavior of salts as a paradigmatic group of chemical compounds. Because salts could be produced experimentally from an acid and a base, and because they could be decomposed again into these two ingredients, the acid and the base were considered relatively stable building blocks, that is, the "immediate constituents" of salts, although both substances could be further decomposed into chemical elements in subsequent analyses. For example, chemists believed that the salt copper sulfate was not composed directly of the elements copper, sulfur, and oxygen but rather of copper oxide and a sulfur oxide.

The concept of the binary constitution of chemical compounds was significantly older than Lavoisier's concept of the chemical element. It developed in the early eighteenth century as part of the modern conceptual system of "com-

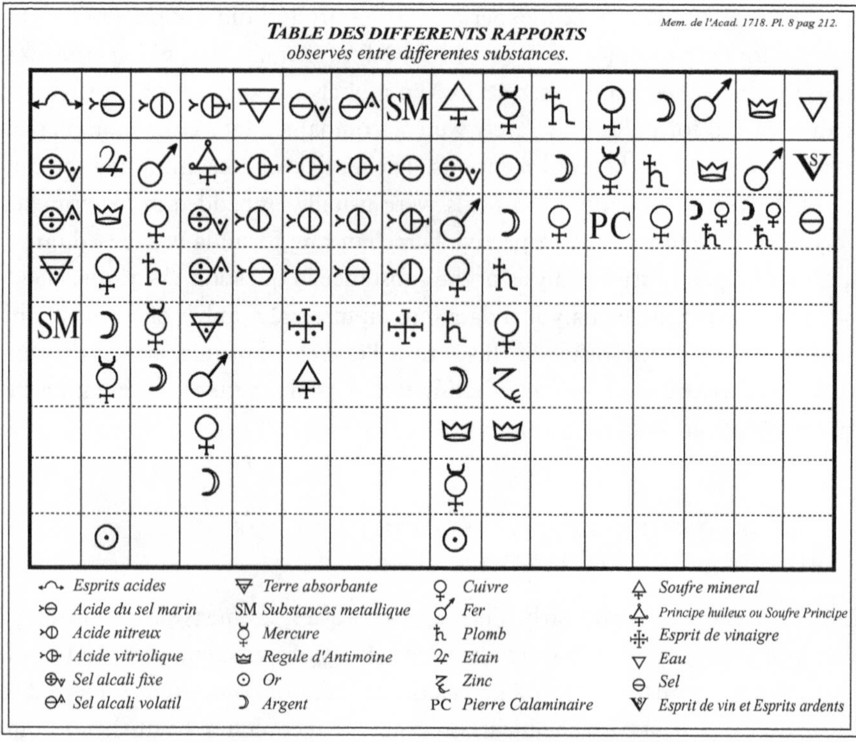

Esprits acides: acids
Acide du sel marin: hydrochloric acid
Acide nitreux: nitric acid
Acide vitriolique: sulphuric acid
Sel alcali fixe: potassium carbonate and sodium carbonate
Sel alcali volatil: ammonium carbonate
Terre absorbante: alkaline 'earth,' i.e., alkaline oxygen compounds or carbonates of light metals like calcium carbonate and aluminium oxide
Regule d'Antimoine: antimony
Pierre Calaminaire: galmei (zinc carbonate)
Soufre mineral: sulphur
Principe huileux ou Soufre Principe: the principle sulphur or oil
Esprit de vinaigre: acetic acid
Esprit de vin et Esprits ardents: alcohols and other combustible and volatile organic compounds

Table 6-1 E. F. Geoffroy's Table of the Different "Rapports" (1718)

pound and reaction," which was formed in the framework of an experimental practice concerned largely with salts and metals.[4] The affinity tables of the eighteenth century constructed after the table by Etienne F. Geoffroy[5] are perhaps the most impressive representation of the concept of binary constitution (Table 6.1). They are made up of multiple columns containing symbols that denote pure laboratory substances, which behaved like stable building blocks in experiments interpreted as reversible chemical syntheses and analyses and displacement reactions. The pragmatics of the table included the presupposition that the combination of two symbols, the one found at the head of a column and the other below, stood for a chemical compound. A chemical compound was thereby represented as a binary entity. The one-to-one correspondence between a symbol and the denoted component represented visually the binarity of constitution. Because an affinity table was a complete map of all pure compounds known at a particular time, it expressed a belief in the general applicability of the concept of binary constitution.

Jacob Berzelius offered the following verbal explanation of the concept of binary constitution in the early nineteenth century:[6]

> In inorganic nature compound bodies of the first order never contain more than two elements; so that we may say that *inorganic nature never contains more than binary combinations, and bodies composed of binary combinations.* All inorganic bodies in which we find more than two elements are evidently composed of binary combinations of these elements, which may be separated from each other without decomposition, and generally they may be united again so as to form the compound substance anew. The double or triple metallic sulphurets, which contain two or three metals, are always to be considered as being composed of as many simple sulphurets. The alkaline sulphurets are to be considered as compounds of a binary compound, the alkali, with an elementary body, the sulphur, and of course constitute only an apparent exception to the law.

Berzelius thus emphasized clearly that in his view not only the salts but all inorganic compounds were always made up of either two elements or two composed immediate constituents. The result was an image of a hierarchical order of binary chemical compounds. For example, Berzelius described sulfuric acid made from sulfur and oxygen as a binary compound of the first order, the salt "sulphate of potash," which was made from sulfuric acid and potash, as a binary compound of the second order, and the double salt alum consisting of sulphate of potash and sulphate of alumina as a binary compound of the third order (Fig. 6.1): "Thus sulphuric acid is a compound of the first order. Sulphate of potash is one of the second, and alum one of the third. For sulphate of potash is composed of two bodies of the first order, the acid and the potash; and alum of two compounds of the second order, namely, sulphate of potash and sulphate of alumina."[7]

Berzelius's electrochemical theory, which was received with undivided acclaim in the first decades of the nineteenth century by the European chemical community, transformed the concept of binary constitution of inorganic compounds into an almost incontestable principle of chemistry. According to this

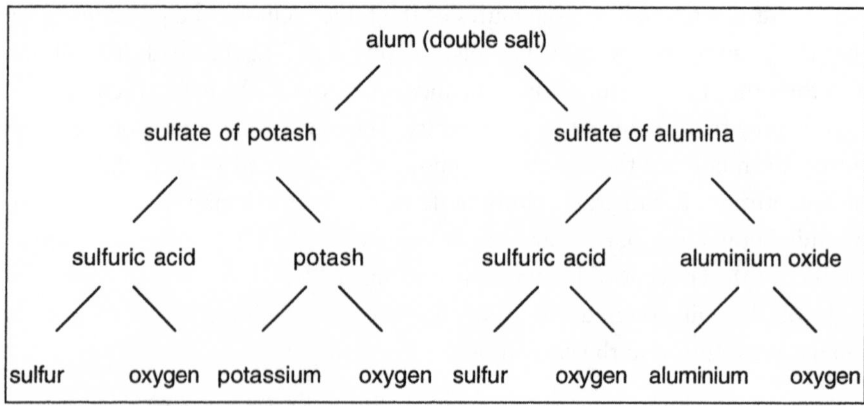

Figure 6-1 The hierarchical order of binary constitution

theory, all inorganic compounds are made up of a positively and a negatively charged constituent. Along with the inorganic oxides, the salts again provided the model substances for this theory. Berzelius wrote, in 1818, "that every compound body, with however many components, can be divided into two parts, one electrically positive and the other electrically negative. Thus, the sulfuric acid natron [sodium sulfate] does not consist of sulphur, oxygen, and sodium, but of sulfuric acid and natron [sodium oxide]; each of which can in turn be decomposed into an electropositive and an electronegative component."[8]

In the historiography of chemistry, the concept of binary constitution has been frequently identified with electrochemical dualism, without mentioning the older historical roots of the former. Some of the historical actors, however, such as Jean Dumas, in fact traced the concept of binary constitution at least as far back as the chemistry of Lavoisier. After he already had abandoned the concept of binarity, Dumas wrote in 1840:[9]

> What is so striking about the *chemistry of Lavoisier* and its nomenclature, which is based on [this chemistry] and expresses it, *is the antagonism of the elements which combine to form binary compounds*; the antagonism of the acids and the bases which combine to form the salts; the antagonism of the salts which combine to form double salts, etc. The chemistry of Lavoisier and its nomenclature thus seem to have predicted and prepared the way for the electrochemical theory, which had nothing to add other than to name one of these antagonistic bodies a positive element and the other a negative.

Here Dumas points out that the new systematic nomenclature introduced by Lavoisier and his allies presupposed the concept of a binary constitution of chemical compounds. In addition, we can see from Dumas's comments that the electrochemical theory only added the electrical dualism of the two building blocks to this earlier concept.

Extension of the Concept of Binarity to Organic Compounds in the 1830s

In the previous chapters, we saw that since the first decade of the nineteenth century, French chemists assumed that some organic compounds, such as alcohol and ether, had a binary constitution analogous to the inorganic salts.[10] Seen in the context of European chemistry, French chemists—including Louis Thenard, Joseph Louis Gay-Lussac, Michel Chevreul, Jean Dumas, and others—took an outsider position. In the first two decades of the nineteenth century, most chemists were convinced that the dichotomy between inorganic and organic substances also extended to their constitution and therefore organic compounds were composed directly of chemical elements.[11] For example, Leopold Gmelin, professor of chemistry and medicine in Heidelberg, wrote in his *Handbuch der theoretischen Chemie*:[12]

> All organic compounds differ from the inorganic in the following: 1) all inorganic compounds can be seen as binary compounds, that is, they are either made up of simple substances, or binary compounds of simple substances, or binary compounds grouped in pairs. . . . All organic compounds must be seen as ternary, quaternary, etc., that is, *at least three substances are immediately unified with one another without previously having formed binary compounds.*

Jacob Berzelius, the leading European chemist of the first decades of the nineteenth century, had occasionally considered it possible before 1833 that organic compounds were also binary compounds and subject to electrochemical dualism; he was fascinated by the idea of a larger unity of nature. After considering various arguments, however, he always returned to the constitutive difference between inorganic and organic compounds.[13] In the context of his atomic theory of 1813, he unmistakably stated that "organic atoms" must be made up of more than two elementary constituents,[14] and a year later he still wrote:[15]

> All organic bodies contain oxygen united to more than one combustible radicle; and chemical experiments on these substances have shown us that these combinations of oxygen with two or more radicles cannot be considered as composed of two or more binary bodies. Consequently organic bodies of the first order contain more than two elements. According to the number of these elements we may call them ternary, quaternary, quinquarny etc. oxides.

By *radicles*, following Lavoisier's nomenclature, Berzelius here meant nonmetallic elements like carbon and hydrogen. According to him, organic compounds always contained oxygen, which in turn was always combined with more than one radicle (*element*). This occurred without any particular arrangement of the elements, so that there was neither a binary constitution nor electrochemical dualism in organic compounds:[16]

> There is still another essential difference between organic and inorganic nature. This difference consists in the electro-chemical modification of the organic products, which does not appear to depend immediately on that which the elements have in organic [inorganic] nature; that is to say, on the original electrical modifications of the elementary substances. We cannot even form conjectures respecting the manner in which nature modifies the electrical properties of the elements in the economy of living bodies, nor of the means which she employs to combine the elements in other proportions, and in other numbers than in inorganic bodies.

Ultimately, however, Berzelius was unsatisfied by the hypothesis that the laws of electrochemistry could not be applied to organic compounds. On the one hand, he was a chemist who emphasized the difference between inorganic and organic bodies; on the other hand, he was also convinced of the unity of nature. In the tension between the two differing beliefs, questions about the extent of the differences between organic and inorganic matter were permanently on the agenda.[17] In his elaborated electrochemical theory of 1818, Berzelius wrote that "every composed body, with however many components, can be divided into two parts, one of which is positive and the other negative."[18] He thus concluded that "in organic chemistry, the same must also be true, and every organic product can be considered electrically divisible into oxygen and a compound radical."[19] In 1818, this was still an isolated remark that he denied in the same text: "In the formation of compound organic atoms of the first order," he wrote, "three or more of these elements [oxygen, water, carbon, nitrogen, etc.] are combined."[20]

Liebig and Wöhler's Experiments with the Benzoyl Radical and Berzelius's Reversal

In the short time span between 1832 and 1833, Berzelius threw all his remaining doubts about the binarity of organic compounds overboard and unambiguously accepted an analogy between the binary constitution of inorganic and organic compounds. An incentive for this was a series of experiments by Wöhler and Liebig on the oil of bitter almonds and its derivatives, the results of which had been published in 1832.[21] In an emphatic letter, Berzelius wrote, "The results you drew from the study of oil of bitter almonds are surely the most important thus far obtained in vegetable chemistry, and promise to shine an unexpected light on this area of science."[22]

Why did Berzelius consider these experiments so significant? The substance for this series of experiments, which the two chemists had received in a purified form from their French colleague, Théophile Pelouze, was an oil extracted from bitter almonds. The oil of bitter almonds was a true organic substance, that is, a "natural" substance extracted from plants. Furthermore, the experiments with this substance and its derivatives seemed to provide relatively clear empirical evidence for the hypothesis that oil of bitter almonds, the benzoic acid made from it (which also could be found in plants), as well as chemically manufactured derivatives, had a binary constitution with the compound organic radical "benzoyl" as a shared organic component.[23]

In the 1830s, chemists understood "organic radicals" as compound immediate constituents of binary organic compounds. They believed that these compound constituents were as stable building blocks as chemical elements in most chemical experiments. Radicals, as Liebig put it in 1838, are compound bodies that "play the role of elements."[24] Because knowledge about such compound constituents of binary organic compounds served as criterion for their new classification, Liebig qualified his statement, stating that only those compound constituents were radicals, which "are found in a series of compounds."[25] Another aspect of Liebig's definition referred to the way in which the radicals were "found" in experiments. A radical was a constituent within a compound that could "be replaced by other simple bodies."[26] The analogy between chemical elements and organic radicals thus included the belief that radicals were veritable laboratory substances that, in principle, could be isolated by chemical analyses.[27]

In their series of experiments, Liebig and Wöhler had produced and identified a total of six new and three already known reaction products of oil of bitter almonds.[28] Except for one reaction product, benzamide, they interpreted all as binary compounds made up of the benzoyl radical with the formula $C^{14}H^{10}O^{2}$ and an additional component, either hydrogen, oxygen, chlorine, bromine, iodine, sulfur, or a complex inorganic component like cyanogen or ammonia. The oil of bitter almonds, for example, was thus termed *benzoyl hydrogen*, the binary constitution of which they represented with the formula $(14\ C + 10\ H + 2\ O) + 2\ H$. Analogously, they represented the binary constitution of benzoyl chlorine with the formula $(14\ C + 10\ H + 2\ O) + 2\ Cl$.

Both formula models, as well as the corresponding verbal interpretations, were embedded in unusually simple experimental results.

In each of their experiments studying the chemical reactions of oil of bitter almonds and its derivatives, Liebig and Wöhler had obtained only one or two reaction products, whereas most reactions of organic substances produce a cascade of reaction products.[29] For example, the reaction between oil of bitter almonds and chlorine yielded only the two reaction products "benzoyl chlorine" and hydrochloric acid. If, instead of chlorine, bromine was used, the two chemists obtained merely "benzoyl bromine" and hydrogen bromide. From benzoyl chlorine and potassium iodide, they produced "benzoyl iodine" and potassium chloride, with lead sulfide "benzoyl sulfur" and lead chloride, and so on. The new systematic names, such as *benzoyl chlorine*, which Wöhler and Liebig gave their products, expressed well their conviction that these substances possessed a binary constitution of the radical benzoyl and a further component. These new names did not precede the interpretation of the reactions; rather, they were only the end results of multiple interpretive steps in which the construction of formula models set the course.

All the reactions studied by Wöhler and Liebig fit relatively easily into the collectively shared building-block scheme of simultaneous analyses and syntheses developed in inorganic chemistry. In the reaction of chlorine with oil of bitter almonds, for example, the reaction product hydrochloric acid could only have come from the added chlorine and the hydrogen of the oil of bitter almonds. Because the second reaction product, "benzoyl chlorine," like the initial oil of bitter almonds, consisted of carbon, hydrogen, oxygen, and also chlorine, it was not hard to conclude that a part of chlorine displaced a part of hydrogen from the oil of bitter almonds and formed hydrochloric acid, and another part of chlorine combined with the remaining constituents of oil of bitter almonds to form benzoyl chlorine. The qualitative composition of the reaction products thus seemed to support the interpretation of the reaction as a decomposition (that of oil of bitter almonds) and two simultaneous syntheses (that of hydrochloric acid and benzoyl chlorine). Similarly, the reactions of benzoyl chlorine with inorganic metal compounds fitted almost seamlessly into the traditional framework of displacement reactions: "If one treats benzoyl chlorine with a metal bromide, iodide, sulfide or cyanide, an exchange of the components takes place, whereby on the one side a metal chloride is produced, and on the other a compound of benzoyl with bromine, iodine, sulfur or cyanogen, which is constituted proportionally to the benzoyl chlorine."[30]

In this special case of organic chemical experimentation, the mere knowledge of the qualitative composition of the initial substances and their reaction products provided ample empirical support for Liebig and Wöhler's assumption that there was an analogy between these reactions and the reactions of inorganic salts and that, therefore, the organic compounds also have a binary constitution.

The additional quantitative analysis of oil of bitter almonds and three of its reaction products and the transformation of the analytical results into chemical formulas confirmed this hypothesis (see my reconstruction in Fig. 6.2). For example, two analyses of samples of oil of bitter almonds yielded 79.438% (79.603%) carbon, 5.756% (5.734%) hydrogen, and 14.808% (14.663%) oxygen. Wöhler and Liebig transformed these analytical results into integral numbers of "atoms": fourteen "atoms" of carbon, twelve "atoms" of hydrogen, and two "atoms" of oxygen.[31] Based on this kind of transformation as well as on the results of the experimental investigation of the reactions of these substances, they then proposed the following formula models for the binary constitution of oil of bitter almonds (benzoyl hydrogen) and benzoyl chlorine in connection with a model for the formation of benzoyl chlorine:[32]

> Benzoyl hydrogen (oil of bitter almonds) consists of
>
> (14 C + 10 H + 2 O) + 2 H.
>
> Due to the effect of the chlorine, the 2 atoms of hydrogen combine with 2 at. of chlorine to form hydrochloric acid, which is released. 2 at. of chlorine take the place of this hydrogen, according to the following formula:
>
> (14 C + 10 H + 2 O) + 2 Cl.

The two formula models of oil of bitter almonds and benzoyl chlorine displayed the binary constitution of the represented substances because of the division of the Berzelian raw formula into two partial formulas, separated by parentheses and the plus sign. The qualitative building-block scheme thus was complemented by a quantitative one, which provided it with additional empirical support.

What was the role of chemical formulas here? Chemical formulas did not illustrate only the "benzoyl radical" $C^{14}H^{10}O^{2}$ and the model for the binary constitution of the two substances, but they were also the essential paper tools for their construction. The "benzoyl radical" was a purely hypothetical entity. There was no laboratory substance denoted by this name. The question of what a benzoyl radical is thus could not be answered, as was usually the case, by pointing to a substance observable in the laboratory, with observable properties and a particular composition discovered by experimentation. Moreover, neither the experimental results of the quantitative analysis nor the experiments studying the chemical reactions of the substances were on their own sufficient

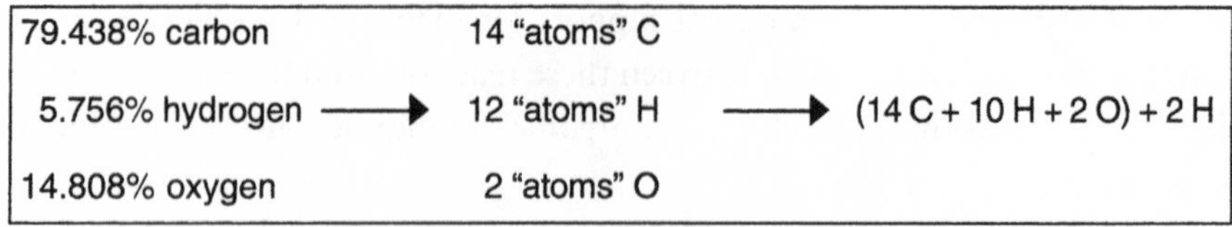

Figure 6-2 Liebig and Wöhler's construction of the formula model of binary constitution for benzoyl hydrogen (my reconstruction)

for a clear definition of the benzoyl radical. From studying the reactions, it could be concluded with only a certain probability that there was a stable organic building block in all derivatives of oil of bitter almonds. The quantitative analysis of these derivatives then showed that this building block was made up of carbon, hydrogen, and oxygen. Because the analytical results provided only a percentage composition that *varied* depending on the second component of the derivative, however, these experimentations were not sufficient for defining the "benzoyl radical" as a qualitatively and *quantitatively stable* building block.[33] For example, the analytical results for oil of bitter almonds were 79.438% (79.603%) carbon, 5.756% (5.734%) hydrogen, and 14.808% (14.663%) oxygen; for benzoic acid, they were 69.155% (68.902%) carbon, 5.050% (5.000%) hydrogen, and 25.795% (26.098%) oxygen; benzoyl chlorine yielded 60.83% carbon, 3.74% hydrogen, 11.01% oxygen, and 24.42% chlorine. Only transforming these analytical results into integral numbers of "atoms" *and* their arrangement within interpretive formula models of binary constitution consisting of two partial formulas—such as (14 C + 10 H + 2 O) + 2 H and (14 C + 10 H + 2 O) + 2 Cl—made it possible to define clearly the radical 14 C + 10 H + 2 O or $C^{14}H^{10}O^{2}$ as a stable building block identical in all derivatives and with a precisely defined quantitative composition.

Liebig and Wöhler's experiments and their hypothesis about the binary constitution of benzoyl compounds were accepted without hesitation not only by Berzelius but also by most European chemists. Their work was a catalyst for the general acceptance of the concept of binarity in organic chemistry. This can be seen, for example, in the fact that Berzelius not only accepted the benzoyl radical and the formula models of binary benzoyl compounds, but he also simultaneously agreed to Dumas and Boullay's formula models of the binary constitution of alcohol and its derivatives. In doing so, he only made a small change, transforming the formula for bicarbonated hydrogen $C^{2}H^{2}$ (with C = 6) into the formula $C^{4}H^{8}$ (with C = 12), which, according to Berzelius, did not denote the laboratory substance "bicarbonated hydrogen" but rather the hypothetical radical "etherin."[34] He also suggested a new way of abbreviating the binary formula models for the etherin and benzoyl compounds, in which the benzoyl radical $C^{14}H^{10}O^{2}$ was represented by the symbol Bz and the etherin radical $C^{4}H^{8}$ by Ae. This was intended to make "the concept of constitution that one wants to express . . . immediately clear to the eye of the reader."[35] In a vehement critique of his two French colleagues a year later, however, Berzelius rejected the formula models of the constitution of alcohol and its derivatives and offered revised formula models.[36] The year 1833 was thus a turning point that not only led to the collective adoption of the general concept of binarity for organic compounds but also introduced a phase of passionate controversies about the particular models of binary organic compounds and their classification.

Liebig's Defense of the Concept of Binary Organic Compounds and the Manifesto of 1837

By 1834, Justus Liebig, who had become one of the leading European chemists, wholeheartedly agreed with the extension of the concept of binarity from inorganic to organic compounds. This can be seen, for example, in his interest in a problem arising from the concept of binary constitution of organic compounds. Why did so many organic compounds, supposedly analogous to the salts in their constitution, behave differently in chemical experiments? Unlike the salts, they could neither be produced from their immediate constituents nor decomposed into these components by bases or acids.[37] This problem was particularly pressing because the reversibility of composition and decomposition constituted the empirical bedrock of the concept of binary constitution. Liebig did not really solve the problem but explained it away. He argued that in the case of inorganic compounds, there was no complete certainty about the existence of binary constitution. Furthermore, he added a theoretical reason for the failure of his experiments in which he had attempted to decompose organic compounds into the two postulated immediate constituents. Borrowing from the affinity theory of Claude Louis Berthollet that claimed that the compositions and decompositions of the paradigmatic inorganic compounds were caused not only by chemical affinity but also by cohesion and other nonchemical forces, Liebig claimed, "Organic chemistry includes a series of compounds in which the affinity outweighs all other forces that we see active in the formation of compounds or the decompositions in inorganic chemistry; for this reason we cannot decompose them [organic compounds] with our usual methods."[38]

In 1834, Liebig so strongly believed in the analogy between the constitution of organic compounds with inorganic salts and oxides that he dared to present bold theoretical speculations. Whereas chemists of the older generation, for example Fourcroy and Berzelius, usually explained the easy decomposition of organic compounds in the chemical laboratory by their weak affinity, Liebig now claimed just the opposite.

For Jean Dumas, who led the French side of the controversy on radicals, extension of the concept of binary constitution to organic compounds already had become a truism in 1835; in the fifth volume of this textbook, dedicated to organic chemistry, he wrote:[39]

> *All of current chemistry is based on the perspective of an antagonism between bodies that agrees admirably with electrical phenomena.* In supposing that the force producing compounds is identical with electricity, one can explain so many facts of chemistry that it is quite natural to assume that compound formation almost always takes place between two bodies of opposite electricity, be it between simple or compound bodies. All theories of mineral chemistry are based on this general conception . . . *all of my efforts are dedicated to transferring [this idea] to organic chemistry.*

The concept of the binary constitution of organic substances was also clearly expressed in a paper that Dumas and Liebig wrote for the British Association for the Advancement of Science in 1837.[40] The task of this manifesto was to describe the current state of organic chemistry, but the two chemists took the opportunity to outline *a plan général* for research in organic chemistry, combined with an emphatic invitation to young European chemists to visit the chemical laboratories in Paris and Giessen and to participate in the collective project. The two chemists described the intellectual side of their project as follows:[41]

> It is easy to understand that with the fifty-four elements known today, with the assistance of a small number of laws of chemical combination and by producing all possible binary compounds or salts not only can all known compounds in the inorganic kingdom be created, but in addition a very large number of analogous compounds. But how can one apply such concepts successfully to organic chemistry?

"Elements," "laws of chemical combination," "binary compounds"—these were key concepts of inorganic chemistry that now were to be made useful for organic chemistry and, as we are also told, in principle were already applied.[42] In particular, the task was to elucidate all compound organic radicals, which, like chemical elements, formed the characteristic immediate constituents of organic compounds. The second related task, much more difficult and requiring a collective solution, was the "natural classification" of organic compounds, and here Dumas and Liebig made it clear that they understood "natural classification" to mean grouping substances based on the knowledge of their composition and their common radical.

MODELS OF CONSTITUTION PRIOR TO 1833

Dumas and Boullay as well as Liebig and Wöhler had embedded their models of the binary constitution of alcohol and ethers and of binary benzoyl compounds, respectively, in experiments studying the synthesis and decompositions of these compounds. By 1833, chemists began to construct models of binary constitution underpinned exclusively by the quantitative analyses of the organic compounds. Chemists first would transform the analytical data into integral numbers of "atoms" or volumes and into a Berzelian raw formula, which they then transformed into a formula model containing two partial formulas representing the two immediate constituents of the compound. This made model constructing much more hypothetical. According to the methodological standards established in inorganic chemistry, analytical data alone were not sufficient to support empirically claims of knowledge about the binary constitution of a substance. In that chemical area, experimental investigations of chemical reactions of the substance at stake were always included. Furthermore, as we have seen above in the case of "benzoyl compounds," weight percentages of a group of chemical derivatives vary depending on their different elemental components, and this makes it impossible to identify a stable compound constituent based on analytical data alone. Hence, it is not surprising that many of the formula models of the binary constitution of organic substances constructed from the late 1820s onward were highly controversial.

The models of constitution at the center of research on the constitution of organic compounds and of the controversy about organic radicals during the 1830s were primarily those of alcohol, ethers, and other alcohol derivatives. Around 1830, German, French, and British chemists began to repeat and modify Dumas and Boullay's 1827 and 1828 experiments with alcohol, ethers, and other alcohol derivatives. They also reconsidered Dumas and Boullay's formula models of the constitution of these substances. In doing so, all chemists used Berzelian formulas as paper tools for constructing the intended models. Before 1833, they were often undecided about which of the formula models to prefer and suggested various possible formula models in their publications.

This changed dramatically after 1833, when Berzelius opened the controversy with the publication of a fundamental paper entitled "Considerations on the Composition of the Organic Atoms."[43] After Berzelius retired from professional experimental life, he had to rely on experiments of his younger colleagues. In the following, I analyze the experiments and subsequent construction of models of constitution that Berzelius chose as the most relevant for his attack on Dumas and Boullay. These were Liebig and Wöhler's experiments with sulfovinic acid in 1831 and 1832 and the subsequent experiments with the same substance by Gustav Magnus; the experiments by Pelouze with alcohol, phosphoric acid, and a phosphovinic acid analogous to sulfovinic acid; and experiments by Liebig with wood spirit and acetal. I mention repeatedly how Berzelius used the results of these chemists.

The Case of Sulfovinic Acid

Sulfovinic acid, which had been found and identified in 1820 as a by-product in the manufacture of ordinary ether, was an important link for establishing models of the constitution of alcohol and alcohol derivatives. In 1827, Dumas and Boullay had assumed that sulfovinic acid consisted of the two immediate constituents hyposulfuric acid (S^2O^5) and sweet wine oil (H^3C^4). A year later, they had transformed their formula model into 4 C^2H^2 + SO^3 (with the "atomic weight" C = 6). This had been necessary for ordering sulfovinic acid into the class of compounds of bicarbonated hydrogen.[44] Around the same time as Dumas and Boullay, Henry Hennell and Georges Simon Serullas had carried out similar experiments with sulfovinic acid. Hennell believed that sulfovinic acid was a compound of bicarbonated hydrogen and sulfuric acid; but, in contrast to Dumas and Boullay's 1828 model, he thought the acid was only half saturated with the bicarbonated hydrogen.[45] Serullas also assumed that sulfovinic acid included sulfuric acid; but, because his analyses had shown a higher hydrogen and oxygen content than Dumas and Boullay's, he accordingly thought that sulfovinic acid consisted of ether and water-free sulfuric acid.[46] Figure 6.3 reconstructs and compares these alternative models of the binary constitution of sulfovinic acid.

In their first experimental report on the constitution of sulfovinic acid in 1831, Liebig and Wöhler referred to all four of these studies. Their quantitative analysis of the barium salt of sulfovinic acid provided a still higher water content than Serullas had established.[47] After representing their analytical results in weight percentages—54.986% barium sulfate, 19.720% sulfuric acid, 12.370% carbon, 3.060% hydrogen, 9.864% oxygen—they wrote:[48]

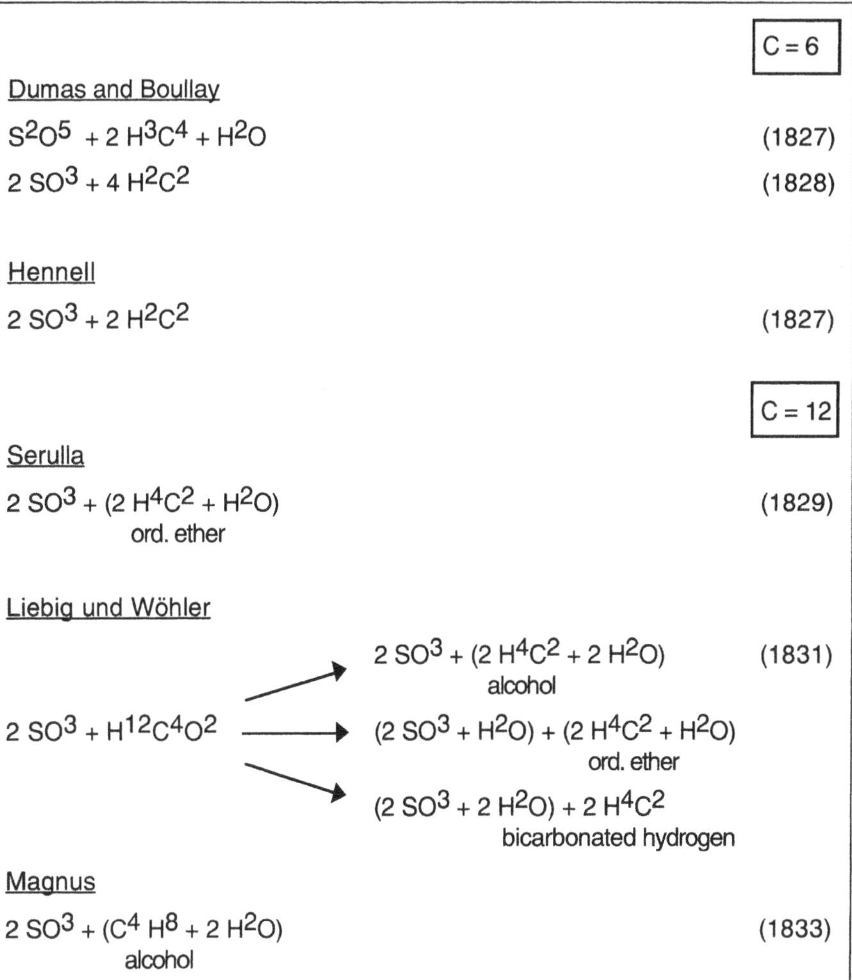

Figure 6-3 Different models of binary constitution for sulfovinic acid (my reconstruction)

From this composition, we see that sulfovinate of baryta contains

2 atoms of sulfuric acid,
1 – baryta,
4 – carbon ⎫
12 – hydrogen ⎬ 2 atoms alcohol,
2 – oxygen ⎭

One can thus consider sulfovinic acid as a compound of water-free sulfuric acid with alcohol or, more likely, as a compound of sulfuric acid hydrate with ether.

Liebig and Wöhler transformed the results of the quantitative analysis into a model of the binary constitution of sulfovinic acid from sulfuric acid and alcohol, assembling the "atoms" of alcohol by a bracket (they used the "atomic weight" C = 12). In their verbal summary, however, the two chemists noted an alternative model: Sulfovinic acid could also be composed of ordinary ether and sulfuric acid hydrate. This resulted from the assumption that alcohol and ordinary ether differed only in the higher water content of the former, as Dumas and Boullay's theory stated; it was therefore conceivable that water could be removed from the constituent alcohol to leave ordinary ether and then combined with sulfuric acid as hydrate water (or water of crystallization). In a reprint of the experimental report that appeared in the first volume of Liebig's *Annalen der Pharmacie*, they suggested a third possibility: "Sulfovinic acid can thus be considered a compound of water-free sulfuric acid with alcohol, or, more probably, *a compound of sulfuric acid hydrate with olefiant gas.*"[49]

With this third variant, Liebig and Wöhler had suggested all possible binary constitutions of sulfuric acid resulting from the successive separation of the formula for water from the partial formula of alcohol as one of the two constituents of sulfovinic acid (Fig. 6.3). Again, Dumas and Boullay's formula models of the constitution of alcohol and ordinary ether were taken a certain precondition.

In 1832, Liebig and Wöhler did not decide which of these alternatives they preferred; but a year later, Gustav Magnus, who would later have a brilliant career as a physicist, attempted to make that decision. He transformed Liebig and Wöhler's analytical results into a Berzelian formula and remarked that these two chemists had assumed that the barium salt of sulfovinic acid is made of "$2\ SO^3 + BaO + C^4H^8 + 2\ H^2O$. They left it open, whether one should consider the acid to be composed of 2 atoms of water-free sulfuric acid, combined with alcohol, or of 2 atoms of hydrous sulfuric acid, combined with hydrocarbon (etherin)."[50]

According to Magnus, because experiments with naphthalene had shown that water-free sulfuric acid could combine with organic substances, it was permissible to assume in the analogous case of sulfovinic acid "that the two atoms of water belonging to its composition are not combined with sulfuric acid but with the hydrocarbon" and that sulfovinic acid should therefore be considered a binary compound of alcohol (consisting of the hydrocarbon and water with the formula $C^4H^8 + 2\ H^2O$) and water-free sulfuric acid (see Fig. 6.3).[51]

Furthermore, Magnus attempted to test the possible presence of water of crystallization, stated by the alternative model $2\ SO^3 + C^4H^8 + 2\ H^2O$ by heating a salt of sulfovinic acid. This was an experimental technique that was familiar from inorganic chemistry. In the case of the relatively unstable organic compounds, however, it became a precarious enterprise, especially when studying the presence of water of crystallization in a quantitative fashion because it was possible that water could be absorbed from the air.[52] In fact, Liebig and Wöhler already had performed a similar experiment with a salt of sulfovinic acid two years earlier, but without a clear result.[53] Magnus's results were by no means clearer, but he interpreted the impossibility of positively detecting water of crystallization as evidence for the hypothesis that water combines with etherin to form alcohol. With this result, he contradicted quite clearly Dumas and Boullay's model of the constitution of sulfovinic acid, but, at the same time, without denying their claim that one "can consider alcohol and ether compounds of etherin with water."[54]

The papers by Magnus and Liebig and Wöhler about the constitution of sulfovinic acid are epistemologically revealing in several ways. All three chemists constructed models of the binary constitution of sulfovinic acid based exclusively on the results of its quantitative analysis without carrying out the otherwise usual experiments to study its chemical reactions. They transformed the analytical results into the number of "atoms" of the elements, and then they grouped the carbon, hydrogen, and oxygen "atoms" in different ways to represent alcohol, ordinary ether, or bicarbonated hydrogen as an immediate constituent of sulfovinic acid. Whereas Magnus worked with Berzelian formulas, Liebig and Wöhler used a slightly different synoptic form. Because, according to the shared view of the chemists of that time, only additional experiments could have provided the necessary support for deciding on one of the three models of constitution, Liebig and Wöhler had not decided between these models in 1832. Based on analogical reasoning and a technically difficult experiment, Magnus, however, decided in favor of a formula model of constitution that showed sulfovinic acid to be a binary compound of alcohol and sulfuric acid. Although this formula model contradicted Dumas and Boullay's formula model for sulfovinic acid, none of the three chemists thought it raised any major questions about the claim that the substance denoted by the formula C^4H^8 (with $C = 12$; or C^8H^8 with $C = 6$), whether it was termed *bicarbonated hydrogen* or *etherin*, was an immediate constituent of alcohol and ether. A year later, however, referring to Magnus's and Liebig and Wöhler's work on sulfovinic acid in his attack on Dumas and Boullay's models of constitution, Berzelius presented the former's results as clear experimental facts, contradicting Dumas and Boullay's models. He wrote that it was now "proven that the composition of sulfovinic acid can be seen as a compound of equal numbers of atoms of sulfuric acid and alcohol."[55] He also saw an experimental fact of this kind in Pelouze's model for the constitution of phosphovinic acid, which is discussed in the following section.

Pelouze's Model for the Constitution of Phosphovinic Acid

In 1807, the Paris pharmacist Pierre Boullay already had attempted to form ether with alcohol and phosphoric acid instead of the usual sulfuric acid. He had claimed that this alternative procedure also produced ordinary ether.[56] In 1828, Jean Dumas and Polydore Boullay adopted this opinion, and in 1833, Théophile Jules Pelouze also linked his research to Boullay's earlier experiments.[57] In repeating these experiments, Pelouze obtained a considerable amount of ordinary ether and a new acid that he named phosphovinic acid. Because he considered phosphovinic acid to be analogous to sulfovinic acid, the study of its constitution was of particular interest. Pelouze took on this task, again exclusively on the basis of the quantitative analysis of the new compound, without carrying out further experiments on the reactions of phosphovinic acid. The successive transformations of inscriptions were the decisive steps of his model construction.

In his analyses of baryta of phosphovinic acid, Pelouze first determined the weight percentages of phosphoric acid and "baryta" (barium oxide) contained in the salt and then the weight percentages of carbon, hydrogen, and oxygen using Liebig's "Kaliapparatus." Starting from the analytical results given in weight percentages—82.800% baryta of phosphoric acid, 9.166% carbon, 2.266% hydrogen, 5.768% oxygen—he presented the following series of transformations:[58]

> These numbers [the analytical results] give in atoms:
>
> 2 BaO
> Ph^2O^5
> $H^{12.35}$
> $C^{4.06}$
> $O^{1.95}$
>
> and thus approximate as much as one can hope the formula 2 BaO + Ph^2O^5 + 12 H + 4 C + 2 O. Because (H^4C^2 + H^2O) represents one atom of alcohol, one can see the baryta of phosphovinic acid as a one-and-a-half basic salt, in which one at. of phosphoric acid is saturated with 2 at. of baryta and 2 at. of alcohol. According to the two experiments described above, it contains 12 atoms of water of crystallization. Baryta of phosphovinic acid thus has the following formulas in a dry state:
>
> 2 BaO + 2 (H^4C^2 + H^2O) + Ph^2O^5,

in a water-containing state:

$2\ BaO + 2\ (H^4C^2 + H^2O) + Ph^2O^5 + 12\ H^2O.$

Pelouze required three steps to construct his formula model of the binary constitution of phosphovinic acid: $2\ (H^4C^2 + H^2O) + Ph^2O^5$ for water-free phosphovinic acid and $2\ (H^4C^2 + H^2O) + Ph^2O^5 + 12\ H^2O$ for the water-containing acid, according to which water-free phosphovinic acid consisted of phosphoric acid (Ph^2O^5) and alcohol ($H^4C^2 + H^2O$). He first transformed the weight percentages into the numbers of "atoms" of the elements by dividing the weight percentages by the theoretical combining weight of each element; as an exceptional step, he provided the calculated values and rounded off only in the next step of transformation when he constructed the Berzelian formula. Furthermore, in this first step, Pelouze represented the baryta (barium oxide) and phosphoric acid contained in the salt and analyzed separately, by the formula $2\ BaO + P^2O^5$.

In the second step, Pelouze constructed the Berzelian formula for phosphovinic acid, $2\ BaO + Ph^2O^5 + 12\ H + 4\ C + 2\ O$, by rounding off the numbers of "atoms" and newly arranging the symbols in a linear sequence. The third step consisted of regrouping the partial formula $12\ H + 4\ C + 2\ O$ into $2\ (H^4C^2 + H^2O)$ and adding the parentheses to denote the stable immediate constituent "alcohol." In doing so, Pelouze took Dumas and Boullay's model of the constitution of alcohol (see my reconstruction in Fig. 6.4). Pelouze's rhetoric represented his formula model of the binary constitution of phosphovinic acid from phosphoric acid and alcohol as if it were a necessary consequence of the experimental data. Finally, he argued that it could be concluded from this result that the analogous sulfovinic acid must consist also of sulfuric acid and alcohol. The decisive step for this move was in comparing the formula model $2\ BaO + 2\ (H^4C^2 + H^2O) + Ph^2O^5$ for the barium salt of phosphovinic acid with the formula model $BaO + 2\ (H^4C^2 + H^2O) + 2\ SO^3$ for the barium salt of sulfovinic acid, selected from the three alternative models that had been previously presented by Wöhler and Liebig.

In his 1833 paper, Berzelius adopted Pelouze's model for phosphovinic acid stating that Pelouze had "shown" that phosphovinic acid is made up of one "atom" of phosphoric acid and two "atoms" of alcohol.[59] This view was, however, not shared by Liebig, who had received a sample of the barium salt of phosphovinic acid from Pelouze and repeated its quantitative analysis. Liebig obtained different analytical results, in particular a lower proportion of hydrogen and oxygen: 60.875% baryta of phosphoric acid, 29.150% crystal water, 6.578% carbon, 1.195% hydrogen, and 2.212% oxygen.[60] His calculation of the compound's composition in integral numbers of "atoms" resulted in the

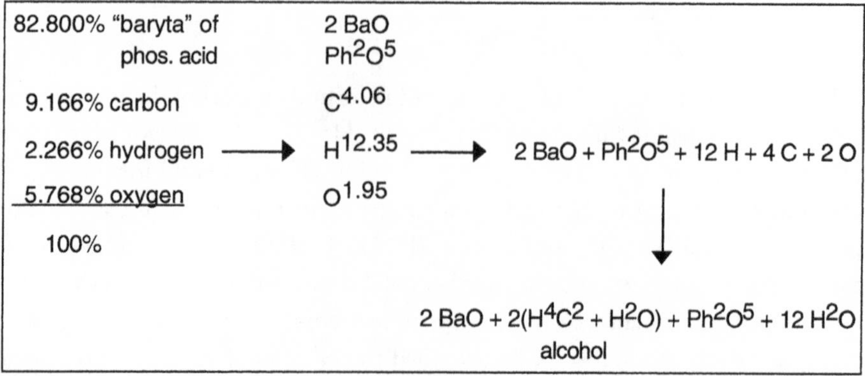

Figure 6-4 Pelouze's construction of the model of binary constitution for phosphovinic acid (my reconstruction)

revised ratio of ten "atoms" of hydrogen, four "atoms" of carbon, and one "atom" of oxygen, instead of Pelouze's formula 12 H + 4 C + 2 O. This new ratio of "atoms" now could no longer be transformed into the partial formula for alcohol ($H^4C^2 + H^2O$ or its multiples); so Liebig concluded "that phosphovinic acid cannot be considered a compound of phosphoric acid with alcohol."[61] Instead, he suggested that it was more probable that ordinary ether was a constituent of phosphovinic acid:

60.875% "baryta" of phos. acid		1 at. baryta of phos. acid	
29.150% crystal water		12 at. water	
6.578% carbon	⟶	4 at. carbon	}
1.195% hydrogen		10 at. hydrogen	} 1 at. ether
2.212% oxygen		1 at. oxygen	}

He also added a second possibility: "The results [agree] just as precisely with the hypothesis that phosphovinic acid is composed of phosphoric acid and olefiant gas or etherin."[62] Although we do not have direct evidence, it is probable that this alternative resulted from removing the partial formula H^2O from the formula for ordinary ether ($2\ C^2H^4 + H^2O$) and grouping it together with the formula for phosphoric acid (see my reconstruction in Fig. 6.5); thus, H^2O no longer signified an immediate constituent of a compound; rather, it signified hydrate water. Both of Liebig's models implied acceptance of Dumas and Boullay's formula models of the constitution of alcohol and ordinary ether as compounds of bicarbonated hydrogen.

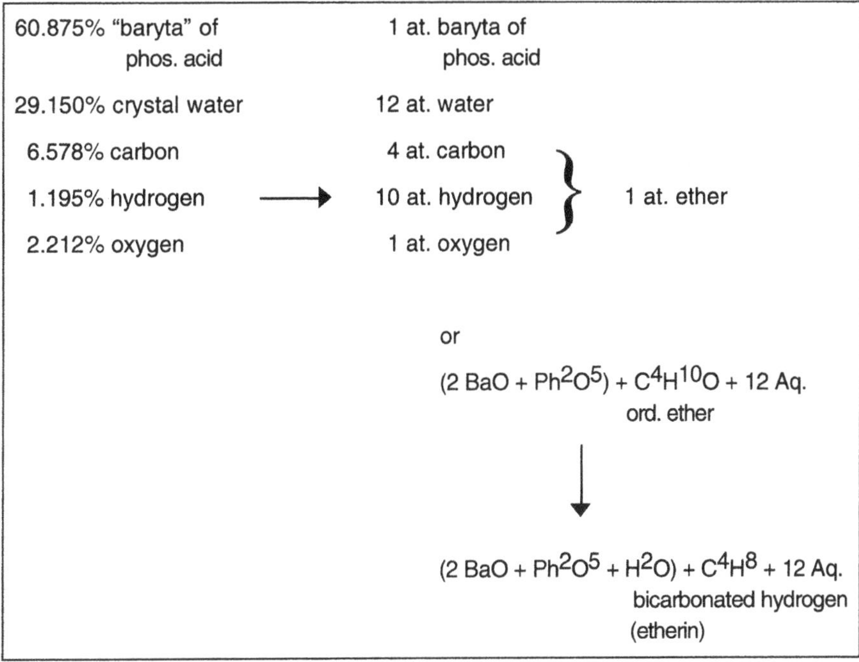

Figure 6-5 Liebig's construction of two models of binary constitution for phosphovinic acid (my reconstruction)

Liebig's Early Formula Models of Compounds Containing Ordinary Ether

Liebig's alternative model of the constitution of phosphovinic acid, which showed it as a binary compound consisting of phosphoric acid and ordinary ether or, alternatively, of hydrated phosphoric acid and bicarbonated hydrogen, is intimately linked to three other models of constitution, published in the same year for "acetal," "wood alcohol," and "acetic ether," all of which were closely related to ordinary alcohol.[63] Liebig had obtained acetal from Johann W. Döbereiner, who had produced it by vaporizing alcohol over platinum black. Liebig's quantitative analysis of this new substance and his transformation of the analytical results into integral numbers of "atoms" came to eight "atoms" of carbon, eighteen "atoms" of hydrogen, and three "atoms" of oxygen, or the formula 8 C + 18 H + 3 O.[64] Based on this result, Liebig disagreed with Döbereiner that the new substance was an ether; instead, he claimed it was a product resulting from the oxidation of alcohol, which he called *acetal*.[65] He supported this claim first with a formula model of the reaction in which acetal was created, second by a formula model of the binary constitution of acetal. In both cases, he used Berzelian formulas in their simplest form, that is, avoiding exponents, parentheses, and dots to represent oxygen.

Based on the Berzelian formulas 8 C + 18 H + 3 O for acetal and 2 C + 6 H + O for alcohol (C = 12), he constructed the formula model of the formation of acetal as follows:[66]

> From 4 at. of alcohol, through the partial oxidation of the hydrogen, 4 at. of hydrogen and 1 at. of water are removed.

8 C + 24 H + 4 O	=	4 at. alcohol
6 H + 1 O	=	4 at. hydrogen + 1 at. water
8 C + 18 H + 3 O	=	1 at. acetal

The general rule for constructing a model of reaction was that the sum of the theoretical combining weights of the initial substances had to be equal to that of the reaction products. This was the case if the formula of alcohol, 2 C + 6 H + O, that according to Liebig represented one "atom" of alcohol, was multiplied by a factor of four. From the balancing of 8 C + 24 H + 4 O with 8 C + 18 H + 3 O, a difference of 6 H + 1 O resulted, a formula that Liebig divided into 2 H + O (the formula for water) and 4 H (the symbol for the oxidizable hydrogen contained in alcohol). The information that, in the formation of acetal, water and hydrogen are removed simultaneously, and the quantitative details of it were thereby exclusively a result of the work with chemical formulas.

Based on the Berzelian raw formula 8 C + 18 H + 3 O for acetal, Liebig then presented the following model of its binary constitution:[67]

1 at. of water-free acetic acid	4 C + 6 H + 3 O
3 at. of ether	12 C + 30 H + 3 O
	16 C + 36 H + 6 O

According to this balancing schema, acetal was a binary compound consisting of acetic acid and ordinary ether. The main task for constructing the model was to compare acetal's formula with formulas of possible constituents and to fit these formulas into an equation. Liebig did not perform any experiment to study the reactions of acetal to support the model. As Pelouze had done in the case of phosphovinic acid, he constructed the model merely by manipulating chemical formulas. The result was again a model showing ordinary ether as one of the two immediate constituents. Liebig, however, mentioned another possibility of a formula model that would support Dumas and Boullay's models. If the formula 4 C + 10 H + 1 O for ether was divided into that of water 2 H + O and bicarbonated hydrogen or etherin C^4H^8, one could also consider acetal to

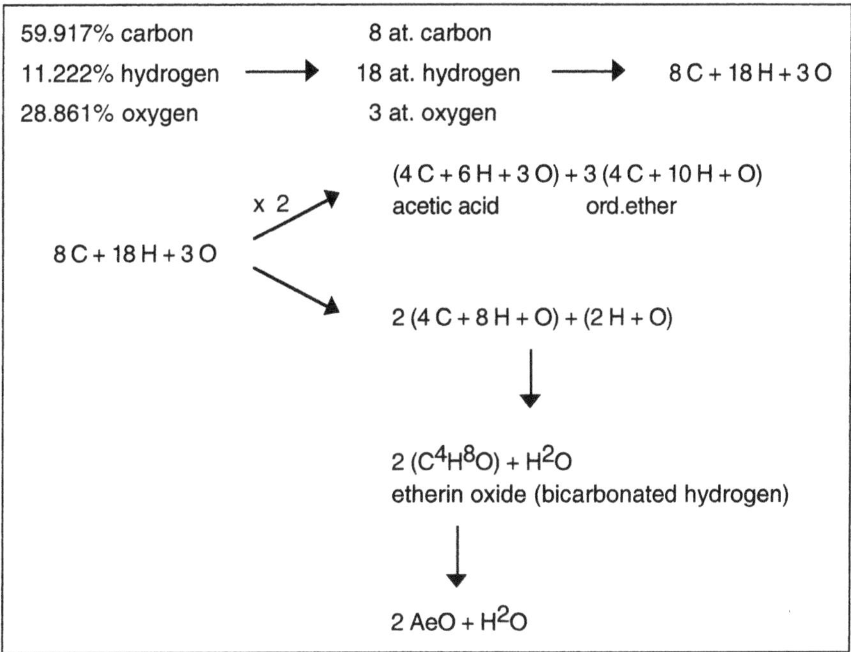

Figure 6-6 Liebig's construction of two models of binary constitution for acetal (my reconstruction)

be etherin oxide combined with water, 2 (C^4H^8O) + H^2O or 2 AeO + H^2O, as Berzelius had suggested a year earlier.[68] The fact that the second model of constitution was based on a Berzelian raw formula that represented one "atom" of acetal (8 C + 18 H + 3 O), whereas the first model was based on a Berzelian raw formula for two atoms (16 C + 36 H + 6 O), shows once again the extent to which the formal possibilities of manipulating Berzelian formulas, their multiplication, and the regrouping of the individual symbols determined the construction of models of constitution (see my reconstruction in Fig. 6.6).

Liebig presented his model of constitution of acetal together with the formula models for the constitution of two other compounds. By distilling wood vinegar, Liebig had obtained a substance very similar to alcohol, which he called, analogous to "spirit of wine," "spirit of wood."[69] He analyzed this substance and transformed the analytical results—54.747% (54.753%) carbon, 10.753% (11.111%) hydrogen, 34.500% (34.136%) oxygen—into integral numbers of "atoms:" two atoms of carbon, five atoms of hydrogen, and one atom of oxygen. Based on this, he constructed the following model of the binary constitution of wood spirit:[70]

According to this composition, wood spirit can be considered a compound of:

	C	H	O
1 atom of ether	4 +	10 +	1
with 1 atom of oxygen		+	1
wood spirit	4 +	10 +	2

Liebig obtained this model of the constitution of wood spirit, in which it is represented as a binary compound of ordinary ether and oxygen, merely by removing the symbol O for oxygen from the formula for wood spirit. Again, he succeeded in constructing a model of constitution showing ordinary ether as an immediate constituent (see my reconstruction in Fig. 6.7).

A third model possessing the same immediate constituent—ordinary ether—was of that of acetic acid ether. Liebig had repeated the quantitative analysis of acetic acid ether, transformed the analytical results—54.47% carbon, 9.67% hydrogen, 35.86% oxygen—into an integral number of atoms—eight atoms of carbon, sixteen atoms of hydrogen, four atoms of oxygen—and then constructed the following formula model of the constitution of acetic acid ether:[71]

		C	H	O
1 atom	of ether	4 +	10 +	1
1 –	of water-free acetic acid	4 +	6 +	3
acetic acid ether		8 +	16 +	4

According to this model, acetic acid ether was made up of ordinary ether and acetic acid (see my reconstruction in Fig. 6.8). Liebig achieved this model by regrouping the formula for acetic acid ether without performing additional experiments to study the reactions of acetic acid ether. It is important to note that in this case the purpose of modeling informed the first transformation step in which the analytical results were transformed into an integral number of atoms. In this step, Liebig did not prefer the smallest integral number of atoms—two atoms of carbon, four atoms of hydrogen, and one atom of oxygen—but a multiple of this, which allowed him to construct a model of constitution that contains the formula for ordinary ether.

In 1833, merely by working on paper with chemical formulas, Liebig constructed formula models of the binary constitution of four different organic compounds—phosphovinic acid, acetal, spirit of wood, and acetic acid ether—which showed these substances to be binary compounds with ordinary ether as a common immediate constituent. It seems clear that in 1833, Liebig wished to pursue the construction of a new class of organic substances that contained not bicarbonated hydrogen (Dumas and Boullay), or etherin (Berzelius), but

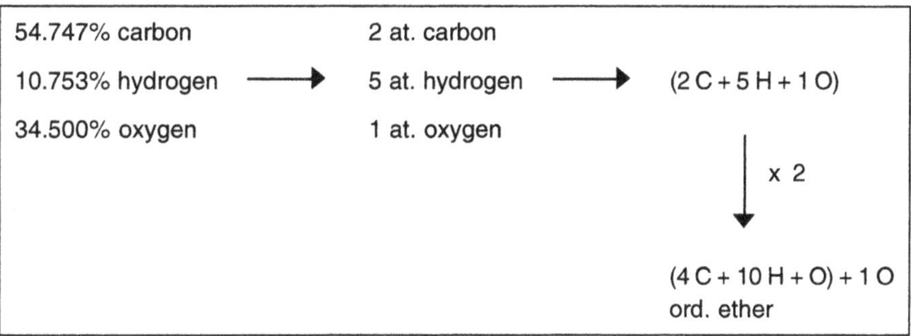

Figure 6-7 Liebig's construction of the model of binary constitution for wood spirit (my reconstruction)

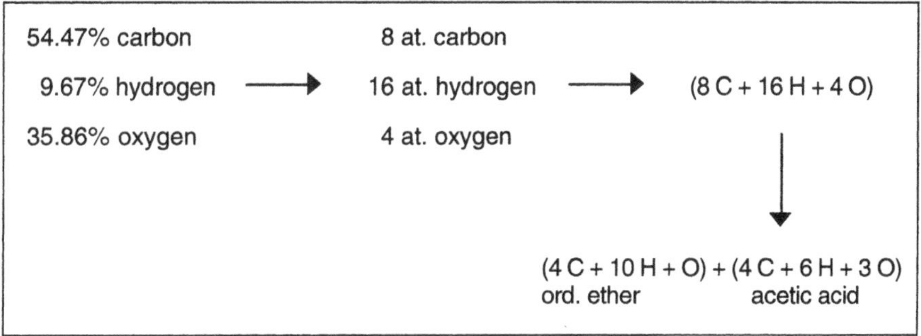

Figure 6-8 Liebig's construction of the model of binary constitution for acetic acid ether (my reconstruction)

the common denominator ordinary ether. Berzelius eagerly took up these formula models, writing that Liebig had "found" that wood spirit, acetic acid ether, and acetal contained ordinary ether as an immediate constituent.[72] He did not give the slightest hint that Liebig lacked definite positive experimental evidence for his models—given the fact that chemists continued to consider that claims about the constitution of chemical compounds ought to be embedded in the study of their chemical reactions. Thus, it is not surprising that the construction of models of constitution by manipulating chemical formulas on paper was occasionally the target of criticism in the ensuing controversy about radicals. If, for example, Liebig objected that Dumas made claims "with the quill in hand,"[73] the latter answered with similar charges. He remarked, for example, that in constructing wood spirit's model of constitution, Liebig "was satisfied with the formula, which in the end proved nothing."[74]

THE CONTROVERSY

Whereas Liebig and other chemists before 1833 had not questioned the two basic models of Dumas and Boullay, according to which alcohol and ether were compounds of bicarbonated hydrogen (C^4H^8 with C = 12, or C^8H^8 with C = 6) and water, Berzelius in 1833 began to attack the models of his French colleagues. He now suggested that alcohol was a compound that contained the radical CH^3, whereas ordinary ether and many other alcohol derivatives were binary compounds of the radical C^2H^5. There is no direct indication in his paper how Berzelius had obtained the new formula models containing the radicals CH^3 and C^2H^5. In a letter to Liebig from the same year, we find an example of his procedure. In 1832, following the models of constitution of Dumas and Boullay, Berzelius had considered the ethers of the hydrogen acids as compounds of etherin C^4H^8 so that, for example, the model of constitution of the ether of hydrochloric acid was $C^4H^8 + \not{H}\not{Cl}$.[75] Now he transformed this model into a model of constitution containing the radical C^2H^5 by two manipulations:[76]

$$C^4H^8 + \not{H}\not{Cl} = \not{C}^2\not{H}^5 + \not{Cl}$$

2 H from the formula for hydrochloric acid was placed in the formula for the hydrocarbon so that from C^4H^8 the partial formula C^4H^{10} was formed, and the latter was divided into C^2H^5 and completed with a line through it, which denoted two "atoms" of the radical C^2H^5. Berzelius gave no explicit explanation for this division, but it is reasonable to assume that it was motivated by his electrochemical theory. The smaller the number of elementary atoms in a compound immediate constituent or radical, the more likely it was that the radical would be stable. The preference for the simpler radical CH^3 instead of C^2H^6, which also would have been possible in his new system, also resulted from this kind of considerations.

It is quite probable that Berzelius constructed additional models of constitution analogous to the way he modeled the binary constitution of hydrochloric acid ether, that is, by reshuffling letters of either the existent formula model or the Berzelian raw formula for the substance. From his own etherin model of alcohol in 1832, $C^4H^8 + 2\ H^2O$, the model containing the radical CH^3 might have been constructed as follows:

$$C^4H^8 + 2\,H^2O \rightarrow C^4H^{12} + 2\,O \overset{:2}{\rightarrow} C^2H^6 + O \rightarrow \not{C}\not{H}^3 + O$$

In the first step of transformation, the symbols had to be regrouped, then the formula obtained was halved, and finally the partial formula C^2H^6 was replaced by the formula ~~CH~~^{3} with a line through it, symbolizing two "atoms" of the radical CH^3, which was specific to alcohol. In the case of the new model of constitution of ordinary ether containing the radical C^2H^5, a simple regrouping of the individual symbols in the summation formula $C^4H^{10}O$ yielded $C^4H^{10} + O$, which again had to be divided in half to arrive at the constitution formula ~~C~~^{2}~~H~~^{5} + O.[77]

At the end of his article, Berzelius assembled the formula models for a total of seven organic compounds in one class, the binary compounds of the organic radical C^2H^5, which Liebig a year later termed the *ethyl radical* (with the formula C^4H^{10}). To make the binary constitution visually clearer, he replaced the partial formula C^2H^5 for the postulated radical with the symbol *Ae*. Again, two "atoms" were represented by a line through the symbol; the symbol for the oxygen was a dot:[78]

~~$\dot{\text{A}}$e~~ = ether
$\dot{\text{A}}$e = spirit of wood
~~Ae~~ ~~Cl~~ = light hydrochloric acid ether
~~Ae~~ ~~Br~~ = hydrogen bromide ether
~~$\dot{\text{A}}$e~~ ~~$\dddot{\text{N}}$~~ = nitric acid ether
~~$\dot{\text{A}}$e~~ $\bar{\text{A}}$= acetic acid ether
~~$\dot{\text{A}}$e~~^{3} ~~$\bar{\text{A}}$~~= acetal.

Earlier in his paper, Berzelius suggested a new terminological distinction between *empirical* and *rational* formulas. This distinction indicates to some extent the generative function of chemical formulas for the construction of models of constitution and radicals:[79]

> In order to express myself with greater ease, I will in the following use two kinds of formulas for the composition of organic bodies. *The one I will call empirical; they follow immediately from a correct analysis, and are immutable. The others I will call rational,* because they are intended to give an idea of the two electrochemically opposite bodies, out of which one can consider the atom to be formed, that is, they are intended to show its electrochemical division. The empirical formula of alcohol is C^2H^6O. *The rational formula varies according to perspective,* for example, $C^2H^4 + \dot{H}$ or ~~CH~~^{3} + O. To determine, however, which of these is the true rational formula is a difficult problem. Only this much can be said: *The rational formula cannot have more than two parts.* But these can be assumed quite differently.

Berzelius here termed the formulas developed directly from the analytical data *empirical*, in contrast to the *rational* formulas, which represent the constitution of the substances by a particular order of the symbols, generating partial formulas, supplemented by parentheses and the plus sign. There were no general methodologic rules for the construction of rational formulas—or *formula models*, as I have called them. Berzelius refers to this fact by commenting that the rational formulas varied "according to the perspective" of the individual chemist. The only positive overall criterion he mentioned for the construction of rational formulas was the concept of binary constitution.

Unlike bicarbonated hydrogen, the "organic radicals" denoted by the formulas CH^3 and C^2H^5 never had been isolated in chemical experiments and identified as laboratory substances. Both were hypothetical entities, and claims about their existence would have needed a particular justification in light of the fact that the chemists of this period understood "radicals" to be veritable laboratory substances. Yet, although Berzelius is today known for his warnings against the overestimation of chemical formulas prior to 1833, and for demanding experimental evidence from his colleagues, he himself was now satisfied with mere work on paper, the only visual referents of his supposed radicals being marks on paper. Nevertheless, Berzelius's rhetoric suggested that his new models were based on experimental facts and compelling conclusions. This style was pioneering for chemists' mode of argumentation in the controversy about radicals after 1833. As a rule, chemists compensated the lack of positive experimental evidence for their own model of constitutions by proving the empirical inadequacy of the alternative models under attack. The historical analysis shows, however, that none of the participants of the controversy could support his own model of constitution with positive experimental results. In all cases, tinkering with Berzelian formulas on paper was the decisive and necessary means of constructing models of binary constitution and radicals.

Continuation of the Controversy

A year later, Berzelius's younger colleague and friend, Justus Liebig, accepted the new theory of the constitution of alcohol, ethers, and other alcohol derivatives.[80] Although Liebig adopted Berzelius's formula models for ether and other alcohol derivatives without any alteration, he no longer represented alcohol as a compound of the radical CH^3, placing it instead in the class of compounds of the radical C_2H_5 or C_4H_{10},[81] obtained by multiplying the former formula. He also disagreed with Berzelius's postulate that all organic compounds in the strict sense, that is, those found in nature, have to be oxides of compound radicals. In this respect, he was willing to compromise with his French colleagues in that he viewed organic compounds both as oxides and salts. After consideration of all the "facts" in support of this or that view of the constitution of ordinary ether, Liebig wrote in the introduction to his paper, he came to the "conclusion" that ether is "the oxide of a compound radical made up of 4 atoms of carbon and 10 atoms of hydrogen." He named this immediate constituent the *ethyl radical*.[82] A little later in this paper, he wrote that, "utterly convinced of the correctness" of Berzelius's theory, he believed that the key to deciding between the controversial models of constitution lay in experimentation.[83] In fact, Liebig now attempted for the first time to find experimental evidence for the existence of the ethyl radical, but his efforts were not crowned with success. Although he did not find evidence for his own models of constitution, his experiments yielded enough material for the emerging controversy.

Liebig based his experiments on the following consideration. If ordinary ether, as Berzelius had claimed, was an oxide of the radical C_2H_5 or C_4H_{10}, which behaved analogously to the metal oxides, it could be expected that it decomposed by adding potassium because potassium was considered an element with a strong affinity to oxygen, which would displace the radical C_4H_{10} and release it as an observable laboratory substance. It turned out, however, that potassium not only reacted extremely slowly with ordinary ether, but it also created a whole series of new products when air was present. Liebig could not identify any of these products as the expected ethyl radical. If the experiment was executed without air contact, the reaction came quickly to a stop without having produced sufficient amounts of reaction products. Although Liebig introduced a series of *ad hoc* hypotheses—for example, that the multiplicity of the new products "complicates the separation of the radical"—in the end he could not help admitting that he had performed "some cursory experiments with ordinary ether and potassium, but without giving any decisive result."[84]

Further experiments, which Liebig referred to only briefly, presumably also were unsuccessful. For example, he tried to decompose the ether of hydrochloric acid, which according to Berzelius was a binary compound of the ethyl radical C_4H_{10} with chlorine. In this case, he expected the formation of potassium chloride and the free ethyl radical. In view of the negative result, Liebig comforted his readers by promising that the future "will decide very quickly the extent to which this supposition is probable."[85] Despite these failed experiments, Liebig's conviction that the ethyl radical was a stable immediate constituent of all kinds of ethers and other organic substances remained unshaken: "I have no doubts that it will be possible to obtain the radical of ether, that is the hydrocarbon compound C_4H_{10}, free of all other bodies."[86]

If Liebig was not able to yield positive experimental results for Berzelius's models of constitution, it was all the more important to find experiments for countering his opponents, Dumas and Boullay. Thus, he employed Pelouze's experiment with phosphovinic acid, without being disturbed that he himself a year earlier had still thought it possible that phosphovinic acid consisted of phosphoric acid and olefiant gas or etherin. Now, following his mentor's rhetoric, he claimed that it "had been proved that these acids [sulfovinic and phosphovinic acid] do not contain olefiant gas, but rather spirit of wine [alcohol] or ether."[87] Furthermore, Liebig presented an alternative interpretation of Dumas and Boullay's experiment with oxalic acid ether, which they had introduced as positive experimental evidence for their models of constitution.[88]

Another of his arguments against Dumas and Boullay's models of constitution rested on reinterpretation of the quantitative analysis of a salt that the Danish chemist William Christopher Zeise had produced in 1831 from potassium chloride and a new organic compound, which he had obtained from alcohol and platinum chloride.[89] Zeise had interpreted this salt as a compound analogous to the double salts, made up of bicarbonated hydrogen, platinum chloride and potassium chloride. Thus, it provided further evidence for Dumas and Boullay's models of constitution. Liebig adopted Zeise's analytical data, but he transformed them into a different "empirical formula," based on his intention to reject Dumas and Boullay's models. The new empirical formula was then a suitable starting point for constructing a new formula model that did not contain the partial formula of bicarbonated hydrogen but rather that of ordinary ether.

Liebig's method is of particular interest not least because it shows that the path from the results of quantitative analysis to "empirical formulas" and formula models of constitution was not a one-way route. Thus, it throws a telling light on how chemists attempted to fit empirical formulas and rational formulas together. Liebig emphasized that although Zeise had analyzed the new compound with unusual and great "precision,"[90] the individual measurements of the repeated analyses with different samples of the compound still varied widely. Zeise had obtained the following analytical data: 51.3326% (51.2540%, 50.4535%, 50.1980%, 52.5659%) platinum, 20.375% (19.744%) potassium chloride, 7.1520% (6.9455%, 6.8861%, 6.6191%, 5.7113%) carbon, and 1.53259% (1.64035%, 1.19870%, 1.04169%, 1.42120%, 1.05310%) hydrogen. He had selected some of these data and transformed them into the following integral numbers of "atoms":[91]

2 at. platinum,

4 at. chlorine,

1 at. potassium,

2 at. chlorine,

4 at. carbon,

8 at. hydrogen.

In contrast, Liebig used Zeise's analytical data to construct an average of the different analytical data for each element. He thus obtained the following average weight percentages: 51.1796% carbon, 20.059% potassium chloride, 6.6623% carbon, and 1.3146% hydrogen. On the basis of this average, Liebig now constructed the following different model of the constitution of the salt:[92]

2 at. platinum,
4 at. chlorine,
1 at. potassium,
2 at. chlorine,
4 at. carbon, }
10 at. hydrogen, } 1 at. ether
1 at. oxygen. }

The particular ordering of the "atoms" in the preceding representation, for example, by dividing the six "atoms" of chlorine into four plus two "atoms," already paved the way for the subsequent construction of the model of constitution. The addition of the bracket further completed the model by visually showing that four "atoms" of carbon, ten "atoms" of hydrogen, and one "atom" of oxygen formed the stable constituent ordinary ether. Immediately after his presentation of the model, Liebig claimed that "we know with certainty that the radical of ether C_4H_{10} is contained in this compound."[93] His "certainty" lay not on additional experiments but rather on the visual image of his model.

At the end of his paper, Liebig grouped alcohol and eighteen of its chemically produced derivatives together as a new class, the binary compounds of the ethyl radical C_4H_{10}. This new class included all known kinds of ether, spirit of wood, and further immediate and more distant derivatives of alcohol (Fig. 6.9). He also tentatively ordered two formula models of "natural" organic compounds into the new class, but not without questioning this again by separating the two formula models with a line from the formula models of the artificial chemical products. At the end of his list, he mentioned additional possible candidates for the new class of binary compounds of ethyl by giving their "empirical formula." In his table, Liebig replaced the formula of the radical C_4H_{10} by the symbol E, making visually more obvious the binary constitution of the classified compounds. In the case of the more complex binary compounds, such as sulfovinic acid, which was analogous to the double salts, the binary constitution was emphasized visually by the added parentheses. For example, the parentheses around the first two partial formulas EO and H_2O in the constitution formula (EO + H_2O) + 2 SO3 for sulfovinic acid meant that this acid consisted of the two immediate constituents ethyl oxide hydrate and sulfuric acid. Furthermore, the formula showed that the constituent ethyl oxide hydrate (EO + H_2O) also had a binary constitution (ethyl oxide and water) and that these constituents for their part were also made up of two substances.

In 1833, Berzelius proposed that organic compounds were analogous to inorganic oxides. In contrast, Liebig's table presented organic compounds as binary compounds, which were either analogous to inorganic salts, as Dumas and Boullay had claimed, or to inorganic oxides. For example, ordinary ether E + O was an example of an oxide; cane sugar (4 CO_2 + 2 EO + H_2O) and grape sugar (4 CO_2 + 2 EO + 4 H_2O), in contrast, were examples of salts. One must "consider the species of sugars as salts and ether as an oxide,"[94] Liebig wrote in a commentary on his table. The formulas for cane sugar and grape sugar showed that the only difference between these two species of sugar was the content of water of crystallization. Dumas and Boullay also had seen cane sugar and grape sugar as salts consisting of water of crystallization, carbonic acid, and—this is a difference—bicarbonated hydrogen. In their system, grape sugar also differed from cane sugar only in the content of water of crystallization, and they also succeeded in presenting a plausible interpretation of alcoholic fermentation in the framework of their system.[95] Berzelius's immediate and vehement protest against the claim that sugars were analogous to inorganic salts was obviously not shared by Liebig.

With the example of Liebig's construction of the model of constitution of Zeise's salt, we have seen that he arrived at his models of constitution containing the ethyl radical by manipulating chemical formulas. It is not necessary to go into additional individual cases because the example shows that Liebig, in principle, proceeded in the same way as Berzelius. The new models of the constitution of alcohol, ethers, and other alcohol derivatives of both chemists and the new classification based on them were the result of work on paper with chemical formulas. Liebig did not succeed in introducing even one

Bezeichnen wir die Kohlenwasserstoffverbindung 4 C + 10 H als das Radikal des Aethers mit E_2 und nennen es Ethyl, so haben wir zum Ausdruck der Zusammensetzung seiner Verbindungen folgende Formeln.

E = Radikal des Aethers = $C_4 H_{10}$
E + O = Aether
E + 2 O = Holzgeist
EO + H_2O = Hydrat (Alcohol)
E + Cl_2 = Chlorur (Chlorwasserstoffäther)
E + I_2 = Iodur (Iodwasserstoffäther)
E + B_2 = Bromur
E + S = Sulfur?
EO + Ox = Oxalat (Oxaläther)
EO + BO_3 = Benzoat (Benzoeäther)
EO + N_2O_3 = Nitrit (Salpeteräther)
EO + $\overline{A}$ = Acetat (Essigäther)
3 EO + $\overline{A}$ = Acetal
(EO + H_2O) + 2 SO_3 = Weinschwefelsäure
(EO + H_2O) + P^2O^5 = Weinphosphorsäure
EO + 2 SO_3 = Aetherschwefelsäure
EO + 2 Cl_2 Pl = Weinchlorplatinsäure
(EO + $\overline{A}$) + $CaCl_2$ = Verbindung von Essigäther mit Chlorcalcium. (a)
(EO + $\overline{A}$) = Essigäther) + 2 ClH = Verbindung im rohen schweren Salzäther. (b)
3 ECl_2 + 2 Cl_2 Cl_5 = Chloräther (c)

Rationelle Formel

Stärke	
Milchsäure	= $C_6H_{10}O_5$
Rohrzucker	= 4 CO_2 + 2 EO + H_2O
Kryst. Milchzucker	
Schleimsäure	= $C_6H_{10}O_8$
Traubenzucker	= 4 CO_2 + 2 EO + 4 H_2O
Gummi	
Mannit	
Brenzl. Schleimsäure	= $C_9H_3O_6$?

Empirische Formel

Stärke	= 12 C + 20 H + 10 O (f)
Milchsäure	= 12 C + 20 H + 10 O
Rohrzucker	= 12 C + 22 H + 11 O (d)
Kryst. Milchzucker	= 12 C + 24 H + 12 O (h)
Schleimsäure	= (12 C + 20 H + 10 O) + 6 O (i)
Traubenzucker	= 12 C + 28 H + 14 O (e)
Gummi	= 12 C + 22 H + 11 O (g)
Mannit	= 12 C + 28 H + 12 O (k)
Brenzl. Schleimsäure	= 18 C + 6 H + 12 O ?

Figure 6-9 Liebig's classification of ethyl-compounds, after Liebig 1834a

single positive experimental result from his studies of the chemical reactions of these substances. The sole link of his models of constitution to experiments was the quantitative analysis on which the transformations into integral numbers of "atoms" and into Berzelian raw formulas were based. The latter were used as tools that were manipulated to construct a model of constitution. At least three conditions determined the modeling, besides of the formulas: the concept of binary constitution, which prescribed that constitution formulas should consist of two partial formulas; the attempt to construct analogies to particular inorganic classes (oxides and salts); and the syntax (additivity) of the Berzelian formulas themselves, which made possible the free addition of individual symbols (for additional conditions see next section).

On the Causes of the Controversy

With Liebig's publication of his own theory of radicals, a social dynamics of controversy was established that would last a number of years and divide the leading European chemists into two camps. On the one side were Jean Dumas and his French supporters, on the other Jacob Berzelius and Justus Liebig with their German and French allies. With his sarcastic remarks, Liebig contributed significantly to an aggravation in tone. Thus, for example, he commented on Dumas's opinion about the constitution of sugar with the following remark: "Surely only the extreme hurry with which new theories in France have been made and proposed since Mr. Dumas is responsible for the fact that these theories lack the completion which one is justified in expecting from such excellent, talented chemists."[96]

We saw in the preceding how the models of constitution that became the subject of controversy were constructed, but we have not addressed the questions of why various chemists preferred one or the other model and why they carried out heated debates about them. In older historical interpretations, the multiplicity of models of constitution and hypotheses about radicals, which from today's perspective are false, often were evaluated as a result of confusion not worthy of further examination.[97] In more recent accounts, however, they have been viewed as a challenging explanandum.[98] F. L. Holmes has pointed out that psychological and social factors played important roles in the controversy: "Liebig's shifting alliances, and the personal and national rivalries that motivated them," Holmes writes, "exerted a strong influence on the course of the most prominent controversy in organic chemistry during the 1830s, that over the various radical theories proposed by the leading chemists in the field."[99]

Psychological and social factors, such as the making of alliances and intrigues, unquestionably contributed to the style of the controversy, its emotional intensity, and its duration. As causes for the choice of the different models of constitution and hypotheses about radicals, however, they are unconvincing because they are not specific enough to explain why, for example, Dumas and Boullay preferred bicarbonated hydrogen as an immediate constituent of alcohol and its derivatives, whereas Berzelius and Liebig preferred the ethyl radical. It easily could have been the other way around. There is no causal relationship between the various social interests and emotions of the historical actors on the one hand and the particular intellectual content of the different chemical models of constitution on the other. Nor can chemists' choices be explained by pointing to a superior experimental foundation for this or that model of constitution. According to the collectively accepted standards, quantitative analysis alone could not provide a sufficient experimental basis for the models of constitution. The study of chemical reactions would have been necessary, but in this respect no chemist was able to present positive experimental results.

To create a historically adequate understanding of chemists' predilection for a particular model of constitution, we need to consider the broader intellectual context that was different for individual chemists. For example, Berzelius closely tied his models of constitution to his encompassing electrochemical theory. When, in 1833, he extended the concept of binarity to organic compounds, he did not make a secret of this connection and the particular significance he gave to it. The possibility of applying his electrochemical theory to organic compounds was a decisive and explicit motive for accepting the concept of the binary constitution of organic compounds. In doing so, some particularities of his theory, which were ignored by many chemists, served as a framework that conditioned his selection of possible models of constitution and "radicals." According to Berzelius, all natural organic compounds were made up of three elements: carbon, water, and oxygen. They always must contain oxygen, which in Berzelius's electrochemical theory played an outstanding role as an "absolutely negative element."[100] This special role of oxygen is reflected in Berzelius's postulate that the organic compounds are oxides and that the organic radical always must play the role of the electropositive partner with the electronegative oxygen.[101] The construction of formula models of constitution was significantly constrained by this additional theoretical principle, which was not shared by other chemists, despite the fact that they accepted other core principles of Berzelius's electrochemical theory.

Furthermore, the models of constitution must be placed in the context of the new classification enterprise, which by the end of the 1820s had become most urgent to chemists. All chemists supported their own formula models of constitution by fitting them together to represent a class of substances. Classification thus constrained the construction of a particular formula model of a new compound. Examples can be seen in Dumas and Boullay's 1828 classification of the compounds of bicarbonated hydrogen. The two chemists again altered already existent formula models, such as that of sulfovinic acid and of the ethers of vegetable acids and nitric acid, to fit them into the newly built class.[102]

Additional issues were linked to the classification of organic substances that exceeded the narrow boundaries of chemistry involving more fundamental beliefs. How different were "organic" and "inorganic" substances? How strong was the border between plant and animal "life" and minerals? How far did the unity of nature extend? Questions like these bordered metaphysics and religious convictions. If the distinction between organic and inorganic things and between life and nonlife was blurred or was done away with entirely, the exceptional place of humans, grounded in Christianity, was at stake as well. For the French, the view of unity of nature outweighed the reservation against demolishing the border between organic and inorganic matter. Already in the first decade of the nineteenth century, Louis Thenard and Pierre Boullay had established the analogy between organic substances and inorganic salts.[103] A few years earlier, Fourcroy went so far as to speculate about the synthesis of organic substances from inorganic ones. This line of reasoning was continued by Joseph Louis Gay-Lussac, Jean Dumas, Polydore Boullay, and others. In keeping with the French tradition, in which alcohol and ordinary ether were binary compounds of bicarbonated hydrogen, by the end of the 1820s, Dumas and Boullay did not hesitate to construct analogous formula models of other alcohol derivatives as well as of natural organic substances, despite the fact that bicarbonated hydrogen was viewed as an inorganic substance. For these two chemists, it was not taboo to group alcohol and its clearly "artificial" derivatives together with "natural" organic substances, such as the sugars in their class of compounds of bicarbonated hydrogen. In 1835, Dumas no longer hesitated to declare the distinction between organic and inorganic to be outdated.[104]

Berzelius had a different view of nature. Although his conviction in the unity of nature slowly grew, first in 1814 with the extension of his theory of chemical portions to organic substances, then in 1833 with the extension of the concept of binary constitution, he never revised his basic belief that inorganic and organic matter constituted different realms. He thus never termed artificially produced carbon compounds that contain elements like chlorine and bromine "organic" substances. Justus Liebig echoed the same conviction in 1841, after he had left organic chemistry and turned to physiological and agricultural chemistry.[105]

The belief that organic and inorganic substances constituted significantly different classes had to be supported by showing a difference in the "composition" and in the "constitution" of the substances because these were seen as the most important criteria for the identification and classification of chemical substances. The composition of organic compounds was in this sense relatively unproblematic because organic substances actually differed significantly from inorganic substances in terms of their elementary composition in that they contain a much narrower spectrum of chemical elements. For constructing models of the constitution of organic substances, however, metaphysical beliefs imposed additional constraints. Why did Berzelius refuse Dumas and Boullay's classification of organic compounds with bicarbonated hydrogen, even though bicarbonated hydrogen met the requirements of his electrochemical theory because it could be considered an electropositive constituent? Berzelius was compelled to do so because bicarbonated hydrogen was commonly viewed as an inorganic substance. Hence, to accept it as the denominator of an entire class of organic substances would have blurred the boundaries between the realm of the organic and the inorganic. Already in 1818, Berzelius commented on the assumption of his French colleagues that alcohol and ordinary ether were made up of olefiant gas (bicarbonated hydrogen) and water with the following skeptical remarks: "Some chemists have proposed views about the composition of organic bodies which differ from the prior opinions which I derived from my studies on this subject. They imagine these to be binary *inorganic* compounds."[106] For Berzelius, "organic" radicals had to be specific organic substances apart from the realm of inorganic ones. When in 1832 he was briefly tempted to accept Dumas and Boullay's constitution theory, this happened only with the crucial modification in which bicarbonated hydrogen became the organic radical "etherin" with the formula C^4H^8.

CHAPTER SEVEN

The Dialectic of Tools and Goals

> Dans tous les cas, il est essentiell de marquer que notre puissance va plus loin que notre connaissance.
>
> —BERTHELOT 1860, 810

Between 1834 and 1840, in the context of the controversy about the binary constitution of alcohol, ethers, and their derivatives, Jean Dumas and other chemists, such as Auguste Laurent, introduced and developed a new and rather far-reaching chemical concept, the *concept of substitution.* The coming into being of the concept of substitution, which is studied in this chapter, exhibits a dynamics that is of broader epistemological relevance. It was closely associated with experiments investigating the chemical reaction of alcohol with chlorine and the construction of interpretive formula models of this reaction. When Dumas set out to study the reaction of alcohol with chlorine, which was to become a model object for "substitution," his goal was to support his own model of the binary constitution of alcohol with new empirical evidence.[1] To reach this goal, he applied chemical formulas as suitable paper tools. Yet the manipulation and visual display of these tools had a suggestive power that propelled Dumas to go a step further than would have been necessary to construct his intended model. In an unintended way, this additional step led to the new concept of substitution. The tools reacted back on the goals.

In this chapter, I study the introduction of the concept of substitution by Dumas in 1834 and its further development until 1840. I show that the development of this new concept depended heavily on the work on paper with chemical formulas, and I argue that because of this constitutive function of Berzelian formulas, it is impossible to reconstruct the historical meaning of *substitution* by any direct reference to experimental evidence, that is, without recurrence to formulas. The new concept of substitution meant not only a minor deviation from the original intentions, concepts, and models of chemists, but it meant a change on a much deeper intellectual level. In its elaborated form of 1840, the concept of substitution was incompatible with the concept of the binary constitution of organic substances. Because most of the leading European chemists after 1834 accepted the concept of substitution step-by-step and tacitly by accepting the associated new significance of the manipulations of formulas and of the resulting formula models, development of the full-blown concept of substitution went hand in hand with the fading away of the older concept of the binary constitution of organic substances.

With the introduction and development of the concept and formula models of substitution between 1834 and 1840 and the associated experimental objects and practices, the process of the transformation of vegetable and animal chemistry into the experimental culture of organic chemistry accelerated. In the second section of this chapter, I discuss one aspect of this transformation, the mutual adjustment of analytical data, Berzelian raw formulas, and formula models, and the reflection of this mutual adjustment in the new mode of justification.

A PERFORMATIVE ACCOUNT OF CONCEPTUAL DEVELOPMENT

The concrete incentive for Dumas's experiments and model in 1834 was a series of experiments performed two years earlier by Justus Liebig.[2] The research topics involved in these experiments originally had little to do with the controversy about the constitution of alcohol and organic radicals. They belonged to a much more traditional, natural historical agenda studying vegetable materials, such as "oils." More specifically, chemists were interested in the possibility of artificially producing "oils" and in the question of whether oils produced in different ways were identical or different species.[3] One of the ways to produce an oil in the laboratory was to introduce chlorine gas into alcohol and heat up the mixture. Several French and German chemists had repeated this experiment—which had a long history going back to the late eighteenth century[4] and maybe even earlier—shortly before Liebig in 1832 took up the question and repeated and modified the experiment.

Referring to the earlier experimental results and different opinions concerning the nature of the chemically created oils, he wrote:[5]

> The erroneous and contradictory assumptions that have been made about the nature of these bodies [the oils] and the decomposition of alcohol by chlorine stimulated me several times to study these problems. However, the variety of the products and compounds that I saw being created, as well as the false way that I took then, were the reasons why I did not continue these studies. Intending to gain precise knowledge of all these processes, I tried to learn above all about the bodies which might be created in the complex action of chlorine on alcohol.

Liebig wanted to solve the problem of the identity of the oil created from alcohol and chlorine by studying the chemical reaction underlying its production. The construction of an interpretive model of the reaction was, however, extremely difficult because many different reaction products were produced in addition to the oil and hydrochloric acid produced at the beginning of the operation. Liebig tried to remove this obstacle by successively varying different parameters of his experiment, which he described in detail. Two variations were of particular importance. The first was to apply carefully dried alcohol and chlorine. Liebig gradually reached this approach. He had noticed that more of the oily substance segregated when water was added. Likewise, less was observed when he used purer alcohol, that is, alcohol that contained less water. When he repeated the experiment with absolute alcohol and carefully dried chlorine, the oily substance could not be observed. He inferred that it was still produced but not segregated.

The second variation concerned the quantity of chlorine passed into alcohol and the apparatus applied for that. Liebig had observed that the transformation process, which at first seemed to be completed when the oily substance had been created, actually continued when the temperature was increased; hence, he continued with the introduction of chlorine into the alcohol. Control of the velocity of the stream of chlorine and control and variation of the temperature during the whole operation—cooling of the mixture at its beginning, moderate heating when the absorption of chlorine became slower, and stronger heating up to the boiling point of alcohol at the end of the operation—were important for the outcome of the operation. To obtain a slow and continuous velocity of the stream of chlorine, Liebig altered the common apparatus used for saturating fluids with gases, replacing the Woulfe bottle by a horizontal tube (Fig. 7.1). The whole operation, which took an extremely long time (between eleven and thirteen days), ended with an unexpected result: the formation of a white crystalline substance that was much easier to isolate than any fluid.

Liebig's interest shifted from the oil to this new substance, which he termed *chloral*. Investigating its chemical properties by testing it with the usual tool kit of reagents, he found another new substance, called *chloroform* by Dumas shortly afterward. Liebig then attempted to construct an interpretive model of the formation of chloroform from chloral and an alkali, based on the quantitative analyses of chloral and chloroform whose results subsequently were transformed into chemical formulas. He did not obtain results that satisfied him, however. In particular, he was not convinced that he had found the right formula for chloral.[6] Even less satisfying were his results concerning the formation of chloral from alcohol and chlorine. In particular, the nature of the oil pro-

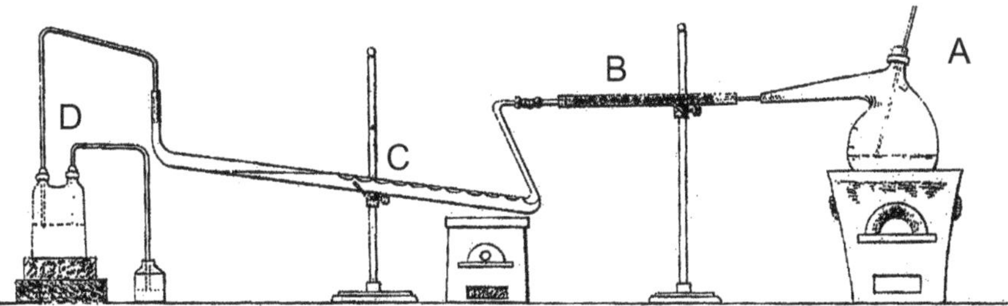

Figure 7-1 Liebig's apparatus for saturating liquids with gases (Liebig 1832): In flask A chlorine is produced, which is dried in B, and passed into alcohol in C. The hydrochloric acid produced is absorbed by water in D.

duced before chloral remained obscure. Because Liebig adhered to the traditional style of studying chemical reactions that prescribed to isolate and analyze *all* the reaction products as a necessary precondition for the subsequent construction of an interpretive model of the reaction, he gave up his original goal.

As far as Dumas was concerned, however, the results published by Liebig in 1832 contained enough information to invoke a new idea: to use these results and to repeat the experiments to support with new experimental evidence his own model of the binary constitution of alcohol, which had been attacked the year before by Berzelius.[7] Dumas repeated both the production of chloral and chloroform and the quantitative analyses of these substances. He obtained different results in the analysis of both chloral and of chloroform. Based on his new analytical results, he constructed new Berzelian formulas for both substances—$C^8H^2Ch^6O^2$ for chloral (Liebig's was $Ch^{12}C^9O^4$; Ch being the symbol for chlorine) and $C^4H^2Ch^6$ for chloroform (Liebig's was C^2Ch^5)—which turned out to be quite suitable paper tools for constructing new models of the two reactions yielding chloral and chloroform. It was in particular the model of the synthesis of chloral, constructed in several steps, that yielded the intended empirical evidence for Dumas's model of the binary constitution of alcohol. I discuss in detail Dumas's stepwise construction of this model in the following section. Then I show how he proceeded from this first formula model, which would have been fully sufficient for his initial purposes, to develop his new concept of substitution.

Dumas's Interpretive Formula Models for the Formation of Chloral

According to Liebig, the explanation for the transformation of alcohol and chlorine into chloral, hydrochloric acid, and other reaction products was extremely difficult because many different reaction products were created, some of which had not yet been identified.[8] Dumas additionally mentioned a hindrance to the identification of all the reaction products, namely, that several of these products could not be isolated: "The complicated reaction of chlorine with alcohol creates a variety of very different substances difficult to isolate."[9] Unlike Liebig, however, Dumas did not give up the goal of representing the complicated reaction. Whereas the former was clinging to the traditional methodology of first isolating *all* the products of a chemical reaction, and hence refrained from modeling the reaction, the latter understood that chemical formulas were suitable tools for modeling the reaction, even without knowing all the reaction products. He followed a strategy already pursued in 1814 and 1815 by de Saussure and Gay-Lussac: to reduce the complexity of the model construction by focusing on the most relevant reaction product—in this case, chloral—and to balance its composition and relative combining weight with those of alcohol. The important difference from the earlier procedure was, however, that Dumas used chemical formulas.

Based on his new formula of chloral, $C^8O^2H^2Ch^6$, and on his older formula for alcohol, $C^8H^{12}O^2$ or $C^8H^8 + H^4O^2$, in a first step, he constructed the following primary model which balances the volumes (indicating the preserved masses) of the initial substances and the reaction products:[10]

The reaction demands

4 vol. alcohol	= 4 vol. hydrogenated carbon = C^8H^8
	4 vol. vapor of water = H^4O^2
16 vol. chlorine	[= Ch^{16}]

It eventually produces

20 vol. of hydrochloric acid	= $Ch^{10}H^{10}$
4 vol. of chloral	= $C^8O^2H^2Ch^6$

Dumas's first interpretive model of the formation of chloral demonstrates that this reaction yields only one byproduct: hydrochloric acid. This can be seen by comparing the formulas of alcohol and chlorine with those of the two reaction products. Because the theoretical combining weights of the elements contained in the initial substances and in the reaction products are balanced, the representation of the reaction is complete. As in the case of the representation of the reaction between alcohol and sulfuric acid yielding ordinary ether, which Dumas had published seven years earlier together with Polydore Boullay,[11] it was now evident that all the other by-products of the reaction were created in independent reaction paths and hence could be ignored in the model of the reaction yielding chloral. The formula model also gives information about the quantity of the produced chloral and hydrochloric acid (in proportions by weight or volumes). Moreover, it shows that in the formula of chloral $C^8O^2H^2Ch^6$ the part $C^8O^2H^2$ of the formula of alcohol $C^8H^{12}O^2$ is preserved, while ten portions (or "atoms") of hydrogen (H^{10}) are displaced by six portions of chlorine (Ch^6). A glance at the formula of hydrochloric acid shows that the ten portions of hydrogen displaced from alcohol recombine with ten portions of chlorine to form $Ch^{10}H^{10}$, representing ten portions by weight or twenty theoretical volumes of this acid.

A year later, in 1835, Dumas further simplified his reaction model by comparing only the formulas of alcohol and chloral:[12]

> The formula of alcohol is:
>
> | | |
> |---|---|
> | | $C^8H^{12}O^2$ |
> | that of chloral is | $C^8H^2O^2Ch^6$ |
> | alcohol has lost | H^{10} |
> | and gained | Ch^6 |
>
> for producing one atom of chloral from each atom of alcohol.

This simplified model brought to the foreground a specific observation: the alleged equivalence of 10 H and 6 Ch. Dumas used this observation for his next, more important formula model, which, in his own eyes, was a successful empirical underpinning of his own model of the binary constitution of alcohol.

In inorganic chemistry, the equivalence of ten proportions by weight or volumes of one element displacing six proportions or volumes of another element was rather unusual. Dumas even argued that it was an "anomaly."[13] In 1828, he had pointed out: "To displace one proportion of any body contained in a compound, as a rule, it is necessary to apply one proportion of the body chosen for displacing."[14] This presupposed, Dumas seized the opportunity to accommodate his first formula model to the "anomaly" by integrating and thus "empirically proving" his own formula model for alcohol $C^8H^8 + H^4O^2$, representing its binary constitution from bicarbonated hydrogen and water. Splitting the formula equation of the reaction between alcohol and chlorine into two parts, he obtained:[15]

$$C^8H^8 + H^4O^2 + Ch^4 = C^8H^8O^2 + Ch^4H^4$$
$$C^8H^8O^2 + Ch^{12} = C^8H^2Ch^6O^2 + Ch^6H^6$$

By virtue of this modified formula model of the reaction, Dumas explained the formation of chloral as a two-step reaction and as an interaction differing in each step. First, four portions of the hydrogen of alcohol contained in water (H^4O^2) are removed by combining with chlorine to form hydrochloric acid, however without a simultaneous chlorine substitution. Then, in the second step, six portions of hydrogen contained in C^8H^8 are removed (to combine with chlorine) and simultaneously replaced by exactly six portions of chlorine. Dumas envisioned this new model of the reaction—which implied the equivalence of one portion of chlorine replacing one portion of hydrogen—as supporting his own model of the binary constitution of alcohol. He argued that the model demonstrates that chlorine acts differently on the hydrogen contained in the two constituents, depending on the different "state" (*état*) of the hydrogen contained in the two immediate constituents of alcohol.[16]

In addition, it could be recognized from the new formula model that a substance with the formula $C^8H^8O^2$ was formed as an intermediary. The model showed not only that this substance was produced before chloral but also that it was consumed in the second step when chloral was created. In an unforeseen and unintended way, this contributed to solving the older problem of the identity of the oily substance. Based on the formula, Dumas believed that the intermediary substance was acetic acid ether.[17] He concluded that the oil, produced before the choral, was a mixture containing acetic acid ether and other substances stemming from the different simultaneous parallel reactions between alcohol and chlorine. In 1835, he wrote that the "oil has many variable and alterable properties . . . it seems to me that a report on the attempts of the analyses done with this impure oil is useless. They result in considerable differences as a result of the state of the material or to the imperfection of analytical methods."[18] Later, in the same year, Liebig happened to isolate experimentally and to identify the substance with the formula $C^8H^8O^2$ as a new chemical compound, called aldehyde.[19]

The Unintended Introduction of the New Concept of Substitution

Dumas's secondary formula model of the formation of chloral was highly successful because it not only plausibly represented the experimental formation of chloral, but it also fitted this representation, first, with the explanation of the identity and the formation of the oil occurring before the creation of chloral and, second, with Dumas's model of the binary constitution of alcohol. The model was a perfect "accommodation."[20] Yet Dumas did not stop with this result. A year later, in 1835, he added a new vision of the reaction that eventually required a complete overthrow of his own binary model of alcohol, whose empirical underpinning had been his main goal in 1834, along with the general concept of the binary constitution of organic substances, which had intellectually framed the controversy about radicals and constitution in the 1830s. Why did Dumas not stay with his successful model? What caused him to do something completely new that served neither his individual goals nor the collectively shared goals of European chemists? Before I suggest an answer to this question, I want to analyze further the meaning and novelty of *substitution.*

In the fifth volume of his chemical textbook, published in 1835, Dumas introduced the term *substitution* to denote the equivalent replacement of one portion (or "atom") of hydrogen by one portion of chlorine in the formation of chloral.[21] He extended this new concept to other reactions of organic substances, such as the reaction of formic acid with mercury or silver oxide to form carbonic acid and the transformation of alcohol with air or magnesium peroxide into acetic acid; and he stated the general "rule" that one "atom" of hydrogen of an organic substance can be substituted by one "atom" of chlorine, bromine, or iodine or half an "atom" of oxygen. In 1835, this rule, together with a second rule claiming that the hydrogen contained in water is always displaced without substitution (which Dumas viewed as the most important rule because it supported his own binary model of alcohol), made up what Dumas called his theory of substitution.[22] In the following presentation, for analytical reasons, I distinguish between this "theory" of substitution, embracing the additional second rule, and the "concept" of substitution expressed in the first rule. Only the latter was broadly accepted by the European chemical community.

Referring to his formula model of the substitution of the hydrogen of alcohol by chlorine, Dumas added the following reflections:[23]

> If alcohol has the formula C^8H^8, H^4O^2 chlorine can withdraw H^4 without replacing it, transforming alcohol into acetic acid ether $C^8H^8O^2$, which is what really happens. From this point in time, *each atom of hydrogen* that has been *withdrawn will be replaced by one atom of chlorine*, and without being concerned here *with the intermediary compounds*, we state that chloral $C^8H^2O^2Ch^6$ is formed.

With this statement, Dumas assumed that there was not only an equivalent substitution of one portion (or "atom") of hydrogen by one portion of chlorine but additionally that this substitution proceeded step by step, creating further intermediary compounds. The assumption of intermediary compounds created by a stepwise substitution—instead of the substitution of 6 H by 6 Ch all at once—was speculative in 1835, but it was in principle testable by experiments.

Six years later, instead of his verbal statement, Dumas modeled the stepwise substitution of one portion of hydrogen by one portion of chlorine and the production of a series of intermediary products as follows:[24]

$$C^8H^8O^2 + Ch^2 = C^8H^7ChO^2 + HCh$$
$$C^8H^7ChO^2 + Ch^2 = C^8H^6Ch^2O^2 + HCh$$

.

.

.

$$C^8H^3Ch^5O^2 + Ch^2 = C^8H^2Ch^6O^2 + HCh$$

In this 1840 formula model of substitution, Dumas fully realized the implications of the concept of substitution. The model visually demonstrates the stepwise substitution process and simultaneously shows that there is no such thing as a stable immediate constituent represented by the formula C^8H^8 and a binary constitution of alcohol.

To understand what was new in this hypothetical model of substitution, we need to compare it with the traditional concept of displacement reactions, which Dumas had applied and extended in the construction of the two interpretive formula models of chloral formation in 1834. The concept of displacement reaction, developed from the beginning of the eighteenth century and embodied in all eighteenth-century affinity tables,[25] referred to chemical substances as the building blocks of chemical compounds, which can be set free in chemical operations by adding another substance with a stronger affinity to the second component of the compound. The displacing substance always belonged to the same class of substances as the displaced one; an acid could be displaced only by another acid, a base only by another base, or, in terms of the electrochemical system, electropositive substances could be displaced only by electropositive ones, electronegative substances by electronegative ones. Alignment of this concept with the concept of equivalence, based on quantitative experimental investigations of the reactions denoted as "displacement reactions," added information about the relative quantity of the displaced and the displacing substance. The mass of the displacing substance had to be "equivalent" to that of the displaced substance. By comparing the measured masses of various displacing substances (such as acids) in relation to a constant mass of a second component (such as an alkali), chemists determined constant "equivalents" of different chemical substances.

When chemists extended the concepts of displacement reaction and equivalence to the reaction of organic substances, as was the case in Dumas's 1834 models of the formation of chloral, they preserved the original meaning of the concepts and their traditional kind of referents. Hence, the significance of Berzelian formulas, applied as tools to construct models that fitted the traditional conceptual framework, was determined by a conceptual framework older than the chemical formulas. In this traditional framework, the assumption that H^6 was replaced by Ch^6 and Ch^6H^6 was formed by this action referred to observable, measurable quantities of laboratory substances—even if these quantities could not be measured in practice. Moreover, as in the case of inorganic displacement reactions, the first 1834 model meant that H^6 was replaced by Ch^6 all at once—in the same way as, for example, an equivalent of hydrochloric acid is replaced by an equivalent of sulfuric acid from its combination with a metal in only one chemical action.

What was new, compared with this traditional concept of displacement reaction, in the hypothetical model of substitution? First, the displacing and the displaced substances no longer belonged to the same class of substances; hydrogen was electropositive, and chlorine, bromine, iodine, and oxygen were electronegative. Second, the formula model of substitution represented no longer unambiguously measurable masses of laboratory substances. Chemical formulas denoted invisible, scale-independent portions of chemical elements and compounds identified by their unalterable relative combining weights. From the theoretical combining weight of chemical portions, however, the measurable combining weights of the substances involved in a reaction, such as chlorine and hydrochloric acid, could always be recalculated. Third—and this was a particularly discerning feature of the new model—the new concept assumed a discontinuous, stepwise replacement of one portion of hydrogen by one portion of chlorine and intermediary reaction products created in each step.

Now I return to the question raised previously: Why did Dumas not stop with his 1834 interpretive models, which were a highly successful accommodation? Why did he introduce the new concept and formula model of substitution? Dumas did not say explicitly what caused him to introduce this new concept; from the way he defined it, namely, by means of chemical formulas, the following assumption is plausible: He followed a suggestion visually displayed by chemical formulas and further triggered by his previous work with this sign system. If a two-step reaction yielding chloral was possible—in which in the second step 6 H was replaced by 6 Ch—was it not also possible that the replacement of 6 H by the 6 Ch was not all at once but step by step? The latter assumption was further entrenched by the work of his assistant, Auguste Laurent.[26] Around the same time that Dumas performed his experiments with alcohol and chlorine, Laurent did analogous experiments with naphthalene isolated from coal tar. In his interpretation of these experiments, Laurent followed a strategy similar to that of Dumas. He performed quantitative analyses of the substances involved in the reaction, transformed the analytical results into Berzelian formulas, and applied these formulas to construct formula models of the reactions. In two papers published in 1835, he reported a series of different chlorinated, brominated, and oxidized products of naphthalene; the formula of the latter was $C^{40}H^{16}$, and the reaction products were represented by formulas such as $C^{40}H^{14}Cl^{2}$, $C^{40}H^{12}Cl^{4}$, $C^{40}H^{8}Cl^{8}$, $C^{40}H^{14}Br^{2}$, $C^{40}H^{12}Br^{4}$, $C^{40}H^{8}Br^{8}$, and $C^{40}H^{14}O$. Laurent concluded from these formulas that a stepwise replacement of one portion of hydrogen by equivalent portions of chlorine took place.[27] Both Dumas and Laurent "derived" the new concept or theory of substitution from chemical formulas, not in a logical sense, but rather from the graphic suggestiveness and maneuverability of formulas and from their actual manipulations on paper.

I want to add further arguments for this thesis by excluding possible historical alternatives. First, it was not possible to derive the concept of substitution from the general theory of chemical portions or any existing atomic theory. These theories made assumptions about the composition of chemical compounds, not about their reactions. All hypotheses concerning chemical reactions had to be introduced from the outside into this kind of theory. Moreover, both a one-step replacement of all atoms of hydrogen by a given number of atoms of the displacing element (in accordance with the traditional concept of displacement) and a discontinuous, stepwise substitution would fit atomic theories and the theory of chemical portions. Second, in 1835, nothing on the phenomenal, macroscopic level of experimentation supported the view that there was a discontinuous, stepwise substitution. The assumption that intermediary products were created in each step of the substitution was mere speculation at that time. Therefore, I am talking about a "hypothetical" rather than an "interpretive" model. What actually could be observed in Dumas's experiments—and in experiments of Laurent and other chemists—was a continuous production of hydrochloric acid (or another acid) during the operation. Third—and this is perhaps the most important argument—nothing pertaining to the goals and interests of Dumas or any other leading European chemist directly contributed to the introduction of the concept of substitution. On the contrary, chemists' goals and interests were quite different, namely, to support empirically and to defend their own models of the binary constitution of organic substances and of organic radicals. It was precisely these shared goals that were undermined by the concept of substitution.

In 1834, Dumas had modeled the formation of chloral with chemical formulas to defend his model of the binary constitution of alcohol. In the introduction to his research report from 1834, Dumas announced that he wanted to render "palpable and manifest even to the eyes of the most biased chemists" that his new interpretive model of the formation of chloral strongly supports his own model of the binarity of alcohol.[28] Yet, ironically, the tools applied to achieve this goal eventually made "palpable and manifest" that he was wrong, as were all of the leading chemists of the time. As can be easily seen in Dumas's 1840 formula model, there was no such thing as a binary constitution of organic compounds containing a compound radical that was as stable as an element. In 1840, most chemists agreed that the concept of substitution was incompatible with any model of a binary constitution of organic substances. Substitution then had become a key concept in the new experimental culture of organic chemistry, referring to an entirely new practice of synthesizing organic compounds. This new concept of substitution was not the consequence of an atomistic research program, of social interests, or of any other form of deliberate intention on the part of chemists; rather, it was an originally unforeseen and unintended result of work on paper with chemical formulas that was coordinated with a new type of experiments performed with organic substances.

Musing about the role of material agency in knowledge production, in particular about the role of inscriptions, Bruno Latour asked, "Everyone agrees that print, images, and writing are everywhere present, but how much explanatory burden can they carry? How many cognitive abilities may be, not only facilitated, but thoroughly explained by them?"[29] The introduction of the concept of substitution is an example that demonstrates that the work with material things, both paper tools and laboratory tools, stimulates and shapes conceptual innovation. It demonstrates that "material agency" can be much more active than exhibiting "resistance"[30] to the goals of human actors. I hasten to add that this is not a mysterious power that material things *per se* have; rather, it is a feature of labor in which human beings interact both intellectually and physically with material things. In the case at stake here, it was in particular a sign system—a material device in the widest sense of extramental entities—applied as tools on paper that developed an unintended dynamics. Like laboratory instruments, paper tools are resources whose possibilities are not exhausted by scientists' attempts to achieve existing goals but that rather react back on the goals and ideas of their appliers. The introduction of the concept of substitution was due to this dialectic of material tools and goals.

Acceptance of the Concept of Substitution in the European Chemical Community

Among the young generation of European chemists, the concept of substitution was met with complacency. In 1835, Dumas and Eugène Melchior Peligot, an assayer at the Paris mint and professor of chemistry in the Conservatoire des Arts et Métiers, coauthored an article on the substitution of hydrogen by chlorine in cinnamon oil, which they viewed as another confirmation of the theory of substitution.[31] In the same year, Auguste Laurent published articles on substitution products from naphthalene.[32] Two years later, the Italian chemist Faustino Malaguti, a political refugee working with Pelouze, examining the reactions between different "ethers" of oxalic acid and chlorine, wrote of the "phenomenon" of substitution.[33] Similarly, Dumas had support from German scientists, such as Johann Christian Poggendorff, the editor of the *Annalen der Physik und Chemie*,[34] and from Justus Liebig.

Liebig fully accepted the concept of substitution, albeit not Dumas's rule that the hydrogen contained in water (as an immediate constituent of organic compounds) was withdrawn without replacement. The reason is obvious: This rule yielded the crucial support for Dumas's own model of the binary constitution of alcohol that was under attack by Berzelius and from 1834 also by Liebig. In 1836, Liebig, together with his French friend Théophile Pelouze, published an article that, among other issues, also addressed the question of substitution. In this article, the two chemists attacked Dumas's model of the binary constitution of alcohol, trying to defeat him with his own means. "By means of the theory of substitution," they wanted to prove that their own model of the binary constitution of alcohol was right and that Dumas's was wrong.[35]

The preceding year, Liebig had accepted Dumas's formula for chloral, $C^4H^2Cl^6O^2$ (with the theoretical combining weight $C = 12$). Furthermore, he had isolated and identified the intermediary $C^4H^8O^2$, which had been postulated by Dumas's secondary formula model, as a new substance called *aldehyde*.[36] Now, with their own model of the binary constitution of alcohol, according to which it was a hydrate of "ethyl" oxide, $C^4H^{10}O + H^2O$, Liebig and Pelouze constructed a new formula model of the reaction between alcohol and chlorine:[37]

1. alcohol	$C^4H^{10}O + H^2O$
	$-H^4$
aldehyde	$C^4H^6O + H^2O$

2. aldehyde	$C^4H^6O + H^2O$
	$- H^6 + Cl^6$
chloral	$C^4Cl^6O + H^2O$

Like Dumas, Liebig and Pelouze constructed a two-step model of the reaction in which first the intermediary substance aldehyde is created and then chloral is formed. They also assumed that in the first step hydrogen (H^4) is withdrawn without a substitution and that in the second step six portions of hydrogen are withdrawn ($- H^6$) and replaced by exactly six portions of chlorine ($+ Cl^6$) to produce chloral ($C^4Cl^6O + H^2O$) and hydrochloric acid (H^6Cl^6). Liebig and Pelouze emphasized that the second reaction step confirmed their own general "rule" of substitution, according to which "substitution operates in equivalents."[38] From their additional remark that the reaction of alcohol and chlorine also yields "chlorinated intermediary compounds," it becomes clear that their rule of substitution fully embraced Dumas's new concept of substitution.[39]

The only difference between Liebig and Pelouze's reaction model and that of Dumas was that the former assumed that chlorine acted differently on the hydrogen contained in one and the same immediate constituent $C^4H^{10}O$ of alcohol, whereas the latter stated that different "states" of the hydrogen were contained in two different immediate constituents, which caused different actions from chlorine. Both models came with advantages and explanatory difficulties. Arguing against Dumas's model, Liebig and Pelouze said that it could not explain why chlorine stopped acting on the hydrogen contained in C^8H^8 after six portions of hydrogen had been substituted by chlorine rather than substituting all eight portions.[40] In the alternative model of Liebig and Pelouze, this specific problem did not occur because all ten portions of hydrogen contained in the immediate constituent $C^4H^{10}O$ were withdrawn and in part substituted; however, this alternative had also a price. First, to construct their reaction model, Liebig and Pelouze had to build an *ad hoc* formula model of the constitution of aldehyde, transforming its raw formula $C^4H^8O^2$ into the binary formula model $C^4H^6O + H^2O$. As a consequence, the formula they obtained in the reaction model for chloral was $C^4Cl^6O + H^2O$, which had to be regrouped into $C^4H^2Cl^6O^2$ to obtain the raw formula for chloral. The two chemists simply stated without explanation that "the water of the aldehyde becomes part of the composition of chloral."[41] Second, in the new model, no explanation was given for why chlorine acted in two steps and acted so differently on the hydrogen contained in $C^4H^{10}O$.

Liebig and Pelouze's acceptance of the concept of substitution and their specific application of it implicitly showed that "substitution" actually did not support any model of the binary constitution of alcohol. Jacob Berzelius was perhaps the first chemist who clearly recognized this as well as the significant consequences for chemical theory. In a letter to Pelouze, he wrote, "It seems to me that the theory of substitution, established by M. Dumas—according to which, for example, chlorine can exchange hydrogen in equal numbers of atoms by taking its place—has a destructive influence on the progress of science."[42] In particular, he criticized the assumption that an electronegative element such as chlorine can substitute for an electropositive element and become part of an organic radical. Rather, he claimed, elements such as chlorine can combine only from outside with an organic radical, in the form of a component of the second immediate constituent of the compound. Berzelius saw his electrochemical theory challenged and even more so his view of the organic world. Elements such as chlorine, bromine, and iodine were not found in the naturally occurring organic substances and hence must not be part of what he viewed as "organic radicals."

Dumas's immediate response to Berzelius's criticism was moderate.[43] In the following years, from 1839 to 1840, however, he fully acknowledged the consequences of the criticism and broke with the concept of the binary constitution of organic substances and with the application of the electrochemical theory to organic compounds.[44] Instead, he introduced his concept of the "type" and the "type theory," which he also called the general theory of substitution.[45] Dumas's type theory amalgamated the chemical tradition of modeling the constitution of organic substances by means of chemical formulas and of classifying organic compounds based on their formula models of constitution with the natural philosophical tradition of atomism. He also stated explicitly that his term of a *type* was a metaphor that belonged to the "language of the naturalists,"[46] thus adding credibility to his claim that his classification was a "natural one." It should be noted, however, that Dumas's practice of constructing type models of constitution and new classes of organic substances based on the shared type was defined by the manipulation of formulas on paper. In practice, his ideas about atoms in the sense of submicroscopically small bodies that have a spatial arrangement within a molecule did not have much relevance.

THE CREATION OF MUTUAL ADJUSTMENTS AND ITS REFLECTION IN THE NEW MODE OF JUSTIFICATION

The story of the coming into being of "substitution" lends itself easily to the discussion of another topic that is both of a broader epistemological and historical interest. It displays features of a laboratory science pointed out before by Ian Hacking and Andrew Pickering,[47] that is, the "self-vindication" of laboratory sciences (Hacking) and their "dialectic of resistance and accommodation" (Pickering). Hacking and Pickering argue that there is no immutable platform of research in the laboratory sciences, such as a set of certain observation sentences or protocol statements, and that laboratory scientists rather use "plastic resources" that they mutually adjust to each other. They further argue that this adjustment is not governed by any explicit, general methodological rule (logical or otherwise) but rather is an interplay of concrete, specific items, going back and forth from one to the other and proceeding in small *ad hoc* steps. Moreover, they state that this dialectic of mutual adjustment is not performed exclusively on a mental level but includes the accommodation of instruments and apparatus applied to study a scientific object and to produce data or other traces of it.

According to Hacking, the fitting together of types of theories, types of apparatus, and types of data analysis creates a "self-vindicating" system that is essentially irrefutable in the sense that "any test of theory is against apparatus that has evolved in conjunction with it—and in conjunction with modes of data analysis."[48] In other words, there is no independently gained and certain platform from which the other activities might be evaluated and justified. Andrew Pickering shifted the attention from problems relating to the justification of representations to the performances of scientists and the actions of instruments. He asserts that in their experimental practice scientists typically ask the following questions: "Does the machine perform as intended? Has an intended capture of agency been effected?" and so on.[49] If the answer is negative, they may, for example, revise their goals, or they may study the apparatus again and modify some of its parts. This interplay, Pickering states, takes "the form of a *dialectic of resistance and accommodation*, where resistance denotes the failure to achieve an intended capture of agency in practice and accommodation an active human strategy of response to resistance, which can include revisions to goals and intentions as well as to the material form of the machine in question and to the human frame of gestures and social relations that surround it."[50]

Chemists' practice of studying organic chemical substances and their reactions and their classifications, analyzed in the preceding three chapters, confirms Hacking's and Pickering's views of the laboratory sciences. Although chemists temporarily took the so-called empirical formulas for granted when they applied them as tools for constructing models of reaction and constitution, they did not view these tools as completely immutable resources. Rather, they were ready to modify the Berzelian raw formula of an organic compound to adjust it to its formula model of constitution and to its formula models of reactions. If this was the case, they either repeated the quantitative analysis of a compound, or they changed the data processing, as Liebig did with Zeise's results of the quantitative analysis of an organic platinum compound.[51] In doing so, they also mutually adjusted the analytical data and the raw formula for a chemical compound. Chemists further mutually adjusted the models of constitution of various substances to fit them together in a substance class, as we have seen particularly in the case of Dumas and Boullay's classification of alcohol derivatives.[52] In short, chemists neither treated the results of the quantitative analysis and the Berzelian raw formulas (or so-called empirical formulas) as a certain empirical foundation for the subsequent interpretive modeling, nor did they possess any ascertained general theory from which they might have deduced more specific chemical models and experimental phenomena.

As to their "empirical basis" for modeling, chemists were well aware of the many sources of error in quantitative analysis as well as of the fact that there was no unequivocal pathway from the analytical results to an "empirical formula" and from them to a "rational formula" (or formula model of constitution). Division of the weight percentages by the theoretical combining weight (or "atomic weight") of each element often did not yield small integral numbers. As a rule, there was always some free play in the construction of an "empirical formula." Rounding up to integers was sometimes as possible as rounding down. Furthermore, if the theoretical combining weight of a compound (later also called *molecular weight*) could not be determined experimentally, as was often the case with neutral compounds, the "empirical formula" was uncertain in a second aspect. It might have been the multiple of the simplest formula.

For example, in his 1832 experimental investigations of the reaction of alcohol with chlorine, Justus Liebig added several question marks to his results. He had analyzed the newly produced substance chloral, and by dividing the analytical results represented in weight percentages by the theoretical combining weight of the elements, he obtained the "empirical formula" $C^9Ch^{12}O^4$. Liebig, however, added that "*since, with minor changes, two other formulas can be calculated from the presented analyses*, it is of importance to learn in which ratio the chlorocarbon and the salt of the formic acid are created."[53] With this remark about the uncertainty of the Berzelian raw formula for chloral, Liebig introduced his description of another experiment, in which he had decomposed chloral by means of an alkali into "chlorocarbon" (that is, chloroform) and formic acid. He weighed the isolated salt of formic acid and constructed a formula model of the decomposition using the formula $C^9Ch^{12}O^4$

for chloral and C^2Ch^5 for chloroform. The measured weight and the weight calculated from the model, however, did not agree. As a consequence, Liebig repeated the analysis of chloral, but without getting more satisfying results. He wrote, "I have repeatedly made the analysis of chloral, but the deviating results are still leaving me in obscurity concerning the true composition of it [chloral]. *I have only preferred the presented formula, since the behavior of chloral cannot be elaborated more satisfactorily by any other.*"[54]

In this remark, Liebig states explicitly that his choice of the so-called empirical formula of chloral was informed by the possibility of applying this formula in the interpretive formula model of the decomposition of chloral by an alkali. Not less explicit is his statement concerning the fitting of the data from the quantitative analysis of chloral with its Berzelian raw formula, which Liebig calls *theoretical composition* here as well as with the interpretive model of the decomposition:[55]

> The investigation of the composition of chloral presented a lot of difficulties. This was not so much the case for the analysis as such, for it can be easily managed by paying attention to the precautionary measures which I mentioned in the context of the analysis of chlorocarbon; what I mean here is the condition for *reconciling the numeric results [of the analysis] with the decomposition products and the theoretical composition.*

Again, Liebig emphasizes that quantitative analysis was not a certain empirical basis for the construction of the Berzelian raw formula of an organic compound and the interpretive modeling of its reactions, such as decompositions. He makes it clear that the possibility of creating fits between the various items increased the probability that the right approach has been found.

When Dumas repeated the quantitative analysis of chloral and chloroform in 1834, he did this with exactly the same purpose as Liebig: to obtain results that fitted his own formula model of the constitution of alcohol. What particularly strikes the reader in Dumas's article is the seeming circularity of the justification of his raw formulas for chloral and chloroform. Referring to the decomposition of chloral ($C^8O^2H^2Ch^6$) into chloroform ($C^4H^2Ch^6$) and formic acid ($C^4H^2O^3$), which he represented by the formula model $C^8O^2H^2Ch^6 + H^2O = C^4H^2O^3 + C^4H^2Ch^6$, Dumas argued: "Its formula [of chloral] perfectly explains this reaction."[56] In accordance with this, he wrote a year later:[57]

> If one studies . . . the action of alkalis on chloral, one sees that this body is transformed into formic acid and chloroform whose formula is represented by $C^4H^2Ch^6$. The formula chosen for chloral explains very well this reaction; for,

assuming that water combines, which is necessary, one has $C^8O^2H^2Ch^6 + H^2O = C^4H^2O^3 + C^4H^2Ch^6$.

Dumas here argued by means of the formula of chloral in favor of his interpretive reaction model, but shortly after, he argued, in turn, for the formula of chloral by means of the successful modeling of the reaction: "The reaction of chloral leads to the formula $C^8O^2H^2Ch^6$."[58] We find the same mode of argumentation with reference to the formation reaction of chloral from alcohol and chlorine. On the one hand, Dumas used the formula of chloral as a tool to construct his interpretive models of the formation reaction of chloral; and the plausibility of these models depended entirely on the formulas, in particular on the balancing equations. On the other hand, in his 1835 balancing schema, he argued as follows, now in favor of the formula of chloral:[59]

The formula of alcohol is:

	$C^8H^{12}O^2$
that of chloral is	$C^8H^2O^2Ch^6$
alcohol has lost	H^{10}
and gained	Ch^6

for producing one atom of chloral from each atom of alcohol; *which, by the simplicity of the relations, makes the formula [of chloral] very probable.*

Dumas's seeming circularity of argumentation was engendered by an analogous pattern of practical performances. As to the latter, Ian Hacking wrote, "We create apparatus that generates data that confirm theories; we judge apparatus by its ability to produce data that fit."[60] If we replace *apparatus* by *quantitative analysis* and *theories* by *paper tools* or *chemical formulas*, both Berzelian raw formulas and formula models, Hacking's observation describes exactly what Dumas and the other chemists did: Their data of quantitative analysis confirmed the Berzelian raw formulas and the formula models of constitution and reaction, and the quantitative analysis was judged by the ability to produce data that fit both the raw formula and the formula models of a substance.

CHAPTER EIGHT

The Historical Transformation Process

> The progress is astonishing, and the growth of our knowledge for the last ten or twelve years is so great, that organic chemistry has become a more extended and comprehensive science than the comparatively much more elaborated inorganic chemistry.
>
> —BERZELIUS 1839, 289

The previous four chapters brought out a historical peculiarity that needs to be discussed in more detail, namely, chemists' emphasis on experiments with a particular set of substances: alcohol, so-called ethers, and other alcohol derivatives.[1] In the 1830s, controversy about organic radicals, alcohol, and its derivatives were in the foreground. The first coherent series of experimental investigations of organic chemical reactions was performed with alcohol and its derivatives. In the new mode of classifying organic substances suggested by Dumas and Boullay in 1828, the bulk of the classified substances were alcohol and its derivatives. Again, it was the experimental investigation and modeling of a reaction of alcohol that contributed to development of the concept of substitution. Whereas the experiments with alcohol, ethers, and other alcohol derivatives were only of marginal importance in eighteenth-century European plant chemistry, they became the focus of attention between 1828 and 1834. What made them so important? Why did chemists focus exactly on this bit of material culture rather than on any other? In this chapter, I argue that during the 1830s, alcohol and its derivatives became "model objects" in European chemists' endeavors to extend the mode of investigating chemical reactions, the concept of binary constitution, and the mode of classification from inorganic to organic chemistry, and I ask why this was the case. I further argue that the intended model objects simultaneously became unintended exemplary achievements in the emerging culture of carbon chemistry. The distinction between intended model objects and unintended paradigmatic achievements is an important step in my reconstruction of the transformation of organic chemistry between the late 1820s and the early 1840s. Further aspects of this transformation process discussed in this chapter are its temporal frame, its overall trajectories, and epistemological questions concerning the relation between discontinuity and continuity, incommensurability, and causes of the transformation.

MODEL OBJECTS AND UNINTENDED PARADIGMATIC ACHIEVEMENTS

The selection and preparation of a particular object (or group of objects) standing for a larger epistemic set of objects and problems tied to them are by no means unique to chemistry. In his *Lords of the Fly*, Robert Kohler has studied a similar topic with respect to experimental biology: "Why do some organisms, like *Drosophila*, become cosmopolitan, 'standard' species of laboratory creatures—cornucopias of productive methods, concepts and problems—while others do not? . . . What qualities made the partnership between fly and fly people so effective?"[2] Quite similar to Kohler's search for answers to this question, "in the process of experimental production, in the biological and technological nature of *Drosophila*, and in the customs and practices of the drosophilists,"[3] in the previous chapters, I described European chemists' experimental production of alcohol derivatives in the first three decades of the nineteenth century and the extension of this production. This story gives a first (although not comprehensive) answer to the question of why it was alcohol and its derivatives that became model objects or "model substances" in the 1830s: They were a set of pure chemical substances ubiquitous in the European chemical laboratories because of their long history of production and reproduction and the social transfer of both the substances and the experimental technology.

That alcohol and its derivatives, at the end of the 1820s, functioned as model substances for the extension of the mode of studying reactions and the concept of binarity and for the new classification of organic compounds was historically contingent. Nothing in the earlier history of these objects and experiments determined the epistemic role they played later. During the first three decades of the nineteenth century, experiments with alcohol, ethers, and other alcohol derivatives constituted an experimental practice located on the periphery of plant chemistry that overlapped considerably with the pharmaceutical trade. This experimental practice had expanded and developed within this early context to become a stable bit of material culture in European chemistry;[4] until the late 1820s, it remained a marginal part of organic chemistry compared with the practice of extracting natural organic materials and the natural historical and pharmaceutical type of inquiry. It was only after a new historical constellation had come into being—in particular, after new paper tools (Berzelian formulas) had been introduced and accepted, which were external to the older culture of plant chemistry—that the role of the earlier experimental practice with alcohol and its derivatives changed quickly.

Discussion now turns to, in Kohler's words, the "chemical and technological nature" of these objects that made them so suitable as model substances. I align the "nature," or material features, of these objects with their epistemic qualities as they existed in a collectively shared intellectual and practical framework of a scientific culture. In doing so, I do not focus on the nature of "things-in-themselves" but on things that were constructed, handled, and conceptualized by human actors and that, nevertheless, stimulated and guided human actions. Thus, the following analysis also shows that the priority of things is never completely suppressed and dominated in the process of the construction of models and paradigms.

Model Substances

The central epistemic role played by alcohol and its derivatives after 1827 depended on the history of these substances and was shaped by it. In the new historical constellation, many aspects of this history—technological, economic, social, and epistemic—together with the intrinsic material properties of these substances became a previously unforeseen set of preconditions for the new role as model substances. Jacob Volhard, a student of Liebig famous for his Liebig biography, pointed to this conglomerate with the following words:[5]

> In the development of organic chemistry, alcohol and its derivatives play a similar role as did oxygen and the oxides in mineral chemistry. Easily accessible, procured in infinite amounts and at low prices, able to [undergo] various metamorphoses, alcohol serves to manufacture innumerable, well-defined derivatives. The composition of the latter, which is easy to study, and their comparatively simple constitution made possible analogical inference to more complex compounds.

According to Volhard, alcohol was "easily accessible, procured in infinite amounts and at low prices"—several causal factors are condensed in this remark. The model function of alcohol and its derivatives presupposed the historical establishment of a technology of manufacturing alcohol, a specific economy of production that guaranteed low prices, and the social distribution of both technology and economy. Model substances must not be exotic things but rather must be easily accessible, or easy to produce, in both a technological and economic sense. Otherwise, more obstacles would emerge by using them than they help to remove.

It is not different with model organisms in experimental biology, which also require a history and technology of breeding and reproduction at relatively low costs and a social network that guarantees distribution and acceptance. In both cases, it is not "nature" alone (or intrinsic chemical or biological properties) that settles the question of the suitability of a particular set of things as model objects, but also the technological, social, economic, moral, and cultural factors. Referring to *Drosophila* in experimental genetics from about 1910 to the early 1940s, Robert Kohler emphasizes that experimental organisms "can be understood as technological artifacts that are constructed and embedded in complex material and social systems of production."[6] At the same time, he also pointed out that these laboratory creatures have a biological nature (or natural history) that shapes the construction of the standard organism in the laboratory. For example, many *Drosophila* species can live in nearly every climate, they have a high fecundity and growth rate, and their natural habits and seasonal cycle are well suited to the needs of the laboratory.[7]

What fecundity and growth rate were for *Drosophila* was the capacity of "various metamorphoses" (Volhard) for ordinary alcohol. Ordinary alcohol was, in fact, a very reactive substance that could be transformed easily into a great number of different organic species with the available means of early nineteenth-century chemistry, thus creating a large and ever-extending tree of derivatives. Already in the first three decades of the nineteenth century, chemists had created a plethora of so-called ethers, olefiant gas, and some other organic compounds, by mixing and heating (or distilling) alcohol and an acid. For example, when Berzelius was preparing the fourth edition of his textbook in 1835, he wrote in a letter to Wöhler that, unlike the chapters about the naturally occurring substances that could be revised with ease, the chapter about alcohol and the different species of ether required an "entire sea of new additions."[8] The group of immediate derivatives of alcohol was enlarged by oxidation products, such as aldehydes and acetic acid, the alcoholates (created from alcohol and alkalis), and the products of halogenation. After 1840, the number of alcohol derivatives grew to hundreds or thousands, depending on how far we follow the branches of the tree of derivatives. If we include other species of alcohol, such as methanol, which was isolated in the 1830s from "spirit of wood," the derivatives of alcohols present a considerable part of the entire nineteenth-century inventory of pure organic laboratory substances. Berthelot was correct when he stated in 1872 that "the synthesis of alcohol becomes the starting point of nearly all of the other formations."[9]

In sum, compared with other pure organic substances available in early nineteenth-century chemistry, such as vegetable acids or alkalis, alcohol could be produced in large quantities and at relatively low prices, and it presented, together with its derivatives, an expanding system of reactions (which was not the case with organic acids or alkalis). This alone was not sufficient for model substances in the 1830s; additional conditions, simultaneously technical and epistemic, were necessary. Volhard hinted at them when he remarked that the composition and constitution of alcohol and its derivatives were relatively easy to study and that analogies with more complex compounds were possible. The point at issue here is that alcohol and its derivatives could be assimilated to the existing chemical order of inorganic chemistry and trigger its extension to the organic domain. This is a general characteristic of model objects. They must somehow fit the existing cultural framework, and they must facilitate the assimilation of additional cases by allowing convincing analogies to be constructed.

Alcohol and the ethers belonged to the first organic substances that had been obtained in a state of relative purity, and they were among the first organic substances successfully subjected to quantitative analysis. The fact that they do not contain any additional element besides carbon, hydrogen, and oxygen facilitated the quantitative analysis. Quantitative analysis, in turn, was an important precondition for performing the new type of experiments that became central in the transformation period of organic chemistry: the investigation of organic chemical reactions. The possibility of fitting the reactions of alcohol and ethers with acids and alkalis into the existing conceptual scheme of decompositions of binary compounds, established in inorganic chemistry, was another crucial feature of these substances. Several common characteristics of the reaction between alcohol and acids on the one hand and inorganic alkalis and acids on the other hand were easy to recognize, even on the level of experimental observation (that is, without constructing a formula model of the reaction). There was, first, the fact, that alcohol actually reacted with many acids and that at least the most important traces of this reaction (that is, the so-called ethers) could be isolated with the available techniques of separation and purification. Many of the "ethers" had a characteristic smell that helped to detect and to identify them. Second, like many salts, the ethers were neutral, and many of them could be reversibly decomposed with alkalis into alcohol and the acid. In short, even on the phenomenological level of experimentation, there was an obvious analogy to the synthesis of inorganic salts from a base and an acid and the reversible decomposition of these salts into the original base and acid. Extension of the familiar chemical order from inorganic to organic chemistry was further spurred by chemists' application of chemical formulas to construct "models"[10] of the reaction and constitution of organic substances. These formula models displayed a binary constitution of alcohol, ethers, and many other alcohol derivatives, in analogy to inorganic compounds, and the formula models of reaction fitted the general concept of displacement reactions previously developed in inorganic chemistry. Via the formula models, chemists also successfully drew analogies between the "artificially" created ethers and "natural" organic substances, such as the sugars and fats.

It is important to remember here the state of affairs regarding the study of organic chemical reactions in the early nineteenth century.[11] Most organic substances were extracted vegetable and animal materials that were (from our later perspective) either in an impure state or were complex organized materials. Their chemical transformations in the laboratory were viewed by chemists as chaotic cascades of decomposition that were too complicated for any reconstruction. Therefore, the very fact that it was possible to create order in some cases was exceptional and remarkable. We have discussed other examples where this was also possible, namely, the compounds of the so-called benzoyl radical.[12] The compounds of the benzoyl radical also took on the function of model substances; they even played a key role in convincing Berzelius that the concept of the binary constitution was of general applicability. There is, however, an important difference between these and the experiments with alcohol and its derivatives: The latter remained interesting, challenging, and in part controversial objects, whereas the discourse on the former was closed after 1832. It was closed because of the absence of controversy, of possible analogies, and of possible extensions of the system. The same was the case with another experiment, which became prominent in the history of chemistry: Wöhler's urea synthesis.[13]

It has been a persistent myth in the history of chemistry that Wöhler's synthesis of urea from ammonium cyanate was a crucial step in the history of organic chemistry that banished vitalism and considerably contributed to the formation of synthetic carbon chemistry.[14] I argue that this experiment was a single isolated event that alone could not significantly challenge the natural historical agenda of early organic chemistry. As pointed out in Chapter 3, there were other, much earlier cases of artificial production of natural organic substances, such as the production of "oils" and of "artificial camphor," which did not cause deep frictions or even a dissolution of the existing chemical culture. The argument that the synthesis of urea played a crucial role in the transformation of organic chemistry was made plausible by a quite contingent reason: It was first achieved in 1828, when the transformation process was already on its way.[15]

In a recent article, Peter Ramberg observed that "there are surprisingly few known contemporary accounts of the immediate reaction to Wöhler's synthesis."[16] Wöhler's contemporaries were impressed less by the event, Ramberg argues, than the generations afterward, who made it the foundation myth of the new subdiscipline of organic chemistry.[17] In fact, even Wöhler himself was cautious as to the broader impact of his experiment. In his experimental report, he remarked that the formation of urea from ammonium cyanate was "*an example* of the artificial creation of an organic, actually of a so-called animal substance, from inorganic substances."[18] In a letter to Berzelius, however, Wöhler added some doubts. Were the cyano compounds really inorganic compounds, or did they preserve the organic nature? "This artificial creation of urea: can one consider it as an example of the creation of an organic substance from inorganic substances? It is curious that the production of cyanic acid (and also ammonia) always requires an originally organic substance; a natural philosopher (*Naturphilosoph*) would say that *the organic has not yet disappeared from both the animal coal and the cyano compounds formed from it, and therefore an organic body can always be reproduced from them.*"[19]

In contrast to the synthesis of urea, and even the experiments with benzoyl compounds, the experiments with alcohol and its derivatives allowed chemists to extend the concept of binarity, the mode of classification of substances, and the mode of investigating chemical reactions from inorganic to organic chemistry. "The future states of scientific culture at which practice aims are constructed from existing culture in a process of *modeling* (metaphor, analogy)," Pickering pointed out.[20] And he added, "Modeling is an open-ended process with no determinate destination."[21] In an unforeseen way, chemists' modeling was to become part of a historical trajectory that transcended the existing scientific culture. It did so by becoming a paradigmatic achievement of the new.

New Paradigmatic Achievements

After 1833, when Jacob Berzelius had unequivocally approved extension of the concept of binarity to organic substances, alcohol and its derivatives became quickly accepted as model substances throughout the European chemical community. This is not to say that chemists gave them this new epistemic status in a methodologically reflective way. Only some decades later can we find explicit testimony of this epistemic shift. For example, in 1857, Julius Schlossberger, professor of chemistry in Tübingen, remarked, "The study of this family [the alcohols, ethers, and aldehydes] did quite essentially contribute to the elaboration of our theoretical views in organic chemistry."[22] Shortly after this, Liebig wrote in an article (entitled "Theory") for publication in the chemical handbook that he coedited together with Poggendorff and Wöhler: "Alcohol and its numerous derivatives, ethers, acetic acid, etc. play a similar role in the new chemistry to that of the process of combustion in the chemistry of the previous century; that is, theoretical chemistry has developed through the explanation of the composition of these bodies."[23]

Similarly, Berthelot wrote in the introduction to his 1860 textbook on organic chemistry, "It is through the study of alcohol that was begun the long series of discoveries that led to these fundamental results [concerning the reactions of organic substances]."[24] When Liebig and Schlossberger particularly emphasized the role played by the experiments with alcohol and its derivatives for the development of "theoretical views" and a new form of chemistry, they referred to the new culture of carbon chemistry. According to Liebig, the importance of experiments with alcohol and its derivatives was analogous to that of the combustion process in the so-called chemical revolution of the late eighteenth century. The two sets of particulars were exemplary achievements for the coming into being of the new within the boundaries of the old.

During the 1830s, the experiments with alcohol and its derivatives and the subsequent modeling with Berzelian formulas became a crystallization point for the development of new conjunctions, concepts, and skills that became characteristic of the emerging experimental culture of carbon chemistry. It was in this concrete context that a stable conjunction was created for the first time between two types of experiments with organic substances—quantitative analysis and the investigation of chemical reactions—and the mode of classifying organic substances. This conjunction was, so to speak, the practical, material scaffolding that engendered—together with the new sign system—the coherence of what I call a "scientific culture." It was, on the one hand, intended as an extension of the concept of binary constitution from inorganic to organic chemistry, along with the new mode of classification based on information about the composition and the binary constitution of organic compounds. On the other hand, it brought with it several unintended consequences: the creation of a plethora of artificial organic compounds not existing in nature outside the laboratory and a shift of interests concerning the most relevant research objects and problems.

In contrast to the extracted "natural" organic substances, the artificial organic compounds produced in the experimental exploration of the chemical reactions of the former grew exponentially from the 1830s onward; they soon outnumbered the organic substances extracted form plants and animals. Chemists' interest in these artificially produced laboratory substances was not exhausted by the fact that they were "traces" of the metamorphoses of the natural organic substances. Rather, these laboratory substances were also stimulating research objects in and of themselves, and they often exhibited composition and properties different from the organic substances found in nature. The many products of substitution created after 1834, containing elements like chlorine, bromine, or iodine, are examples for this newly opened field of research.

The application of Berzelian chemical formulas as paper tools and the ubiquity of this sign system in all types of activities—experimenting, modeling, theorizing, and classifying—are other characteristics of the new experimental culture of organic chemistry that developed in the 1830s. Chemical formulas were a kind of cultural cement—in addition to the conjuncture mentioned already. They were mediators between various kinds of practices that previously were separated and in part absent in the pluricentered culture of vegetable and animal chemistry. Again, it was in particular in the concrete context of the experimental investigation of alcohol and its derivatives where the productivity of Berzelian formulas became obvious to everybody and their application firmly established. "The language games of scientific practice," Hans-Jörg Rheinberger remarked, "suggest that model things are substances, reactions, systems, or organisms particularly well suited for the production of inscriptions."[25] This was exactly the case with alcohol and its derivatives. In this concrete context, many chemists also acquired the new skills necessary to read and manipulate chemical formulas. It took some time and practice to understand the significance of Berzelian formulas and how to work with them creatively. What kind of regrouping of the letters and numbers within a formula was possible according to the syntactic rules established for the Berzelian formulas? What was the pragmatics of their actual application by leading chemists? What kinds of manipulation on paper had a chemical significance with reference to a concrete experiment? Chemists in the 1930s had to deal with these issues. The construction of interpretive formula models of the reactions of alcohol, the modeling of the binary constitution of alcohol and alcohol derivatives, and the regrouping of the formula models of constitution for classificatory purposes together formed the first coherent and extended practical domain where this could be done. Moreover, it was a domain where chemists learned coordination of the new type of work on paper that became characteristic of the new experimental culture of organic chemistry, that is, the fitting together of raw formulas and formula models of various organic compounds, as pointed out at the end of the previous chapter.

THE STRUCTURAL TRANSFORMATION

In 1841, a year after Justus Liebig had concentrated his interests on agricultural and physiological chemistry, he wrote a harsh criticism of the new type of organic chemistry that he had abandoned shortly before. The laborious investigations of the numerous carbon compounds and their products through decomposition by chlorine and other substances, he remarked, "had been useless for science," and he continued:[26]

> *These substances do not exist in organic nature*, they do not play any role that may capture our imagination; . . . *The most remarkable, the most interesting, and the most useful has little to do with numbers*. There are numerous bodies which we still have to study. The behavior of most dyestuffs is nearly unknown; and of the numerous crystalline *substances existing in barks, roots and fruits*: most of them we only know by their names. . . . If we treat an organic body with chlorine, we transform it into an inorganic body insofar as chlorine enters its composition. *Compounds of chlorine of this kind do not exist in organic nature, their knowledge is useless for us*; if we know the behavior of two or three of these bodies all problems are solved. . . . *Hence, all research on the theory of substitution appears to me rather weary and unsatisfying.*

Historians of chemistry have given psychological arguments for Liebig's attitude, saying he was tired of the controversies with the French chemists, and he "decided that the field was an impossible one for leadership."[27] Although there might be some truth to that, psychological factors alone do not explain why Liebig argued for his new course exactly the way he did. I assert that Liebig's attitude was to a considerable extent a reaction to the deep structural transformation of organic chemistry, which he eventually did not approve. By 1840, organic chemistry had become an experimental culture, which created a plethora of artificial organic substances not to be found in nature. This transformation was propelled mainly by the French chemists, but Liebig himself had contributed considerably to it in the 1830s. Now, in 1841, he was all critic. *Substitution*, the key word of the new synthetic organic chemistry, was "useless to science" and "unsatisfying." Numbers and chemical formulas merely distracted chemists from their most relevant research topic: "nature" as it could be encountered outside the chemical laboratory. Chemists ought to redirect their interests to processes and substances that really existed in nature—such as materials contained in barks, roots, or fruits—instead of creating artifacts containing chlorine, bromine, or iodine never occurring in vegetable and animal substances.

Liebig was not alone in his judgment. Frederic Daniell, professor of chemistry at King's College in London, reiterated his remarks in his textbook. After having described the new field of experimental investigations of organic chemical reactions, in particular of "substitutions," as "boundless,"[28] he continued, "that it is much to be wished the labours of chemists were directed to more really useful objects; for things of this kind truly serve no other purpose than to swell the size of our manuals."[29]

Liebig understood that he could not stop "a rolling train," and so he abandoned organic chemistry around 1840. I take this year as a historical marker, demarcating the culmination of a transformation period in organic chemistry, which brought about the new experimental culture of synthetic carbon chemistry. I date the beginning of this period in 1827, the year Dumas and Boullay's article on the formula models of the formation reaction of ordinary ether was published. Demarcations like these are, of course, to some extent arbitrary, for there were no crucial events—single and outstanding inventions or discoveries—that initiated and terminated the historical process. I do not claim that the experimental culture of organic chemistry suddenly emerged and that the older culture suddenly vanished. There were development processes before 1827 that contributed to the transformation process between 1827 and 1840, and there was—at least in many particular aspects—no closure after 1840. Alan Rocke even characterized the ongoing slow transformation after 1840 that spanned more than two decades as a "quiet revolution."[30]

What, then, justifies my particular emphasis on the time period between 1827 and 1840? What changed in these thirteen years, and how should we conceptualize these changes and their causes? I argue that the period between 1827 and 1840 was a time of a particularly rapid and deep transformation on the structural level of the scientific culture of organic chemistry. It was in this period when the older, pluricentered culture of plant and animal chemistry fell apart and the new experimental culture emerged. Not everything that became an emblem of the new synthetic carbon chemistry was achieved between 1827 and 1840. The big taxonomic reforms were still to be performed, Laurent and Gerhardt being the first to undertake such a huge enterprise. These enterprises were accompanied by further proliferation of types of chemical formulas and by theoretical debates. After 1840, the experimental practice investigating substitutions was further extended, yielding entirely new groups of substances, such as organometallic compounds and derivatives of aromatic compounds. Moreover, new and fundamental chemical concepts were introduced after 1840, such as the concept of valence and that of the chemical structure. In this later period, important social changes also took place, in particular the formation of chemical subdisciplines and the specialization of organic and inorganic chemists.[31] All these changes were accompanied by a growing new self-image of the chemists as architects of a new world. I claim that the transformations in the two or three decades after 1840 took place within the new cultural framework formed in the thirteen years before. They were an extension and development of this culture rather than the creation of a new one.

There is more than one explicit hint that the contemporaries were aware of the dramatic change of organic chemistry between 1827 and 1840. For example, in 1836, Poggendorff felt compelled to add the following warning to an overview on the current state of organic chemistry: "There is such a great activity in the chemistry of organic bodies at present that each endeavor of this kind [of an overview] necessarily is incomplete before it sees the light of the world."[32]

Wöhler's 1835 comparison of organic chemistry with a jungle expressed chemists' view particularly well: "Organic chemistry now drives you completely crazy. It appears to me like a jungle in the tropics, full of curious things, an immense thicket, without entrance and exit, which one does not dare to enter."[33] Even the more optimistic statements, such as Wöhler and Liebig's claim that they "had hit a lucid point in the dark field of organic chemistry,"[34] point in the same direction. In a more moderate tone, in 1839, Berzelius described the development of organic chemistry in the past "ten or twelve years" as follows:[35]

> The chemical *study of organic nature has become one of the most interesting topics* in the inquiry of nature ("Naturforschung"). Having been neglected for a long time, as a consequence of undeveloped concepts, it has *become a favorite* study *of the majority of chemists* proportional to the development of the concepts. *The progress is astonishing; and the growth of our knowledge for the last ten or twelve years is so great*, that the organic chemistry has become a more extended and comprehensive science than the comparatively much more elaborated inorganic chemistry.

Berzelius's judgment supports my claim that a dramatic change of organic chemistry occurred in the 1830s and corroborates my temporal framework of that process. In the next section, I describe the trajectories of this transformation.

The Trajectories of the Transformation

If we depict a very rough sketch displaying not more than the major trajectories of the transformation of organic chemistry between 1827 and 1840, the result is the following (Fig. 8.1). There is one trajectory (trajectory 2) that signifies the rapid formation and extension of the experimental culture of carbon chemistry, and there is another one (trajectory 1) that goes back to the eighteenth century, signifying the traditional culture of plant and animal chemistry and the falling apart of its elements simultaneously with the formation of the experimental culture of carbon chemistry. Trajectory 2 does not suddenly begin in 1827 but is a continuation of a much weaker trajectory going back to the beginning of the century, denoting the experiments with alcohol and its derivatives and the development of quantitative analysis from 1807. Whereas trajectory 2 parallels and is closely linked to trajectory 1 before 1827, both trajectories are separated in the period between 1827 and 1840. The former acquires its proper form and gets stronger, whereas the latter splits into several parts, representing, for example, physiological chemistry, medical chemistry, pharmaceutical botany, agricultural chemistry, and so on.

The year 1827 marks the beginning of dramatic changes. The previously marginal substances and experiments move to the center of research to become model substances in chemists' attempt to extend the concept of the binary constitution, the mode of classification, and the mode of investigating chemical reactions from inorganic to organic chemistry. Berzelian chemical formulas figure as indispensable tools in these endeavors. In fact, 1827 is exactly the year when they began to proliferate. In an unintended way, research performed with the model substances simultaneously becomes an exemplary achievement of the new experimental culture of organic chemistry, which is characterized by the creation of new associations and fits—Berzelian formulas playing the role of mediators, that is, of a cultural cement. Shortly after the beginning of the proper form of trajectory 2, in 1834, another kind of unintended consequence occurred and has to be pointed out: the formation of the new concept of substitution and its aligned experimental practice, which was to create a plethora of artificial organic compounds not existing in nature outside the laboratory. Around the same time, quantitative analysis was stabilized into a routine procedure.

As to the first trajectory signifying the fate of the pluricentered culture of plant and animal chemistry, it must be emphasized that its single elements

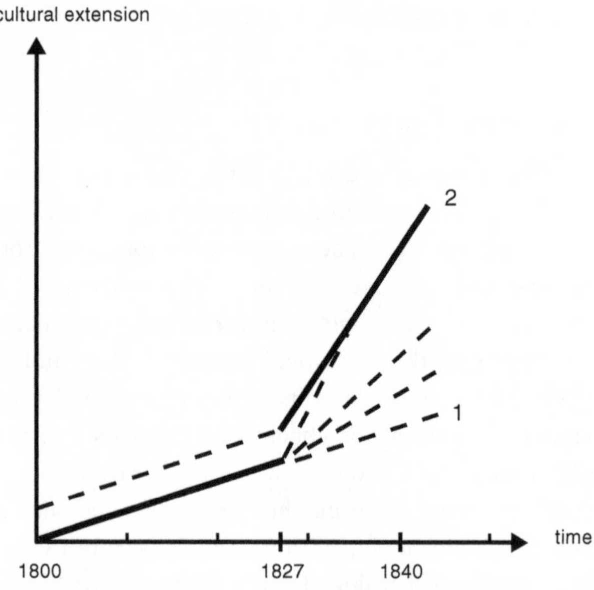

Figure 8-1 Trajectories of the structural transformation of organic chemistry

do not disappear from the scene in the period between 1827 and 1840. There was no such thing as a complete destruction of all the elements of the old culture. Rather, its trajectory is split into different parts. A few of them become part of trajectory 2, signifying preserved traditional elements within the new culture. Whereas extractions, the most important type of experiments in the former culture of vegetable and animal chemistry, remained a type of experiments in the new experimental culture, but lost their central role, physiology and the natural history of "organized" plants and animal materials were completely excluded. They moved to new emerging disciplines and subdisciplines, such as chemical physiology, agricultural chemistry, scientific pharmacy, and pharmaceutical botany.[36]

Incommensurability

In the period between 1827 and 1840, the entire area of research objects within organic chemistry was restructured. Chemists established a full analogy between organic and inorganic substances in all areas of chemical practice—

experimentation, modeling, theorizing, and classification. In this vein, the old dichotomy between the organic and the inorganic was given up. The notion of the organic (and inorganic) nature took on a new meaning. *Organic* no longer meant a particular natural origin of chemical materials from the organs of plants and animals. Moreover, the clear demarcation of the organic substances from inorganic ones by defining a set of characteristic properties and chemical behavior no longer worked. The new concept of the "organic" was defined mainly by the elemental composition of the substances—the fact that organic substances always contain carbon and hydrogen and only a few additional elements. The new meaning was aligned with a new reference. In the experimental culture of carbon chemistry, an "organic" substance was, first, a pure chemical compound rather than any kind of extracted vegetable and animal material. Second, it meant a laboratory substance that might or might not exist in nature outside the laboratory. Synthetic carbon chemistry created a "Leonardo world"[37] of hybrids that were both artifical and natural. The previous distinction between "natural" substances, viewed as "organic," and "artificial" ones, viewed as laboratory derivatives of the former, had to be given up. The two concepts of organic substances were not only incompatible, but they were also "incommensurable" in the sense of Kuhn (and Feyerabend). The point of view for ordering things into the class of the "organic" and for defining its meaning had changed. The incommensurability of the concept of the organic substance (and of the organic nature in general) was naturally continued in the mode of classification. The new mode of classifying organic substances, based on their composition and constitution, resulted in a taxonomic tree utterly different form the earlier one, which was based on observable properties and natural origins.

Thomas Kuhn introduced his concept of incommensurability with respect to different paradigms in the sense of overarching theories.[38] One of his concerns was the problem of theory choice. The parties accepting different theories (or paradigms), he argued, "inevitably see differently certain of the experimental or observational situations to which both have recourse." Their vocabularies have a different meaning and, as a result, "the superiority of one theory to another is something that cannot be proved in the debate."[39] Organic chemists living in the pluricentered culture of plant and animal chemistry and in the experimental culture of carbon chemistry in fact "saw differently certain of the experimental or observational situations," but I use the term *incommensurability* not for pointing out a problem of theory choice and theory justification. Rather, I use it as an analytical category to denote the different modes of identifying and classifying the scientific objects, which make it impossible for us historians of science to compare the two cultures of organic chemistry in any direct way. Drawing on the later work of Thomas Kuhn,[40] Jed Buchwald suggested a new view of incommensurability, which nicely explains what is meant here:[41]

> If scientific kinds do properly represent the sorts of things that scientific schemes deploy, at least since circa 1800, then the notion of *incommensurability* takes on a thoroughly new, and (we shall presently see) pragmatically significant character. But instead of specifying what *incommensurable* corresponds to, we must instead specify what *commensurable* corresponds to. If two schemes are commensurable then their taxonomic trees can be fit together in one of the following two ways: (1) every kind in the one can be directly translated into a kind in the other, which means that the whole of one tree is topologically equivalent to some portion of the other; or (2) one tree can be grafted directly onto a limb of the other without otherwise disturbing the latter's existing structure. In the first case one scheme is subsumed by the other. In the second, a new scheme is formed out of the previous two, but one that preserves intact all of the earlier relations among kinds. If neither case holds then you are in the previously fuzzy realm of the *incommensurable.*

With respect to the taxonomy of organic substances constructed in the pluricentered culture and the experimental culture of organic chemistry, none of the two conditions of commensurability outlined by Buchwald existed.

The thoroughly new way of classifying things in the experimental culture of carbon chemistry was caused by a variety of factors (see later). It was a difference of "culture" rather than a difference of an overarching general paradigm (or theory). Whereas Kuhn's paradigm is an umbrella, which covers each element and practice of a science, my concept of a scientific culture assumes a fragile unity of heterogeneous elements.[42] There is "coherence" between these elements; otherwise, the notion of a culture would be nonsense. Yet the elements of the culture are neither fully determined by the system (network, associations, and so on) nor by any fundamental principle, such as a unifying theory or a paradigm.[43] As a consequence, I do not assume that the incommensurability is total. It concerns only those sites of the two cultures that were involved in the identification of the organic substances and their classification. Another consequence is that there was also not a deep gulf between the two cultures, but overlapping zones existed.[44] There was continuity within the discontinuity.

Continuities Within Discontinuities

I have described above the transformation of organic chemistry between 1827 and 1840 as a transformation on the structural level of scientific culture rather than a transformation of one single outstanding element. If we focus on each single element in isolation from the other elements, the view that there was a slow "metamorphosis of a tradition" might be more suggestive.[45] The traditional extraction of plant and animal substances was preserved in the new culture and further refined, although these experiments lost their former privileged status. The overall social network also remained largely untouched in the transformation period between 1827 and 1840. On the one hand, the formation of an institutionalized subdiscipline of organic chemistry and the division into specialized organic and inorganic chemists took place only from the 1850s onward. On the other hand, a transnational social network already existed in the time prior to 1827;[46] its most important knots were traveling and mutual visits to chemical laboratories (particularly at Paris, Giessen, and Stockholm), correspondence, exchange of laboratory substances and instruments, specialized chemical journals, reference to the research at other places—in particular, the historical overviews in experimental reports—and publications in general, including textbooks and other monographs.

Furthermore, there were few innovations in the type of chemical instruments and apparatus. Ernst Homburg observed that between 1780 and 1860 the scale of chemical apparatus changed. The rise of analytical chemistry went hand in hand with the introduction of the test tube and small-scale experiments as well as with the development of specific skills tied to them.[47] These developments overlapped with the transformation at stake here but without having a deep, specific impact on it. From the perspective of the "big scientific instrumentation revolution" between 1920 and 1950,[48] which introduced the new types of radiation-based physical instruments into chemistry, the introduction of small-scale furnaces, glass vessels, flasks, tubes, and so on appears as minor alterations. Even if we focus on the relationship between science and extrascientific practice, continuity is stronger than discontinuity in the period between 1827 and 1840. If we conceive of the application of chemical formulas in the experimental and classificatory practice of chemistry as its "theorization," this was not done at the expense of the practical application of research products.

Yet, if we look at organic chemistry before the late 1820s and after the early 1840s from a comparative view that focuses on the kinds of objects that were investigated, the concept of organic matter and the corresponding material referents in the laboratory, the mode of classifying organic substances, the type of experiments, the mode of representation, and the self-image of chemists, then the overall picture no longer shows continuity but rather shows discontinuity. Continuity and discontinuity do not mutually exclude each other in an absolute way; rather, it depends on the perspective, emphasis, and methods whether continuity or discontinuity comes to the fore. If we conceive of the scientific culture of plant and animal chemistry not as a homogeneous unity but as a form of life constituted by heterogeneous elements that cohere without creating perfect fits and that might (or might not) develop their own trajectories of transformation, as proposed in Chapter 2, it is not so surprising that the decline and disintegration of this culture implied the preservation of several single elements.

The Dynamics of the Transformation

During the first three decades of the nineteenth century, awareness increased among European chemists of vexing problems with the classification of organic substances.[49] Jean Dumas and Polydore Boullay's new mode of classifying several organic species (alcohol, alcohol derivatives, and sugars) based on formula models of their composition and binary constitution was generally viewed as a step in the right direction. This new mode of classification required not only the quantitative analysis of the substances but also, and most importantly, experimental investigation of their reactions. For the concept of the binary constitution, which was part of the stable, collectively shared intellectual framework in inorganic chemistry and now for the first time extended to a large group of organic substances, was closely tied to this kind of experimental investigation.[50] It was part of the chemical order of inorganic chemistry, that is, of the ensemble of concepts and practices shaping experimentation and the classification of substances and reactions. The extension of this chemical order also required the construction of models of reaction and constitution by means of Berzelian formulas. When Justus Liebig and Jacob Berzelius, as the leading chemists outside France, accepted the extension of the concept of binarity to organic substances in 1832 and 1833, a decisive step in the formation of the European experimental culture of carbon chemistry was taken. The controversy about the binary constitution of alcohol and its derivatives and on organic radicals additionally stimulated investigations in this new research field, thus triggering the extension of the series of experiments, the subsequent modeling and classification, as well as the social transmission of these items. This endeavor was inevitably linked to the increase in new artificial organic substances not existing in nature outside the laboratory and the further stabilization of a new style of research, which was to become characteristic of the experimental culture of carbon chemistry.

Various causal factors for the formation of the experimental culture of carbon chemistry are mentioned in this condensed story: first, a social factor that is simultaneously an intellectual one, that is, the threefold agreement among European chemists that the traditional classification of organic substances needed revision; that the chemical order of inorganic chemistry was to be extended to organic substances and their reactions; and that, apart from experiments, Berzelian formulas were suitable tools to achieve this shared goal. Second, the social dynamics of the controversy considerably contributed to the communal transfer of the new style of research. Third, two new kinds of experiments were necessary to get information about the composition and constitution of organic substances: quantitative analysis and the study of chemical reactions. Fourth, a new paper tool was available to represent chemical reactions and to construct models of the constitution of organic substances: Berzelian chemical formulas. Of course, these immediate causes presupposed further social, material, and cultural conditions—in short, a broader context—which I have to omit here for pragmatic reasons.

Whereas historians of chemistry have acknowledged the importance of quantitative analysis for the development of organic chemistry, they have largely ignored the role played by chemical formulas. Therefore, I have particularly stressed the generative functions of this new sign system. I claim that without Berzelian formulas (or similar paper tools representing the theory of chemical portions), investigations of chemical reactions of organic substances would have remained marginal in organic chemistry; and consequently, the extension of the chemical order from inorganic to organic chemistry would have been impossible. Berzelian chemical formulas were a necessary causal factor for the development of the new experimental culture of carbon chemistry as it was formed after 1827. Without them, it would not have existed. The availability of Berzelian formulas was, however, completely contingent. This sign system was developed outside organic chemistry with purposes at first restricted to inorganic substances.[51] Nothing in organic chemistry contributed to their introduction by Berzelius in 1813. On the contrary, the entire domain of organic chemistry questioned the general validity of the theory of chemical portions embodied by Berzelian formulas.

Many historians of science emphasize the efforts and achievements of individuals as the decisive causal factors for scientific development. For example, with respect to organic chemistry of the first half of the nineteenth century, Berzelius and Liebig have been regarded as the main heroes. I do not deny that individuals play a role—in fact, I have repeatedly referred to the "leading chemists" of the 1830s. Individuals may even play a pivotal role in particular historical constellations. With respect to the historical subject at stake here, however, there was no such constellation. F. L. Holmes, whose conception of "investigative pathways" does focus on individuals, therefore had to concede:[52]

> Even as Liebig and Dumas sought power and influence to lead the development of a growing scientific field, and as Berzelius struggled to retain the influence he had once held, these three, as well as their less eminent colleagues, *were all being swept along in a moving investigative stream that none of them could control. Each of them had to strive not only to direct but to adapt to changes that he could neither arrange nor anticipate.*

Holmes does not further analyze his metaphor of the "moving stream," but it is suggestive to think of it in terms of processes on the emergent level of a scientific community and culture.

Holmes's metaphor and his additional remark that each of the individuals had "to adapt to changes that he could neither arrange nor anticipate" point to another aspect of the process. There was no research program that would have anticipated and intended the structural changes that were actually taking place. Nor can any account in terms of preexisting intentions or interests, individual or social, explain these transformations. At first glance, it may appear that the French did have a research program that outlined the transformations. The French gave up the dichotomy between the organic and the inorganic much earlier than other chemists and pointed to analogies between the organic and inorganic; they were first to extend the concept of binarity from inorganic substances to an entire group of organic ones; and they extensively applied Berzelian formulas to investigate organic chemical reactions, to construct models of reaction and binary constitution, and to classify organic substances. Neither Dumas nor Liebig nor any other chemist who was actively involved in the new practice that engendered the formation of the experimental culture of synthetic carbon chemistry actually wanted to study artificially produced organic compounds at the expense of the extracted natural organic substances, however. No one anticipated or intended the scale of synthetic carbon chemistry that developed quickly after 1827. The exponential growth of artificial substances was an unintended consequence of the experimental study of the reactions or "metamorphoses" of the natural ones. It was the practical research imperative that required the study of the artificial reaction products, and it was the challenging presence of these material things that spurred a shift of interests.

Moreover, nobody foresaw the entire range of the structural transformations of organic chemistry after 1827. New associations between laboratory tools and paper tools were built—in inorganic chemistry the chemical order largely worked without the practical application of chemical formulas, even to a large degree without the quantitative analysis—and completely new concepts with an allied experimental practice were developed, such as the concept of substitution. Whereas the extension of the chemical order of inorganic chemistry to the organic domain actually was intended, first by the French, then by all leading European chemists, the subsequent transformations that eventually transcended the existing chemical order were neither anticipated nor intended. An explanatory historical reconstruction that emphasizes chemists' goals and interests works to some extent for the first stage of the transformation process, that is, extension of the chemical order from inorganic to organic chemistry, and yet it is unable to explain the subsequent events. To explain these events, it is necessary to take into account that material objects, both experimental objects and signs systems, may spur and direct human intentions and actions.

In his *Toward a History of Epistemic Things*, Hans-Jörg Rheinberger points to the material dynamics of experimental research in molecular biology by comparing it with the construction of and movements in a labyrinth: "An experimental system can readily be compared to a labyrinth, whose walls, in the course of being erected, in one and the same movement, blind and guide the experimenter. In the step-by-step construction of a labyrinth, the existing walls limit and orient the direction of the walls to be added. A labyrinth that deserves the name is not planned and thus cannot be conquered by following a plan."[53] In my previous reconstruction of the introduction of the concept and practice of "substitution," and of the construction of new networks and fits, I followed a similar approach; but in this case, I have emphasized the role of another material item: paper tools. My reconstruction of the dialectic of goals and tools, of intentions and unintended consequences, mediates the goals, concepts, and interests of human actors with the material elements of scientific practice.

As to the causes of the falling apart of the pluricentered culture of plant and animal chemistry, it should be noted that they overlapped with, but were not identical to, those engendering the new experimental culture of synthetic carbon chemistry. It is suggestive that the larger transformations of natural history in the first half of the nineteenth century and the formation of life sciences also affected the early natural historical agenda of organic chemistry.[54] The classification in contemporary botany and zoology highlighted the hidden "type" or plan of construction. This could not be ignored by chemists, and the fact that they did pick up the term *type* in the 1840s indicates that there were relations of this kind.[55] Furthermore, the exclusion of scientific objects, in particular of physiological processes and pharmaceutically relevant organic mixtures, from carbon chemistry was not a necessary, intrinsic consequence of the formation and expansion of synthetic carbon chemistry. The former might have coexisted with the latter as well. Again, the goals and interests of the leading chemists cannot explain the exclusion of these two research objects. Berzelius and Liebig explicitly disagreed with this exclusion, and Dumas was ambiguous. In his textbook on organic chemistry from 1835, Dumas, on the one hand, restricted research in organic chemistry to pure organic compounds,[56] but, on the other hand, he remarked that the vegetable and animal physiology and the chemistry of "organized substances" cannot be completely omitted.[57] In his memorandum on organic chemistry from 1837 coauthored by Liebig, we can further read that "the mysteries of vegetation and of animal life" and the "modifications of matter which take place in animals or plants" will be unveiled by organic chemistry.[58] In fact, Dumas himself, after 1840, devoted most of his research not to substitution and the creation of substitution products, but—like Liebig—to chemical physiology and questions related to naturally found organic substances.[59] Regarding the exclusion of physiology and organized materials, institutional constraints may have played a pivotal role; but we must also think of the general trend toward specialization in nineteenth-century sciences. The simultaneous multiplication of chemical subdisciplines and the formation of experimental physiology[60] and academic pharmacy very probably was not without consequences for carbon chemists' self-conception and prospects.[61]

Another possible driving force of the transformation of organic chemistry is commercial needs. Was the emergence of carbon chemistry in any way spurred by industrial interests in the application of synthetic carbon compounds? We know that the synthetic dyestuffs industry that emerged in Western Europe from the 1850s onward was closely linked to the new form of synthetic carbon chemistry.[62] But these events occurred two decades later than those at stake here, and the role that synthetic carbon chemistry was to play for the emerging chemical industry in the second half of the century cannot be viewed as a cause for its support decades earlier. In his reconstruction of the formation of experimental physiology in mid-nineteenth-century Germany, Timothy Lenoir has discussed an analogous case. He has argued, "It cannot be maintained that, for Prussia at least, state interests in supporting experimental physiology derived from the promise of improved medical therapy, in spite of the potential market for it. Furthermore, the economic benefits of supporting physics, in whatever applied forms, became a realistic motivating factor only in the 1880s and

1890s, with the demand of industry for precision mechanics and the associated expansion of technical education."[63] Another fact rejects the assumption that industrial needs played a specific role for the transformation of plant and animal chemistry into synthetic carbon chemistry: Extracted plant and animal materials did have enormous practical use, in particular as pharmaceuticals. As to practical applications before the mid-nineteenth century, there was no significant difference between the synthesized carbon compounds and the extracted natural ones. For example, chloral and chloroform, synthesized by Liebig in 1832, were applied a couple of years later as sleeping tablets (chloral) and narcotic (chloroform),[64] but the pharmaceutical application of synthesized substances like these did not take place at the expense of extracted natural ones. Both groups of organic substances, the artificial and the natural, were applied before 1850 whenever it was possible. Even after 1850, natural-products chemistry remained commercially and industrially important, whereas many, even most, synthesized carbon compounds had no practical application whatsoever.[65]

Remarks on Context and Methodology

In preceding chapters, I have not aspired to give a dense historical description, let alone an exhaustive explanation, of the patterns of the transformation of organic chemistry between 1827 and 1840. For example, I have omitted in my sketch of trajectories the refinements of the technique of extraction, which has been emphasized by F. L. Holmes,[66] because I am convinced that it did not have the same relevance for the formation of the new experimental culture of carbon chemistry as quantitative analysis, the experimental study of chemical reactions, and chemical formulas. This brings us to the methodological question of how a more comprehensive broader picture of the transformations might be achieved. F. L. Holmes has suggested, "If we wish to grasp the larger picture we must show how individual investigative pathways are linked into a network of interacting investigations that comprise the moving front of a shared set of research problems, or a scientific subfield."[67] He immediately added, "It is an open question whether we can assimilate this new level of complexity *without losing the degree of resolution at the level of individual pathways* essential to understand how the experimenter engages nature."[68] I do not want to address in any detail the question of what kind of a broader picture we would get in this way—would we see any pattern or trajectory in it, and how many volumes would it fill?[69] Rather, my point here is a more theoretical one: that scientific culture is emergent, that is, not identical with the sum of individual elements. This is an aspect of scientific culture that has been rightly emphasized by the sociology of scientific knowledge.[70] The conception and historiographical methods of "investigative pathways" have proved highly productive for studying the practical performances and achievements of individuals or local groups of scientists,[71] but it is not scale invariant. For the broader picture, a different approach is needed that, in one way or another, focuses on the communal level of science.

In this book, I have attempted such an approach, but this approach, for pragmatic reasons, required concentrations and exclusions. I concentrated on the practice and culture of European organic chemistry, and I largely excluded the simultaneous transformations of the broader nineteenth-century scientific landscape and its even broader social and cultural context. The result is far from being a global picture, but I hope it is a larger piece for such a picture. Even with respect to my depiction of the two cultures of organic chemistry and the transformation process during the 1830s, I have to add some proviso. Many questions had to remain open. Why did so many chemists of the younger generation turn to the investigation of substitution and of the artificially produced organic substances, as has been described by Liebig and Daniell,[72] at the expense of the study of extractions and naturally occurring organic substances? What career expectations were linked to this choice? Was the establishment of large "research groups" involved in these choices?[73] Did the types of scientific institutions and policy play a role, or did any elements of the larger social and cultural system, such as a new "materialist ideology," as has been described by Timothy Lenoir with respect to the emergence of experimental physiology in mid-nineteenth-century Germany?[74] Future investigations will have to study such questions.

CHAPTER NINE

Paper Tools

> An instrument of labour is a thing, or a complex of things, which the labourer interposes between himself and the subject of his labour, and which serves as the conductor of his activity.
>
> —MARX 1906, 199

The analyses of the foregoing chapters have demonstrated that beginning in the late 1820s, chemists applied Berzelian formulas as productive tools on paper throughout all organic chemistry. Formulas were applied as tools for the construction of interpretive models of organic chemical reactions in the form of balancing schemata, such as formula equations. In that function, they not only enabled chemists to represent recombinations of the parts of reacting substances, but they also filled gaps left open by experimental inscription devices. Exaggerating a bit, one might say that the Berzelian sign system became a reified part of the experimental inscription devices. On the level of interpretation, chemists also used Berzelian formulas to construct models of chemical reaction that incorporated additional assumptions. Furthermore, Berzelian formulas were suitable tools for modeling the constitution of organic substances. This was a precondition for another function, their application as tools for constructing new taxonomies of organic substances. In this taxonomic context, Berzelian formulas also solved the vexing problem of identifying organic species because they represented the composition of each compound like blueprints that clearly distinguished different organic species.

Referring to sign systems in chemistry, both words and different types of formulas, Mary Jo Nye pointed out: "Since the words we use and the symbols we manipulate are instrumental in producing new knowledge as well as in expressing what we already know, the course of chemical research in the nineteenth century and early twentieth century became as closely linked to the conventions of language and imagery as to the instruments chemists built and the network of social circumstances that influenced the problems they chose to study."[1] The various generative functions of Berzelian formulas, however, have been largely ignored by historians and philosophers of science. Many historians of chemistry conceive of Berzelian formulas as illustrations of an atomic theory that was represented much more clearly by verbal language or by Daltonian diagrams. Others have claimed that they were more or less superfluous presentations of empirical findings, equally well expressed by the stoichiometric laws. In both cases, Berzelian formulas are viewed as passive signs illustrating preexisting knowledge. The French philosopher François Dagognet, one of the few philosophers actually to have studied Berzelian formulas, writes of them: "The first mode of writing, which merely transcribes speech by way of letters and vocal symbols, hardly offers any advantages (in comparison with oral chemistry, which it perpetuates). . . . This stenography will occupy or invade chemistry during the first half of the nineteenth century until that moment (rather near) when its insufficiencies will become obvious. At the beginning of the nineteenth century its main proponent was Berzelius, who established the rules of its application."[2] According to Dagognet, Berzelian chemical formulas were mere surrogates for names, that is, signs whose function was exhausted by denoting chemical substances. Whereas he ascribed a generative role in knowledge production exclusively to structural and stereochemical formulas, he conceived of Berzelian formulas almost as a hindrance to progress. His assertion that this sign system was a kind of stenography, a shorthand for names, was the outcome of an ill-balanced focus on formal semiotic aspects of Berzelian formulas, to the exclusion of any historical analysis of the actual application of this sign system by the historical actors.

In contrast to Dagognet and others, I argue that beginning in the late 1820s, Berzelian chemical formulas became enormously productive as tools on paper, *or paper tools*, particularly in organic chemistry. Like Dagognet and other semioticians, I take the materiality of chemical formulas into account. My analyses in the foregoing chapters were not restricted to formal semiotic aspects, however, but included the scrutiny of the historically situated pragmatics of this sign system. Whereas names must be immutable to function as a notational system, chemists manipulated and transformed Berzelian formulas for the purpose of superimposing additional "long-term concepts," such as the concept of binary constitution and constructing concrete formula models displaying the general concept as applied to the particular case.[3] These mutable inscriptions were particularly suited for creating "chains of inscriptions" as part of the enterprise of modeling invisible research objects, closely related to experimentation and classification.[4] In the construction of interpretive models of chemical reactions, analyses could be simulated by deleting letters and numbers, syntheses by adding them; in their formula models of binary constitution, chemists regrouped letters and numbers, and they newly arranged partial formulas within a formula model of binary constitution. In the practice of classification, these formula models were mutually adjusted to each other, for example, by multiplication or division. In all these manipulations, the original function of Berzelian formulas in denoting a particular substance and its composition was preserved while additional reference and meaning were added.[5]

In this last chapter, I first summarize and discuss the way chemists applied Berzelian formulas for representing chemical reactions and constructing simple interpretive models of them. Representations and interpretations of organic chemical reactions had confronted chemists with nearly insurmountable obstacles from the late eighteenth century onward. The fact that chemists were able to overcome these obstacles by using Berzelian formulas is not only historically relevant—for it was one of the most important steps in the formation of the experimental culture of carbon chemistry—but it also bears features that are of broader epistemological interest. In the second section of this chapter, I discuss the notion of chains of inscriptions as Bruno Latour, in particular, has applied it to science studies and extension of this notion by that of "paper tools." I argue that this extension is necessary for the purpose of explaining the historical actors' preference of a certain type of signs as well as for the historical reconstruction of the reference and meaning of a sign system and its changes over time. I further argue that the introduction of new scientific objects and the concepts bound up with them can be explained to a large extent by examination of the dialectic between the goals of this historical actors and their tools, both laboratory tools and paper tools.

FILLING IN THE GAPS OF INSCRIPTION DEVICES

Chemical reactions are invisible processes. Although they are accompanied by observable effects like change of color or smell, the interaction of the substances themselves cannot be observed. In the eighteenth and nineteenth centuries, the most important traces of these invisible transformations were the substances produced in a reaction, the so-called reaction products. Representations of chemical reactions relied mainly on conclusions drawn from the "composition" of reaction products and the comparison of their composition with that of the initial substances. These conclusions were drawn within a collectively shared conceptual framework formed at the beginning of the eighteenth century that included chemical categories like chemical reaction, reaction products and initial substances, composition, constitution, affinity, and others.[6] In this conceptual framework, chemists conceived of chemical transformations as recombinations of the components of the transformed substances into the products of transformation, the components being substances that preserve their integrity while recombining. Furthermore, eighteenth-century and nineteenth-century chemists assumed that these recombinations were directed by an unobservable chemical force between pairs of substances, called *chemical affinity*.

The representation of a particular experimentally investigated chemical reaction traditionally required the experimental separation and isolation of all the reaction products, which, as a rule, were mixed together in the reaction vessel. This was done, for example, by means of distillation in case of liquids or by means of precipitation or crystallization in case of solid products. After its isolation, a particular reaction product had to be further *purified*, that is, separated from admixed traces of other substances (such as untransformed initial substances, other reaction products, or the solvent used). This was done, for example, by its repeated distillation (*rectification*) or by its repeated dissolution and precipitation or crystallization. After this series of experimental procedures, each of the purified reaction products was submitted to another series of experiments in which their composition, qualitative and often quantitative, was investigated. At the end of these processes, which might be compared with the elimination of background noise, data gathering, and data processing in experimental physics, the material traces were transformed into marks on paper, or *inscriptions*. Before the late 1820s, these inscriptions consisted of the systematic names of the reaction products, expressing their elemental composition, the names of their constitutive elements, and the weight percentages of those elements. These inscriptions then were taken as a starting point for constructing an interpretive model of the reaction, which in the eighteenth and early nineteenth centuries was done through ordinary language and by means of tables, such as eighteenth-century chemical affinity tables.

After chemists realized, at the end of the eighteenth century, that organic substances are made up largely of the same components, namely carbon, hydrogen, and oxygen, they considered it necessary that the investigation of organic chemical reactions included, first, information about the quantitative composition of the substances, and, second, about the masses of the reaction products and the consumed initial substances. The construction of a sufficiently distinct building-block model displaying the recombinations of the components of the initial substances into the reaction products was possible only if some information about both kinds of quantitative aspects of an organic chemical reaction was available. Questions concerning the precision of this building-block model and the included quantitative information changed during the first two decades of the nineteenth century. At the beginning of the century, chemists were satisfied with approximate data, in particular with an estimation of the masses of the transformed parts of the initial substances and those of the reaction products. This had already begun to change in the first decade of the nineteenth century, when leading chemists set out to improve quantitative analysis and to measure the masses of substances involved in a chemical reaction.[7]

Beginning in 1807, because of improvements in the technology of quantitative analysis, chemists progressively mastered the determination of the quantitative composition of organic compounds. Yet experimental measurement of the masses of reacting substances turned out to be impossible. In hindsight, the reason for this can be easily understood. As a rule, organic reactions yield a cascade of reaction products; given the technological conditions of nineteenth-century chemistry, it was impossible to isolate all these reaction products in a quantitative way. It was also impossible to separate quantitatively the masses of the initial substances, which remained untransformed in the reaction vessels. Part of the mass of a substance would adhere to the walls of the vessels, another part would get lost while changing a vessel, another one during weighing, and so on. Even worse, it was often impossible to isolate and identify all the reaction products qualitatively, or it was an open question as to whether all the reaction products had been found.

Furthermore, the identification of organic substances by presenting their quantitative composition in terms of percentages of weight of the components often did not yield unambiguous data. Repeated experiments with different samples of one and the same substance yielded analytical data that deviated from each other to greater or lesser degrees. There were many reasons for this, among others the inaccuracy of a chemist's manipulations, an accidental contamination of a sample, or more principal mistakes in the analytical procedure. Therefore, data comparison involved judgments that did not follow any explicit methodological rules.[8] For example, chemists would omit the analytical data from one experiment if they deviated too much from others, or they would compare the analytical data with possible Berzelian formulas. Moreover, the inscriptions stemming from quantitative analysis often did not provide clear boundaries between different organic compounds or species.

In sum, there were three obstacles to chemists' investigation of organic chemical reactions. Figure 9.1 schematically describes these obstacles. Step 1 represents the first experimental manipulations performed in studying a chemical reaction, that is, the mixing together of two initial substances A_1 and A_2

in a reaction vessel, and modification of the target, for example, by heating up the mixture. The immediate traces of this material modification were the reaction products, which might have been observed by a change of color or smell, a spontaneous precipitation, or such. If these reaction products were liquids, which was often the case for organic chemical reactions, they were all mixed together. The traces of the reaction were unordered and unreadable—a situation that might be compared with what physicists call *background noise.*

What was background noise for experimental physicists was *impurity* for chemists. Step 2 of the diagram represents the experimental attempts to isolate and purify the various reaction products and the obstacles in this decisive step of trace gathering. In many cases, it could not be ascertained whether all reaction products had been found and isolated, that is, whether the trace gathering was complete. This was particularly the case if only a small amount of a particular reaction product was yielded. Very often, chemists were also uncertain as to whether they had sufficiently purified an isolated reaction product.[9] Further, and most importantly, there was an insurmountable obstacle in this second step: the fact that, as a rule, the masses of the reaction products and of the untransformed remainders of initial substances were not measurable. Finally, in the third step, that is, the quantitative analysis transforming material traces into inscriptions, a further obstacle occurred. The analytical data obtained from a particular reaction product were often not sharp enough to identify and distinguish it clearly from other organic compounds.

In this situation, Berzelian chemical formulas filled a threefold gap left open by experimental inscription devices. First, a chemical formula was an exact "blueprint" of a pure organic species that clearly denoted its composition and distinguished it from other organic compounds or species. If chemists obtained uncertain analytical data, they would attempt to fit the data with possible Berzelian formulas of the chemical compound as well as with the formula model of its constitution, which might be supported by additional experimental investigations of the chemical reaction of the compound. Thus, "fuzzy" inscriptions were transformed into sharp ones. Second, because a Berzelian formula signified, among other things, the theoretical combining weight of the signified compound (today, *molecular weight*), it carried information about its reacting mass in a theoretical form. If an isolated reaction product was already known and its Berzelian formula more or less accepted, the formula was taken as a substitute for the actual measurement of the mass. This was perhaps the most important and crucial function of Berzelian formulas in the investigation of organic chemical reactions. Here the formula fulfilled a "wholesale" function otherwise performed by a laboratory tool or experimental inscription device,[10]

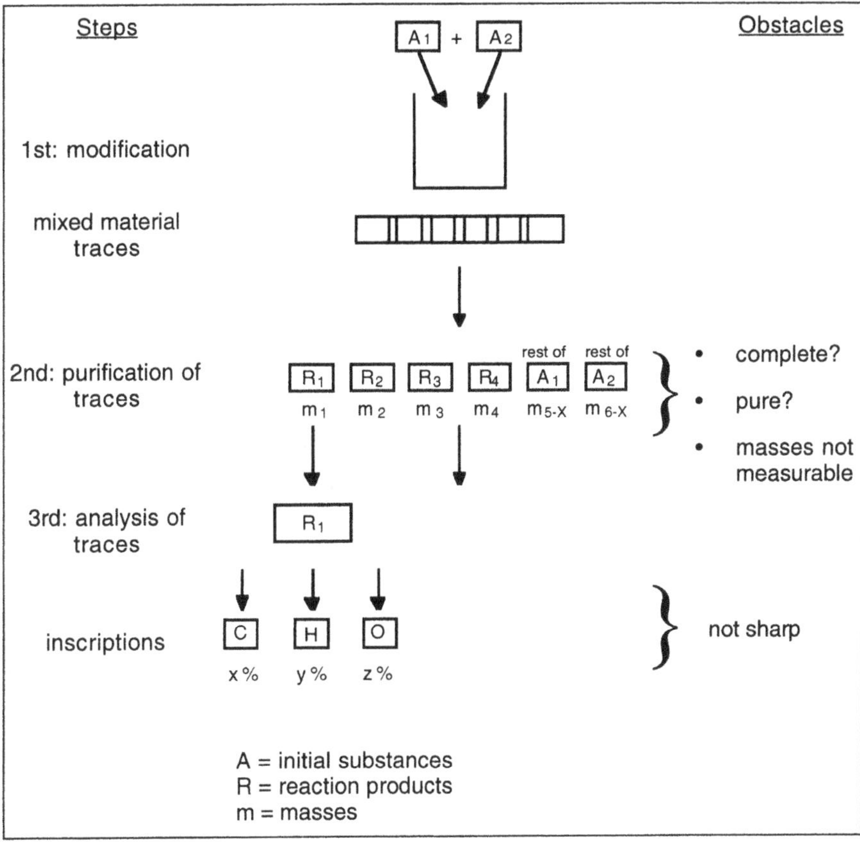

Figure 9-1 Obstacles in studying organic chemical reaction

where *inscription device* refers to the entire sequence of processes represented in steps 1, 2, and 3 of the diagram and the apparatus and manipulations involved in them.

Third, the construction of schemata of balancing the masses of initial substances and reaction products by means of Berzelian formulas often solved the problems of the incomplete qualitative isolation, purification, and identification of reaction products. A simple formula equation balanced the masses of the initial substances and reaction products and thus conveyed information about all the substances involved in a chemical reaction, the qualitative and quantitative composition of these substances, and their reacting masses. It made both a qualitative and a quantitative representation of the recombinations of a reaction possible, even if not all the reaction products were known, isolated, or purified. The reason for this is the following: Formula equations enabled chemists to single out different reaction paths and to establish a distinction between independent parallel reactions and successive reactions, without any overlapping. Thus, chemists could be confident that they would succeed in investigating an organic chemical reaction, even if the experiment yielded a chaotic cascade of reaction products that could not be completely transformed into purified traces of the reaction.

Formulas enabled chemists to eliminate their particular type of background noise. "When the background could not be properly circumscribed, the demonstration had to remain incomplete, like Michelangelo's *St. Matthew*, in which the artist was unable to 'liberate' his sculpture from its 'marble prison.' In physics the analogous process of 'liberation' of an effect from the background is linked to theory." Galison makes this observation with respect to experimental studies of the magnetic field, the turbulence in the cloud chamber, and neutrino collisions.[11] The same kind of "liberation" of experimental effects, linked to theory and the signs embodying it, is to be found in organic chemistry from the late 1820s onward. Without the application of Berzelian formulas, "the demonstration [of the invisible object chemical reaction] had to remain incomplete" as well. In fact, in the time before Berzelian formulas were applied, we find only very few investigations of chemical reactions in the domain of organic chemistry.

In the three functions described previously, paper tools took on tasks, wholesale or complementary, that are otherwise characteristic of laboratory tools and inscription devices. Paper tools and laboratory tools coproduced pure material traces and analytical data of the invisible scientific object "chemical reaction." In this coproduction, impurity was transformed into purity, vagueness into exactitude, incompleteness into completeness. From the very beginning, experience was intertwined with theory and other elements of the broader scientific culture, such as the mathematical worldview and algebraic notation. It was not in the form of hypotheses derived explicitly from a theory or any other kind of propositional statements that theory entered experimentation but via a reified sign system and its skilled manipulations. Theory was used as "part of the craft" of the experimentors.[12]

If Berzelian formulas were indispensable tools for experimental investigations of organic chemcial reactions, filling gaps left open by inscription devices, they were no less productive in the construction of interpretive models of chemical reactions. Schemata of balancing the masses of the initial substances and the reaction products using Berzelian raw formulas, such as linear formula equations, were not only a device for checking whether all the relevant reaction products had been found and identified in the experiment and for singling out different reaction paths; they were also the simplest interpretive models of a reaction. In that additional function, they were a generative tool for simulating possible reaction products not discovered in the experiment, as was often the case for the by-product water. Moreover, chemists used Berzelian formulas for modeling various steps and other "mechanisms" of a reaction.[13] The function of formulas and entire formula models as tools for playing through various possibilities of different regroupings of the components of the initial substances into the reaction products depended on the fact that Berzelian formulas visually displayed the recombining building blocks and were easy to manipulate on paper, their additivity being the only syntactic constraint of these manipulations.

As an example of these two paper tool functions of Berzelian formulas in the process of data gathering and processing and the subsequent interpretive modeling, I have reconstructed Dumas and Boullay's formula model of the formation of ordinary ether and sulfovinic acid in Chapter 4 and in Chapter 7 the building of three subsequent formula models of the reaction of alcohol with chlorine, which led to Dumas's concept of substitution. For example, Dumas and Boullay's formula model for the formation of sulfovinic acid, which was a by-product of the production of ordinary ether, was:[14]

$$2\,SO^3 + 4H^2C^2 = S^2O^5 + 2\,H^3C^4 + H^2O$$

In the model, the masses of the initial substances and of the reaction products are balanced, as can easily be seen by comparing the numbers belonging to a particular letter to the left and to the right of the equal sign; thus, reconstruction of the reaction is complete. The necessary information about the masses of the substances involved in the reaction is provided exclusively by Berzelian formulas. It was not and could not be measured on the experimental level. Hence, Berzelian formulas complemented the experimental inscriptions obtained in quantitative analyses. Furthermore, the formulas give exact quantitative information about the composition of the four substances involved in the reaction, thus clearly distinguishing them from other organic species. Based on this, Dumas and Boullay solved a conundrum that had occurred in 1820, namely, that ordinary ether and sulfovinic acid were produced simultaneously. The formula model attributes the production of sulfovinic acid to a reaction path that is completely independent of the reaction path yielding ordinary ether. It shows that the reaction of the two initial substances leading to sulfovinic acid yields only one additional reaction product (water) and that consequently other products observed in the experiment can be ignored. Formula models like these reinforced the construction of a new ordering of experimental traces, transforming what had been merely occasional, *ad hoc* reasoning about distinct reaction paths into a collective style of reasoning. From the 1830s onward, the distinction between independent parallel reactions and subsequent reactions instituted a stable order in chemists' investigations of the comparatively chaotic transformations of organic substances.

The formula model also visually displays the recombination of the building blocks of the two initial substances, sulfuric acid (SO^3) and bicarbonated hydrogen (H^2C^2), into the reaction products sulfovinic acid ($S^2O^5 + 2\,H^3C^4$) and water (H^2O). In this case, Dumas and Boullay had not observed the second reaction product, water, in the experiment. The information that water was a second reaction product was a new and unforeseen outcome of reshuffling and balancing the formulas of the two initial substances. Attempts to regroup $SO^3 + H^2C^2$ into the formula of sulfovinic acid, $S^2O^5 + 2\,H^3C^4$, first show that the initial formulas have to be multiplied to balance the masses of the initial substances and reaction products. From this, the formula $2\,SO^3 + 4\,H^2C^2$ follows, as well as the further information that 2 H and 1 O are left over, which Dumas and Boullay combined to H^2O, representing the second hypothetical reaction product water.

CHAINS OF INSCRIPTIONS AND PAPER TOOLS

In their construction of models of chemical constitution and reaction, chemists manipulated and repeatedly transformed Berzelian formulas. They thus created series or chains of inscriptions that embodied additional reference and meaning. The following example from chemists' construction of models of binary constitution in the 1830s illustrates this:

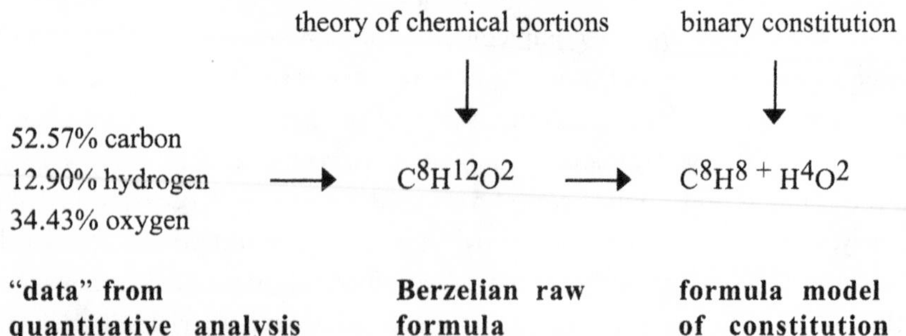

The reference and meaning of the first inscription, which stems from the quantitative analysis of alcohol by Jean Dumas,[15] were embedded in analytical instruments and experimental manipulations as well as in the conceptual network of chemical compound, reaction, analysis, synthesis, and so on formed at the beginning of the eighteenth century and enlarged and modified later by Lavoisier and his collaborators and by chemists developing stoichiometry. It denotes the elemental composition of the chemical compound alcohol using the names of the elements and representing the weight ratio of the elements in weight percentages. In the second step, the analytical data are transformed into a Berzelian raw formula. The Berzelian raw formula preserves the reference and meaning of the first inscription but adds something new, namely, the theory of chemical portions. According to this theory, the formula signifies that the observable, elemental components of alcohol—carbon, hydrogen, and oxygen—consist of invisible, scale-independent portions with an invariant and characteristic relative combining weight, or "atomic weight." From these theoretical combining weights, the quantitative information given by the first inscription can always be recalculated.

In the third inscription, the formula additionally signifies the binary constitution of alcohol from the two substances bicarbonated hydrogen and water. It also denotes the weight ratio of these two immediate constituents in terms of their invariable theoretical combining weights. The additional reference and meaning are due to a simple manipulation of the sequence of signs in the second inscription, resulting in two partial formulas separated by the plus sign. Again, the third inscription preserves the reference and meaning of the former two inscriptions while adding new information. This is achieved without considerably enlarging the sign system. In the final result, an enormous amount of information is stored on a very small piece of paper. In Dagognet's words, the inscription becomes more compact or denser and more effective with each step of transformation.[16]

The Materiality of Inscriptions

The notion of "chains of inscriptions" as introduced by Bruno Latour[17] focuses historical analysis on both the performative aspects of knowledge production and its material conditions, that is, the material medium, the visual form and maneuverability of inscriptions, and the actual manipulations of inscriptions in scientific practice. It presupposes the basic insight of semiotics that the type of sign system shapes our perception of and reflection on the world as well as reference and meaning in general. Studying chains of inscriptions allows us to reconstruct historically the principal steps of scientists' construction of models, concepts, and theories. The formation of abstractions "can be broken down into small, unexpected and practical sets of skills."[18]

Thus, scientific achievements, which traditionally have been viewed as results of spontaneous, individual creativity of the mind, become analyzable, transparent performances. Moreover, individual achievements can be linked to collectively available resources and to the toolbox of signifiers in the broader culture.

What, for example, was the source of the multiplicity of those diverse models of chemical constitution that were fiercely debated in the controversy on organic radicals during the 1830s?[19] Dumas suggested the formula model $C^8H^8 + H^4O^2$ for alcohol (with the "atomic weight" of $C = 6$), Liebig's binary model was $C_4H_{10}O + H_2O$ (with the "atomic weight" of $C = 12$), and Berzelius divided the raw formula in half and suggested $C^2H^6 + O$ (with the "atomic weight" of $C = 12$). My historical analyses of the experiments to which these three chemists referred in support of their respective models have demonstrated that none of them was superior in terms of experimental evidence. Given the assumption that chemists expressed preconceived ideas through their different models, it is difficult to ascertain how these ideas could emerge and what spurred their differences. If we reconstruct the formation of these models by regrouping letters and numbers of Berzelian raw formulas to form two separate partial formulas that fitted the collectively shared general concept of binary constitution, however, these different models become quite intelligible alternatives. It can be shown that the differences between the various models of constitution were primarily the result of different modes of combining and manipulating Berzelian formulas. New possibilities for combination emerged as chemists tinkered with formulas on paper. In Latour's words, "Most of what we impute to connections in the mind may be explained by this reshuffling of inscriptions."[20] It also becomes obvious to what extent the highly disputed particular models of individual scientists depended on resources introduced by other individuals and distributed in the chemical community via institutions such as scientific journals.

Redressing the Balance Between the Material and Mental Dimensions of Inscriptions

The emphasis on the materiality of sign systems and the possibilities and constraints set by their visual form, syntactic rules, and maneuverability raises questions concerning the role of human actors. In Latour's concept of chains of inscriptions, human agency takes a back seat. When interest focuses mainly on the abstract epistemological question of "how many cognitive abilities may be, not only facilitated, but thoroughly explained" by material agency (such as inscriptions),[21] explorations of the explanatory capacity of material agency that go hand in hand with the elimination of notions referring to mental processes are suggestive. By contrast, the historical reconstructions contained in this book also take into account the historical actors' goals, concepts, models, theories, and beliefs. They combine a semiotic analysis of Berzelian formulas and of their manipulations and transformations with the analysis of model building, embedded in its turn in a discussion of beliefs about organic matter and of stabilized long-term concepts, such as binary constitution and displacement reaction, to which the particular models were assimilated. To explain how new reference and meaning accrue to signs, these reconstructions also conceptualize chemists' manipulations of inscriptions in terms of a dialectic of tools and goals.[22]

Redressing the balance between the material aspects of inscriptions and the mental dimensions of human agency was necessitated by concrete questions and puzzles that emerged in my historical analyses. Why did the European chemical community accept Berzelian formulas during the 1830s and 1840s, after this sign system had been largely ignored for more than a decade? Why did they give preference to Berzelian formulas rather than to Daltonian diagrams or to ordinary language? Why were Berzelian formulas applied in the very domain where they most lacked experimental evidence? What were the additional reference and meaning of "higher order" inscriptions—what I term *formula models*—compared with "first-order" inscriptions, such as analytical data? How and why did the additional reference and meaning of higher-order inscriptions change over time, for example, in case of the introduction of the term *substitution*?

It is impossible to answer questions like these without invoking the beliefs, concepts, theories, models, and goals of human actors. Without reconstructing the layers of reference and meaning of Berzelian formulas, which requires additional analyses of texts representing chemical concepts and theories, the acceptance of this sign system could at best be explained as a fashion, a pedagogic decision, or a strategy of gaining academic reputation. As long as we fail to take into account that after 1833 European chemists shared the goal of assimilating organic compounds and their reactions to the chemical order of inorganic chemistry, the reason chemical formulas were manipulated in the particular way they were and the historical reference and meaning of higher order inscriptions—that is, formula models of constitution and reaction—remain in the dark. This shared goal of European chemists and the resulting necessity of constructing particular models of binary constitution and of organic chemical reactions and new classifications based on them explain why European chemists so quickly accepted Berzelian formulas.

Berzelian formulas were much more suitable tools for this goal than Daltonian diagrams and ordinary language. Their compositional syntax and semantics, their graphic suggestiveness, and their simple maneuverability made it easy to play through various possible models, adapting the traditional concept of binary constitution and types of reaction to organic substances. Chemists would regroup letters and numbers of a Berzelian raw formula to build a binary model of chemical constitution; they would add up letters and numbers or eliminate them to construct a new raw formula of a possible reaction product within a model of a chemical reaction; or they would multiply or divide formulas to fit them into a reaction model or into a class of substances represented by formula models of constitution. All this reshuffling of the letters and numbers could be done without any syntactic restriction besides additivity.[23] This was a great advantage of Berzelian formulas over Daltonian diagrams, which imposed additional syntactic and semantic constraints that were alien to the goals of the historical actors.[24] Compared with the additivity of language or names, Berzelian formulas had the dual advantage of being short inscriptions that can be taken in at a glance and of being inscriptions that preserve their original meaning when they are transformed.

In keeping with the traditional intellectual framework, formula models at first referred to *components* and *immediate constituents* of chemical compounds in the sense of pure substances that can be isolated in experiments and stored and observed in laboratory vessels. Yet the graphic suggestiveness of Berzelian formulas also set a new agenda after chemists had learned to read them at a glance and acquired skill in manipulating them. The visual image of Berzelian formulas gave chemists the idea of wedding their traditional concepts of chemical compound, composition, constitution, and types of reactions (all linked to laboratory substances) to the new theory of chemical portions. In the concrete performance of constructing particular models, the tools began to exercise a reverse effect to the chemists' goals and tacitly to alter them. This dialectic of tools and goals becomes particularly obvious in the introduction and development of *substitution*, a concept and correlative scientific object that not only had been unforeseen and unintended in the early 1830s but was incompatible even with the original goals of European chemists.

Jean Dumas, who introduced the concept of substitution in 1834, had set out to construct a formula model of the formation of chloral that fitted with the traditional concept of binary constitution and the traditional types of reaction. His main goal was to defend his own model of the binary constitution of alcohol. Yet the tools applied to achieve this goal led to a shift in Dumas's goals. After 1834, similar applications of Berzelian formulas by other European chemists tacitly undermined the traditional concept of binary constitution, which the European chemical community finally abandoned in the 1840s. At the same time, *substitution* became a hallmark of the new experimental culture of carbon chemistry, referring to an entirely new practice of synthesizing organic compounds. The new object, concept, and practice of substitution were not the consequences of an atomistic research program or of any other form of deliberate intentions on the part of chemists but were originally unforeseen and unintended results of experiments and coordinated work on paper with chemical formulas. At first, this result was embodied in a formula model of reaction, and only afterward was it translated into ordinary language and the new term *substitution.*

Paper Tools and Laboratory Tools

From the perspective of chemists' goals and the historical alteration of these goals, Berzelian formulas can be seen to be paper tools applied to achieve human goals. *Paper tool* is a category that serves to focus the historical analysis and reconstruction on the performative, cultural, and material aspects of representations, without disregarding the goals and other intellectual preconditions of human actors. Although *paper tool* is a metaphor, I claim that on many levels paper tools are fully comparable to physical laboratory tools and instruments and that both kinds of tools contribute to the creation of reference and meaning, or *representations* in that sense.

First, paper tools are material devices in the broadest sense of being exterior to mental processes, visible and maneuverable. Unlike laboratory instruments, they do not interact physically with the object under investigation; this important distinction from laboratory instruments needs to be emphasized. Even so, paper tools are visible marks that can be manipulated to create representations of scientific objects—sometimes even to coproduce inscriptions with laboratory instruments—and to explore their relationships with other objects as they appear in the light of concepts and theories embodied by the paper tools. A laboratory instrument must be suitable to the object under investigation. So, too, with paper tools; their syntax and semantics must fit with scientists' objects of inquiry. The algebraic notation of Berzelian formulas was particularly suited to their application as paper tools for experimentally investigating scientific objects that were mundane in inorganic chemistry of the time but missing in organic chemistry, namely, chemical reactions and the constitution of organic substances. Berzelian formulas conveyed a building-block image of the reactions and of the constitution of organic compounds without embedding this image in the conceptual network of philosophical atomic theories, that is, without invoking simultaneously the idea of submicroscopically small mechanical bodies, which was rejected by many chemists. These building blocks could be manipulated on paper to align them with particular experimental traces, and additivity was the only syntactic rule that constrained manipulations. This was exactly what chemists needed for building particular interpretive models of their scientific objects. Additional constraints alien to chemists' goals were avoided. Second, the material form and maneuverability of both laboratory tools and paper tools have a strong impact on the outcome of experimentation and representation. Similarly, as the technical design of a laboratory instrument and the possibilities and constraints of its manipulations and interaction with the material target shape the kind of traces or data gained in an experiment, the syntax of paper tools—their visual form, rules of construction and combination, maneuverability—shapes scientists' production of chains of representation. On the one hand, the syntax of Berzelian formulas made possible simple chemical models that fitted the existing chemical order and that allowed its extension and modification. On the other hand, they conveyed the intended building-block image of chemical constitution and reactions without simultaneously representing chemical forces and the dynamics of a chemical reaction, which were parts of chemists' reasoning in the eighteenth century and the early nineteenth century. As a consequence, during the 1830s, chemists' style of reasoning about reactions changed tacitly. For more than a decade, arguments invoking affinities and repulsions receded into the background, whereas formula equations came to the fore as a means of justification. Moreover, chemical formulas—Berzelian formulas and even more so the structural and stereochemical formulas developed from them—reinforced a thoroughly mechanical image of chemical compounds and reactions that was not intended by chemists of the time. Hence, many verbal comments were necessary that corrected this image.

Third, speaking of tools raises the question of "tools for what," relating the tools to the goals of their users. Goals, understood as a social category, relate to tools of experimental cultures, laboratory and paper tools, as material and transportable things that potentially can be shared by the entire scientific community. Tools are an amalgam of working stability, adaptive flexibility, and creative openness. They consist of elements with which their users are familiar and allow unprecedented manipulations in new concrete settings that produce new specific effects. When Berzelius introduced the quasi-algebraic notation familiar from mathematics and physics into chemistry, he used it as a creative sign system that forged the difference between "atoms" in the philosophical tradition and discrete "chemical portions." In their new epistemically fluid environment situated between the traditional vocabulary of philosophical atomic theories and the vocabulary of stoichiometric and volumetric experiments, which nourished chemists' intuitions about discrete quantitative units of substances, the new sign system became a medium to express these chemical intuitions and a tool to construct and define a new, refined chemical concept.

In the late 1820s and 1830s, Berzelian formulas became paper tools in a different context and for different goals. In the new context of organic chemistry, the previous novelty, that is, the theoretical concept of chemical portion, was no longer in itself epistemically relevant compared with the objects of research to which formulas were applied—the binary constitution and reactions of organic substances. Berzelius's 1833 introduction of the term *empirical formulas* and its distinction from *rational formulas*, which was quickly accepted by the European chemical community, witnesses this process of reification. For example, in constructing interpretive models of binary constitution, or *rational formulas*, chemists handled letters of *empirical formulas* as formal building blocks to build two partial formulas representing the two immediate constituents of a given compound, and these partial formulas alone had a representational function. Applied as a material resource for building interpretive chemical models that would have been otherwise difficult or impossible to build, Berzelian "empirical" formulas were taken for granted. In a similar way, scientific instruments, which are objects of research and material adaptation in their first state of development, may become unquestioned technological implements in a subsequent context of application. In the course of research, stabilized technologies may be opened up again to scientific inquiry, for example, if a former noise or impurity comes into the epistemic horizon of a possible meaningful mark or inscription. The same happened repeatedly to Berzelian formulas, for the first time in 1834, when Dumas introduced the new chemical concept of substitution. Examples like this demonstrate that paper tools, like laboratory instruments, are resources whose possibilities are not exhausted by scientists' attempt to achieve existing goals but rather whose applications generate new goals.

NOTES

Introduction

1. The literature in this field has grown enormously. I mention only a few representative monographs and book collections: Buchwald 1995; Galison 1987, 1997; Gooding, Pinch, and Schaffer 1989; Hacking 1983, 1992a; Heidelberger and Steinle 1998; Le Grand 1990; Latour 1987; Latour and Woolgar 1986; Lenoir and Elkana 1988; Pickering 1992; Pickering 1995; Rheinberger 1997.

2. See Galison and Warwick 1998.

3. See Lenoir 1994 and 1998.

4. See Morgan and Morrison 1999.

5. See Pickering 1995.

6. See Warwick 1992.

7. Latour 1987, 235.

8. Pickering 1995, 113.

9. See Galison and Warwick 1998, 288.

10. Galison and Warwick 1998, 288.

11. Hoffmann and Laszlo 1989, 23.

12. This book focuses on tools on paper rather than any other medium for historical reasons only; but many of the following items can be generalized for all kinds of representational tools, including three-dimensional models and computers. The productive function of three-dimensional molecular models as representational tools has been studied by Francoeur (see Francoeur 1997). For an excellent study on the role of computers as "tools to theory" in twentieth-century biology, see Lenoir 1999. A recent paper by Francoeur and Segal studies the first attempt at computer modeling of molecular structures in the mid-1960s by Cyrus Levinthal (see Francoeur and Segal n.d.). Recently, *Science in Context* published a collection of papers studying the functions of various kinds of models as well as computer simulation (see Sismondo and Gissis 1999).

13. Outside the history of physics, Morgan and Morrison's work on models has particularly emphasized the parallels between ordinary physical instruments and representational tools, such as models: "We claim that what it means for a model to function autonomously is to function like a tool or instrument. Instruments come in a variety of forms and fulfill many different functions" (Morgan and Morrison 1999, 11). Hans-Jörg Rheinberger views the entire experimental arrangement in molecular biology as a "graphematic activity." "A written table or a printed curve," he writes, "is only

the last step in a series of transformations of a previous graphematic disposition of pieces of matter, which is given by the experimental arrangement itself" (Rheinberger 1998, 296; see also Rheinberger 1997, 102 ff.). Additional papers studying this issue with respect to the life sciences are included in Lynch and Woolgar 1990. The semiotic aspects of various laboratory sciences and other empirical scientific enterprises have also been repeatedly studied by Bruno Latour, who—in keeping with the French semiotic tradition—emphasizes the material dimension of inscriptions and "chains of inscriptions" (Latour 1987, 1990, 1999).

14. Buchwald 2000, 207.

15. Ibid., 212.

16. See Cartwright, Shomar, Suárez 1995.

17. See Galison 1997, 2.

18. Galison 1998, 392. On Feynman diagrams as representational tools see also Kaiser 2000.

19. Pickering 1995, 101.

20. Ibid., 145.

21. See Warwick 1992, 633.

22. Hughes 1998, 340.

23. For the notion of "dense signs," see Dagognet 1973, 118.

24. For these two notions, see Goodman 1976, 154.

25. For this notion, see Hacking 1999.

26. See Hacking 1992a.

Chapter 1

1. See Berzelius 1813a and 1814a.

2. Before the 1830s, Berzelius himself used his formulas only occasionally, for example, in Berzelius 1815, 1973 (1820), 1821. In France, Jean Dumas belonged to a group of chemists who began to apply Berzelian formulas in their research papers from roughly 1826. Dumas also used them in the fifth volume of his textbook dealing with organic chemistry, which appeared in 1835 (Dumas 1828–1846, vol. 5). In Germany, Johann Wolfgang Döbereiner, professor of chemistry at Jena, used Berzelian formulas as early as 1821 in his textbook *Zur Pneumatischen Chemie* (Döbereiner 1821–1825). Also in 1821, Heinrich Rose accepted formulas in his translation of Berzelius's *Von der Anwendung des Löthrohrs in der Chemie und Mineralogie* (Berzelius 1821). In comparison, a contemporary English translation dismissed the formulas (see Crosland 1962, 278).

3. Quoted after Thackray 1972, 123, emphasis mine.

4. For the debate among British chemists on chemical notation see Alborn 1989, Brock 1986, Crosland 1962, 277 ff., and Thackray 1972, 116 ff. Crosland, quoting from the 1822 English translation of Berzelius's book on the blowpipe, in which Berzelius had used his formulas, gives a telling example of the British attitude. The English translator J. G. Children dismissed the formulas with the following argument: "These formulae I have omitted in toto. . . . I have taken this liberty because these signs and

formulae are little known in England, and consequently without an elaborate and tedious explanation would be perfectly unintelligible to most of my readers. I will candidly own too, that thinking them rather calculated to perplex than facilitate our progress, I do not wish to see them used in this country, and therefore I could not honestly add my mite towards their introduction, by adopting them" (quoted in Crosland 1962, 278). It should also be mentioned that British chemists did not accept the Daltonian diagrams.

5. Crosland 1962, 319.

6. Brock 1986, 33.

7. In general, historians of chemistry have not paid much attention to Berzelian formulas. They are mentioned in overviews, but mostly to characterize them as precursors of structural and stereochemical formulas. See Brock 1993, 154 ff.; Caven and Cranston 1928, 37–43; Cavanna and Rocchietta 1959; Crosland 1962; Ihde 1964, 111–116; Mounin 1981; Nye 1993, 89–102; Partington 1936; 1961–70, 4:158 f.; Priesner 1989; Walden 1927; Walker 1923; Weininger 1998. For studies on structural and stereochemical formulas, see Dagognet 1969, 1973; Hoffmann and Laszlo 1989, 1991; Krätz 1973; Laszlo 1993; Mason 1943; Nye 1993, 96–102; Ramberg 1995; Ramsey 1974a, 1974b, 1981; Suckling et al. 1978.

8. I use the term *model* for all kinds of representations that fit particular phenomena with general concepts, theories, and classifications. See, e.g., Cartwright 1983; Morgan and Morrison 1999.

9. The first edition from 1826 did not contain chemical formulas.

10. Webster 1839, 33.

11. Ibid., 34 f., emphasis mine.

12. See Dagognet 1973, 118.

13. For a more detailed history of these laws, which also contains an analysis of their contents, see Rocke 1984.

14. Quoted after Rocke 1984, 29.

15. See Thomson 1808.

16. See Wollaston 1808.

17. See Berzelius 1811a, and 1811b.

18. See Richter 1968.

19. See Dumas 1828–1846, 1:XXV.

20. Ibid., XXIV.

21. It should be mentioned that experimental stoichiometry was confined to inorganic compounds, more precisely to those inorganic compounds paradigmatic in Lavoisier's chemistry, such as salts, oxides, and sulfides. The formulation of general stoichiometric laws and their extension to all kinds of chemical compounds were underdetermined by experiments. Furthermore, the assumption that the stoichiometric laws were true for all kinds of inorganic compounds required the exclusion of alloys, solutions, and glass from the group of chemical compounds. This violated the earlier definition of chemical compounds based on affinity considerations. For problems of application of stoichiometric laws to organic compounds, see below.

22. See Gay-Lussac 1809.

23. Ibid., 218, 233.

24. Ibid., 234.

25. This has been pointed out before by Christie (Christie 1994).

26. See, e.g., Crosland 1962, 270 ff. Crosland's view was criticized by Wightman, who pointed out that Berzelian formulas had an additional theoretical significance, which is ignored if they are characterized as shorthand for names (Wightman 1961, 265 f.). It was shared by the French philosopher and semiotician Dagognet (Dagognet 1969, 1973), whose view will be discussed in more detailed below.

27. See, e.g., Alborn 1989.

28. See, e.g., Knight 1993, 340.

29. For an excellent study of these events and for an argument showing why the assignment of unique, invariant combining weights was a theoretical enterprise, see Rocke 1984.

30. Thomson 1813, 32.

31. *Analyses of Books* 1815, 302.

32. See Berzelius 1813b, 443.

33. Berzelius 1813a, 277.

34. See Berzelius 1813b, 450.

35. For Berzelius's theory and its relation to Daltonian atomism, see also Eriksson 1992; Lundgren 1992; Rocke 1984, 66 ff.; and Russell 1968.

36. Berzelius 1813a, 359, emphasis mine.

37. Berzelius 1814a, 51 f., emphasis mine.

38. Berzelius 1814d, 122, emphasis mine.

39. Ibid., 123, emphasis mine.

40. See Rocke 1984, 13.

41. See, for example, Thomson 1813, 34.

42. More recently, Kim suggested the term *stoichiometric atoms* for these particular chemical entities (Kim 1992b). In the following, I always use the term *atom* with quotation marks when I take over the actor's term with the meaning of scale-independent chemical portions; and I use it without quotation marks if atoms in the sense of particles of the microworld are denoted, as was the case in Dalton's atomic theory or in Berzelius's atomistic foundation of his electrochemical theory.

43. See Dalton 1808–1827, 2:555 ff.

44. Berzelius 1813b, 450. See also Berzelius 1820, 52.

45. Dagognet 1973, 118.

46. Ibid., 113 ff.

47. Ibid., 124.

48. Berzelius 1814a, 52.

49. Berzelius 1813a, 358.

50. Berzelius 1973 (1820), 30. Berzelius's monograph on the theory of proportions appeared in 1818 in a Swedish version and 1819 in French (see Berzelius 1819). In the following, the translations are from the 1820 German translation of the Swedish and French editions. Unless noted, this and all subsequent translations from German or French into English are my own.

51. Berzelius 1814b, 326.

52. Liebig 1834b, 26.

53. Berzelius 1814b, 327.
54. Berzelius 1814b, 325, emphasis mine.
55. Berzelius 1814c, 157.
56. Of course, this would have been possible if any integer had been allowed, regardless of magnitude.
57. I have divided Berzelius's analytical data by the "atomic weights" used by Berzelius at the time (Berzelius 1814a, 362).
58. Berzelius 1815, 98.
59. Ibid., 265.
60. Ibid., 271.
61. See, for example, Haugeland 1998; Mitchell 1987.
62. See Peirce 1931–1958, 2:227 ff.
63. See Galison 1997.
64. In his first introduction, Berzelius wrote the numbers above the letters; only a year later, he imitated the algebraic notation using superscripts. Some chemists, like Liebig, never used superscripts. From 1834, Liebig used subscripts arguing that he wanted to avoid mathematical confusion. See Liebig 1834a.
65. "Constitution" was a basic chemical category of the time. Chemists believed that all inorganic compounds had a binary constitution. For a detailed explanation of the concept and for its application and extension, see Chapter 6.
66. The sign Po^2 stood for potash or potassium oxide.
67. See, for example, Whewell 1831.
68. Berzelius 1814a, 51.
69. Berzelius 1833a.
70. Goodman 1976, 154.
71. Arnheim 1969, 136, emphasis mine.
72. It should be mentioned that these explicit building-block models were not used by chemists before the 1860s. My reconstruction is drawn after Berthelot 1864.
73. Mitchell 1987, 47.
74. Dagognet 1969, 186.
75. Ibid., 186.
76. Ibid., 187.
77. Ibid., 194.
78. Dagognet 1973, 122.
79. Ibid., 124.
80. Dagognet 1969, 186.
81. There is no historical evidence that chemists of the 1870s and 1880s were actually using the kind of diagrams depicted by Dagognet. As a rule, in the last third of the nineteenth century, chemists represented the spatial orientation of atoms by tetrahedral diagrams.
82. Ibid., 195, emphasis mine.
83. Ibid., 196.
84. Quoted after Ecco 1976, 195.
85. Mitchell 1995, 5.
86. See, for example, Ecco 1976; Gombrich 1969; Mitchell 1987.

87. See Goodman 1976, 161.
88. Ibid., 160.
89. Ibid., 226.
90. Ibid., 170, 173.
91. Ibid., 154.
92. Whewell 1831, 440, emphasis mine.
93. Ibid., 440, emphasis mine. In the secondary literature, Whewell is sometimes presented as a critic of Berzelius's formulas. His criticism referred only to Berzelius's inconsistent application of algebraic notation. In particular, Whewell objected that in Berzelian formulas, "Those combinations of elements which are supposed to be the most intimate, are represented by writing the symbols of the ingredients in the way which in algebra denotes multiplication" (Whewell 1831, 441).
94. See Meinel 1999.
95. Quoted after Brock 1993, 139, emphasis mine.
96. See Dalton 1808–1827, vol. 1.
97. See Thackray 1970.
98. Dalton 1808–1827, 1:212.
99. Ibid., 219; and Dalton 1814, 175.
100. See Lüthy, n.d.
101. Berzelius 1813b, 446.
102. Dalton, 1814, 174.
103. On this problem, see also Russell 1968, 262 f.
104. Dalton 1808–1827, 1:216.
105. Quoted after Thackray 1972, 96.
106. See Thackray 1972, 119.
107. Quoted after Thackray 1972, 120, emphasis mine.

Chapter 2

1. These changes sometimes have been described as a "stage of rapid growth" (Holmes 1971, 129), an "explosive development" (Kim 1992a, 69) of organic chemistry, or as the "emergence of organic chemistry as a subfield of chemistry" (Holmes 1993, 123). Although these descriptions refrain from conceptualizing the transformation process, normative judgments are nonetheless involved in its characterization as emergence of a "new science" (Brooke 1992, 180) or a "coherent science," which can be clearly distinguished from "cookery by trial and error" in the first third of the nineteenth century (Levere 1993, 158). In most older accounts, normative tones dominate, for example, referring to a "rudimentary stage" of organic chemistry in the early nineteenth century (Ihde 1964, 161). In addition to the articles by F. L. Holmes, two monographs questioned the still popular notion of the sudden emergence of organic chemistry in the nineteenth century: Reinhard Löw's *Pflanzenchemie zwischen Lavoisier und Liebig* (Löw 1977) and Evan Melhado's work on the background to the chemistry of Jacob Berzelius (Melhado 1981). My claim that the transformation of organic chemistry between the late 1820s and 1840 was on a deep structural level implies the asser-

tion that it embraced *many* aspects of the chemical culture, in particular, experimentation, model building, classification, theory formation, and the mode of representation. In this respect, the following suggestions have been made by historians of chemistry. Alan Rocke has asserted that Berzelius's theory of radicals was a decisive breakthrough, if not "the" innovation of the organic chemistry of the 1830s (see Rocke 1993, 15, 50). John H. Brooke and Trevor Levere highlight explicit and formal methodological principles of analogy as the decisive innovation in the organic chemistry of the 1830s (see Brooke 1971, 1973, 1987, 1992, and Levere 1993, 158 f.; 1994, 111 f.). Other than theories and methodological principles, one particular belief or conviction, or rather its "overcoming," is often considered to have brought about the changes in organic chemistry in the 1830s and 1840s: vitalism. From this perspective, Wöhler's urea synthesis in 1828 is seen as the decisive event (Ihde 1964, 163). In contrast to this focus on theories, beliefs, and methodology, William Brock has pointed out that technologic problems or successes have been largely ignored, especially the problems related to obtaining pure substances and quantitative analysis (Brock 1993, 208). Frederic L. Holmes has also focused on the role of experimental improvements, such as improvements in extraction methods and analytic techniques (Holmes 1971, 1989). Mi Gyung Kim has added that "experimental investigations served as the locus for coordinating and appropriating various theoretical languages" (Kim 1992a, 71). Kim's promising approach, in which various theories, experimental practices, and classification in organic chemistry of the first half of the nineteenth century are seen as one coherent "research tradition," has not yet been put into practice (Kim 1992a, 1992b).

2. See Rocke 1993, 2 f.

3. See Holmes 1989, 62.

4. See Holmes 1971, 129–148; 1989, 61–84.

5. On Lavoisier's experiments in plant chemistry, see Holmes 1985.

6. Among the older works studying plant and animal chemistry as a chemistry of substances the following should especially be mentioned: Kopp 1966b; Hjelt 1916; Graebe 1991. The tradition in the chemical historiography of representing plant and animal chemistry of the early nineteenth century as a science exclusively concerned with the study of chemical substances was continued in the standard works of the 1960s. See Partington 1961–1970, 4:233; Ihde 1964, 161. In the newer monographs by Löw (Löw 1977) and Melhado (Melhado 1981), in contrast, the research topics of plant and animal chemistry, which are not substance chemical, are granted some more attention.

7. Fourcroy 1801–1802, 1:8 f.

8. Ibid., 8.

9. Fourcroy 1801–1802, 7:4 f., emphasis mine.

10. Ibid., 36–60.

11. Ibid., 61–110.

12. Ibid., 111–370; 8:3–106.

13. Ibid., 7:6–35.

14. Ibid., 8:107–256.

15. Ibid., 257–324.

16. Ibid., 7:57.

17. Ibid., 58.

18. Ibid., 3.
19. Ibid., 17.
20. Ibid., 29, emphasis mine.
21. Thomson 1804, 4:367, emphasis mine.
22. "Of the food of plants," ibid., 382–396.
23. "Of the sap of plants," ibid., 397–413.
24. "Functions of the leaves during the day, and during the night," ibid., 414–433.
25. "Of the peculiar juices of plants," ibid., 433–439.
26. "Of the decay of plants," ibid., 439–441.
27. Ibid., 473, emphasis mine.
28. Berzelius 1833–1841, 7:255–604. On Berzelius's animal chemistry, see Rocke 1992.
29. Berzelius 1833–1841, 6:3, emphasis mine.
30. For an excellent analysis of Berzelius's and Liebig's concept of life force, see Lenoir 1989. Lenoir has pointed out that Berzelius's and Liebig's work "reinstated the importance of conceiving *Lebenskraft* as the expression of a complex interrelation of material parts incapable of further analysis but inseparable from the order and arrangement of matter" (Lenoir 1989, 160).
31. Fourcroy, 1801–1802, 7:5.
32. Ibid., 51.
33. For a more detailed explanation of the notion of "immediate constituent," see Chapter 6.
34. Later in the century, the French chemist Michel Eugène Chevreul referred to the analogy between acids and bases as immediate constituents or complex principles of inorganic salts and the concept of the immediate principle used in plant and animal chemistry in the following words: "If one considers the composition of a salt in general, it is obvious that, as Lavoisier established, it is constituted by the union of an acid and an alkali, and not by the elements of an acid and an alkali. . . . Accordingly, it seems consistent to say that acids and alkalis are the two immediate principles of salts. The same is true of sugar, rubber, tissue, etc., in relation to a plant, and of the fibers, albumen, cellular tissue, etc., in relation to an animal; one must consider these substances as immediate and characteristic principles of the plant or animal from which they come, while oxygen, nitrogen, carbon and water are their more distant or elementary principles" (Chevreul 1823, 4 f.).
35. The problem of purity of animal and plant substances has sometimes been raised in the secondary literature. See, for example, Ihde 1964, 162; Partington 1961–70, 4:233; Brock 1993, 208.
36. In inorganic chemistry, since the early eighteenth century, chemists conceived of "pure chemical compounds" as single chemical compounds that were free of traces of other substances. Thus, a clear difference in meaning was implied between "chemical compounds" (and later "chemical elements") and "mixtures" of substances. The behavior of substances in chemical operations was decisive for the formation and meaning of these concepts in the early eighteenth century. Mixtures could often be separated mechanically, whereas chemical compounds could be separated only through typical chemical operations like heating, distilling, precipitating, etc. Above all, "chemical

compounds," such as acids, bases, and salts, proved to be relatively stable, homogeneous entities in series of experiments, which chemists interpreted as "analyses" and "resyntheses" of compounds. Mixtures, in contrast, separated into multiple substances during this kind of reversible chemical operations. Hence, the experimental investigation of the chemical reactions of substances was an enormously powerful method for distinguishing between mixtures and pure compounds and to give the notion of an immediate constituent an unambiguous chemical meaning. Based on this, at the beginning of the nineteenth century, a further experimental and simultaneously conceptual criterion for the presence of a chemical compound in the sense of a pure substance was added. Chemical compounds needed to have a constant quantitative composition that could be shown in the quantitative analysis of the substances. On the formation of the conceptual network distinguishing between chemical compounds and mixtures, and including concepts such as chemical analysis and synthesis, see Klein 1994a, 1994b, and 1996.

37. "Composés organiques," Fourcroy 1801–1802, 7:4, 40.

38. "Matières organisées," ibid., 96.

39. See Thomson 1804, 4:359. Quantitative physical criteria of identification, like a constant melting point, boiling point, etc., were rare at the beginning of the century. There are in part technical reasons for this, like the facile decomposition of heated organic substances, in part reasons of habits. These measurements did not belong to the everyday repertoire of a chemist of the time.

40. Thomson 1804, 4:242 f., emphasis mine.

41. See Thomson 1804, 4:206.

42. Ibid., 358.

43. See Fourcroy 1801–1802, 7:175 f.: "La sève, le muqueux, le sucre, l'albumine végétale, l'acide végétale ou les acides végétaux, l'extractif, le tannin, l'amidon, le glutineux, la matière colorante, l'huile fixe, la cire végétale, l'huile végétale, le camphre, la résine, la gomme résine, le baume, le caoutchouc, le ligneux, le suber." On the history of individual plant substances and the description of their properties and applications, see Löw 1977, 113 ff.

44. See Thomson 1804, 4:473 f.

45. Webster 1826, 545.

46. For the sake of completeness, it should be mentioned that the definition of plant and animal substances sometimes was accompanied by additional considerations. Some early nineteenth-century chemists, such as Fourcroy, adapted the results of Lavoisier's experiments and his interpretation of them, which differed from the preceding discussed basic orientation toward natural history. Fourcroy remarked that Lavoisier's analyses had introduced the "first exact conception of organic compounds" (Fourcroy 1801–1802, 7:40). Furthermore, because elemental analyses showed that, as a rule, plant substances were made up of carbon, hydrogen, and oxygen, Fourcroy and a few other chemists also thought that the differences in the properties of these substances might be based on the differences of the proportions of the elements (ibid., 55). Fourcroy occasionally also returned to another idea of Lavoisier's, that is, the possible binary constitution of plant substances (ibid., 53 f.; on Lavoisier's ideas about binary constitution, see Melhado 1981, 124 f., 139 f.; and Holmes 1985, 402). At the beginning

of the nineteenth century, this kind of reasoning saw little attention from the European chemical community. Furthermore, at the time, it had no effect on the actual practice of identification and classification of plant and animal species. This did not change until the late 1820s, as the concept of the binary constitution of organic substances began to take shape in concrete formula models.

47. A discussion of this problem can also be found in Melhado. See Melhado 1981 and 1992.

48. Thomson 1804, 4:358, emphasis mine.

49. Melhado 1981, 120 f.

50. Melhado indicates that around 1800 the differentiation between species of vegetable acids was controversial, both in general and in concrete cases. See Melhado 1981, 121.

51. See Lavoisier 1789, 129 f.

52. See Chevreul 1823, 2 f.

53. Melhado argues plausibly that Lavoisier only assumed the existence of a few organic substances, in contrast to Berzelius, who, due to his activities in natural history, was convinced of the multiplicity of organic species. "The thought never occurred to Lavoisier that the diversity and wealth of organic chemistry would ever outstrip his simple outline of names" (Melhado 1981, 141). Lavoisier, in his treatise, distinguished between ten possible forms for the composition of organic substances (see Lavoisier 1789, 127).

54. Berzelius 1812, 76 f., emphasis mine.

55. Ibid., 88.

56. Ibid. Two years later, Berzelius argued still more explicitly against the first approaches in France of a classification according to the composition of organic species. Referring to acids, he wrote, "Consequently it is impossible to determine in organic bodies from the elements and proportions of which they are composed whether they are acid or not" (Berzelius 1814b, 328).

57. Berzelius 1815, 179.

58. See Berzelius 1814b; 1815, 260–275.

59. See the table of results in Berzelius 1814c, 157.

60. Berzelius 1973 (1820), 50 f.

61. See Berzelius 1815, 271 and 265.

62. Ibid., 47.

63. Melhado 1992, 165.

64. See Berzelius 1833–1841, 6:107.

65. Ibid., 396–409.

66. Berzelius 1843–1848, 4:92.

67. Ibid., 591–723.

68. Ibid., 5:213–307.

69. Ibid., 5:320–586 and 587–940.

70. Fourcroy 1801–1802, 7:51.

71. See Holmes 1971, 1989.

72. Thomson 1804, 4:363 f.

73. Fourcroy 1801–1802, 7:51.

74. Ibid., 46.
75. Ibid., 49.
76. Ibid., 46.
77. Diderot and d'Alembert 1966, 8:228, emphasis mine.
78. Von Baer 1824; quoted after Lenoir 1989, 78.
79. See, for example, Liebig's experimental investigations of the reactions between alcohol and chlorine, discussed in Chapter 7.
80. Thomson 1804, 4:442, emphasis mine.
81. Fourcroy 1801–1802, 7:61 f.
82. Ibid., 54.
83. For example, Berzelius wrote, "From the peculiar and special modification of the electro-chemical properties of these elements, organic bodies in general constitute but feeble compounds, which often begin to undergo decomposition as soon as they escape from the influence of the organ in which they were produced. Almost all organic bodies are decomposed by the united influence of air, water, and heat. Their elements resume their ordinary electro-chemical modifications, and there finally results a number of binary or inorganic combinations" (Berzelius 1814b, 329).
84. At the time, "chemical synthesis" meant not only the synthesis from elements but any formation of chemical compounds (see Fourcroy 1801–1802, 1: 61ff.).
85. Fourcroy 1807, 242.
86. Thenard 1817–1818, 3:3 f.
87. Webster 1826, 454.
88. Fourcroy 1801–1802, 1:62, emphasis mine.
89. Daniell 1843, 602 f., emphasis mine.
90. Dumas 1828–1846, 5:1.
91. Ibid., 2.
92. Ibid., 77 f.
93. Ibid., 78.
94. Berthelot pointed out the impressive number of more than ten thousand organic compounds in his 1872 *Traité Élémentaire de Chimie Organique* (Berthelot 1872, V). See also Rocke 1993, 2.
95. Dumas 1828–1846, 5:3.
96. In addition, Dumas wrote that organic substances "are identified if they possess the property of regularly crystallizing or forming crystallizable compounds, as well as if they possess the property of boiling at a constant value" (Dumas 1828–1846, 5:1). A regular crystal form and a constant boiling point were empirical criteria for the difference between a pure chemical compound and mechanical substance mixtures as well as organized substance complexes. These criteria, however, were not generally applicable: The first does not apply to fluid or gaseous compounds, the second does not apply to compounds that already disintegrate before reaching a boiling point.
97. Ibid., 2.
98. Ibid., 78.
99. Gerhardt, 1853–1856, 1:3.
100. Schlossberger 1860, 2, emphasis mine.
101. Kekulé, 1861–1882, 1:11.

102. Butlerov 1868, 5.

103. Wallach 1901, 1: 603–605.

104. Laurent 1854, IX.

105. Dumas and Boullay's new mode of classification is discussed in detail in Chapter 5.

106. The many attempts at classification in the experimental culture of organic chemistry after 1830 cannot be discussed here in detail. See Fischer 1973a, 1973b, 1974; and Kapoor 1969a, 1969b.

107. Butlerov 1868, 82 f., emphasis mine.

108. Kekulé 1861–1862, 1:10 f., emphasis mine.

109. Beginning in 1847, chloroform was used as a narcotic; chloral or chloral hydrate were used as sleeping aids beginning in 1869. See Volhard 1909, 1:255.

110. For details on these experiments, see Chapter 7.

111. Gmelin 1843–1870, 4:38, emphasis mine.

112. Liebig 1841, 203 f.; see also Chapter 8.

113. Kekulé 1861–1882, 1:8.

114. Gerhardt 1853–1856, 1:3.

115. Berthelot 1860, 811 f.

116. Butlerov 1868, 2 f.

117. Schorlemmer 1874, 1 f.

118. Cooke 1874, 292, emphasis mine.

119. Kekulé 1861–1882, 1:7, emphasis mine.

120. Berthelot 1860, 811, emphasis mine.

121. For similar applications of the term *scientific culture*, see Latour and Woolgar 1986, 55; Latour 1987, 201; Jardin and Spary 1996; Pickering 1995, 3; Rheinberger 1995 and 1997, 140. Contemporary studies in the history of science, which use a "cultural history" approach, often apply the concept of scientific culture in the sense of collectively shared patterns of interpretations, value complexes, ideologies, mentalities, and so on. Instead, I do not put the accent on the mentalistic connotations of "culture" but rather on the social, material, and symbolic resources of a scientific community.

122. Books like Auguste Laurent's *Méthode de la Chimie* (1854) or Charles Gerhardt's *Traité de la Chimie Organique* (1853–1856) are evidence for the laborious attempt at integration and abstraction which the new taxonomies of organic chemistry required.

123. On the emergence of stereochemistry, see Ramberg 2001; on that of French theoretic organic chemistry, see Nye 1993, 142 ff.

124. See Nye 1993, 4 f.

125. These experiments are scrutinized in Chapter 3.

126. See Knight 1978, 246. On this issue, see also Nye 1993, 65 ff., 85 ff., 269; 1996, 121 ff.; and Kim 1992a, 70 f; 1995, 158.

127. Knight 1992, 128.

128. See, for example, Carrier 1993; Douglas and Hull (eds.) 1992; Hacking 1993; Kuhn 1987, 1989; Goodman 1978.

129. Buchwald 1992, 40.

130. Bachelard 1973, 5 f.

131. Ibid., 34 ff.

132. Ibid., 38, emphasis mine.

133. I have characterized this conceptual system as the system of "chemical compound, affinity and reaction." See Klein 1994a, 1994b, 1996.

134. Such a distinction is made, for example, by Ian Hacking. See Hacking 1992b.

135. See Kuhn 1987, 1989.

Chapter 3

1. With the exception of so-called ordinary ether, produced from alcohol and sulfuric acid, most of the reaction products were esters in current terminology.

2. Althoug several of these experiments have been described in the secondary literature (see Kapoor 1969a, Priesner 1986), historians of chemistry have not assigned a prominent role to the early nineteenth-century experiments with alcohol and ethers. The reason for this is perhaps that there was no important discovery tied to them.

3. See also Chapter 9.

4. See Liebig 1831b.

5. For this notion, see Pickering 1995.

6. See Kopp 1966a, 4:299 ff.; Priesner 1986.

7. Fourcroy and Vauquelin 1797, 203, emphasis mine.

8. Boullay 1807b, 242 f., emphasis mine.

9. Thenard 1817–1818, 3:278.

10. Dumas and Boullay 1827, 306.

11. See Thenard 1807e–f.

12. See Thenard 1807a, 80.

13. Thenard 1807d, 154.

14. Although it was clear that it did not belong to mineral chemistry, its status in relation to the other two domains was unclear. Alcohol was considered a plant substance, which was also created spontaneously or "naturally" through fermentation; but the ethers were exclusively products of the chemical laboratory, and thus highly unlikely to occur in plants. The experimental study of ether formations thus could not answer questions about physiological processes or the constituents of plants.

15. See Fourcroy and Vauquelin 1797.

16. See Fourcroy and Vauquelin 1797, 207.

17. See Lavoisier 1789, 147.

18. See Fourcroy 1801–1802, 8:156 f.

19. For example, a few years later de Saussure claimed that alcohol contained nitrogen. See de Saussure 1807.

20. Fourcroy and Vauquelin 1797, 210.

21. See Deiman et al. 1794, 182.

22. Fourcroy and Vauquelin 1797, 209 f.

23. Ibid., 211.

24. Again, explanation with affinities played a role. Because of the increased temperature, sulfuric acid would no longer withdraw water but would attach itself to the alcohol, thus preventing its evaporation.

25. For example, see Laudet 1799.

26. Rose is an example of a German chemist who accepted the theory rather quickly. See Rose 1800.

27. See Dabit 1799.

28. This hypothesis was adopted by Gay-Lussac in 1820. See below.

29. See Fourcroy 1801–1802, 8:175.

30. See Thenard 1807a–d.

31. See Thenard 1809a–c.

32. See Boullay 1807a–c, 1811.

33. See de Saussure 1807.

34. For the broader historical context of quantification in eighteenth- and nineteenth-century sciences, see Frängsmyr, Heilbron, and Rider 1990; Wise 1995.

35. See Thenard 1807a, de Saussure 1807.

36. For other components, such as sulfur and phosphor, different analytic methods were used.

37. See Gay-Lussac and Thenard 1811, 2:268 ff. The following overview is based on William Brock's (Brock 1997, 48 ff.) and Alan Rocke's account (Rocke, 2000).

38. See Berzelius 1814b, 1815.

39. See Liebig 1831b.

40. Rocke 2001, 38. I thank Alan Rocke for a prepublication copy of his book manuscript.

41. For details of the apparatus, see Rocke, 2001, 36 ff.

42. See Thenard 1807a, 73.

43. Ibid., 98. In his repetition of the quantitative analysis, Thenard obtained very different results: 14.49% nitrogen, 28.65% carbon, 48.52% oxygen, and 8.54% hydrogen (Thenard 1807 f, 367). These differences indicate the enormous technical problems that confronted chemists in their early attempts at precise quantitative analyses of organic substances.

44. Ibid., 96.

45. See Guyton de Morveau, Vauquelin, and Berthollet 1807.

46. Thenard 1807a, 102 f., emphasis mine.

47. Ibid., 86.

48. Thenard 1807b, 135.

49. Ibid., 132.

50. Ibid., 133 f.

51. See Thenard 1807e.

52. See de Saussure 1807. Thenard mentioned that he had begun his series of experiments on February 21 (1807), finishing them on May 19 (Thenard 1807e, 347). De Saussure read his article at the Paris Academy of Sciences on April 6.

53. Thenard 1807e, 346.

54. Ibid., 356 f.

55. Boullay 1807c, 90 f.

56. Ibid., 100.

57. See Thenard 1807 f., 361.

58. See Boullay 1811.

59. See Thenard 1809a. The lecture was given on November 23, 1807, but the article was published only in 1809.

60. Thenard explained the function of the mineral acid as facilitating the combination of the alcohol and the vegetable acid by "condensing" the alcohol. In an additional series of experiments, he systematically varied the quantity of the admixed mineral acid to study its function (ibid., 19 f.).

61. Ibid., 12.

62. Ibid., 20 f.

63. See Thenard 1809b.

64. See Thenard 1807d.

65. See Thenard 1809b, 24.

66. Ibid., 26.

67. Ibid.

68. Ibid., 31.

69. Ibid., 39.

70. As to the latter view, see Chapter 6.

71. See de Saussure 1807, 316.

72. Ibid., 348 f.

73. Ibid., 349.

74. Ibid., 350.

75. Ibid., 351.

76. See de Saussure 1814.

77. See Deiman et al. 1794, 178.

78. Ibid., 180.

79. Fourcroy 1797, 69 f.; emphasis mine.

80. See Deiman et al., 1794, 189.

81. See Fourcroy 1797, 66.

82. See de Saussure 1811.

83. See Robiquet and Colin 1830, 207.

84. See Berthollet 1816. Berthollet had already performed similar experiments in 1785, but with the different goal of studying the effects of "oxidized muriatic acid" (in 1816, "chlorine") on organic substances (Berthollet 1785).

85. See Liebig 1832 and Dumas 1834. Liebig's and Dumas's experiments are discussed in Chapter 7

86. See de Saussure 1814, 284. In 1811, the result was 14.34:86.43, or 1:6.14 (de Saussure 1811, 68).

87. In 1807, the result was considerably different: 43.5% carbon, 38% oxygen, 15% hydrogen, and 3.5% nitrogen.

88. de Saussure 1814, 286.

89. For this notion, see Latour 1987 and 1999.

90. de Saussure 1814, 297.

91. Ibid., 304. The term *essential water* meant the water chemically combined in alcohol, in contrast to the water mixed with it.

92. See Chapter 1.

93. See de Saussure 1807, 351.

94. De Saussure only remarked that carbon could be found at the end of the distillation process, explaining that it was due to the decomposition process at high temperature (de Saussure 1814, 305).

95. See Gay-Lussac 1809. See also Chapter 1.

96. Gay-Lussac 1815, 312 f. Gay-Lussac did not directly measure the density of olefiant gas but calculated it over a number of steps.

97. Ibid., 313.

98. Ibid., 315 f.

99. Ibid., 316.

100. See Boullay 1807a, 242.

101. See Thenard 1807b, 133 f.

102. See Berzelius 1973 (1820), 49 f.

103. On this belief of chemists, see Chapter 6.

104. Gay-Lussac 1820, 78.

105. Evidence for this can be found, for example, in Dumas and Boullay 1827, 295. See also Partington 1961–1970, 4:349.

106. See Dabit 1799, 289.

107. See Sertürner 1819, 54.

108. Gay-Lussac 1820, 79.

Chapter 4

1. This work is now being undertaken by Alan Rocke and Melvyn Usselman. See Rocke 2000 and 2001, 11 ff. For an overview, see also Brock 1997, 48 ff., and Chapter 3 of this book.

2. Dumas and Boullay 1827, 294.

3. The substance termed sulfovinic acid was most likely ethyl sulfuric acid (CH^3-CH^2-O-SO^3H), also called *ethyl hydrogen sulfate*. Ethyl sulfuric acid is produced as a by-product of ether production with concentrated sulfuric acid and alcohol, according to the formula equation $CH^3\text{-}CH^2\text{-}OH + H^2SO^4 = CH^3\text{-}CH^2\text{-}O\text{-}SO^3H + H^2O$, whereby an equilibrium is created between ethyl sulfuric acid and the secondary product "ethylene," which corresponds to the substance called olefiant gas. The substance termed *sweet wine oil* was identified in 1879 by P. Claesson as a mixture made up of polymerized ethylene and ethyl sulfate (see Partington 1961–1970, 4:349).

4. Dumas and Boullay 1827, 296.

5. Ibid., 297.

6. Ibid., 301. The "percentages" did not exactly refer to 100 parts by weight of the compound, but to 100.04 in the first case, 100.57 in the second.

7. Ibid., 301. This representation, in which, as Berzelius suggested, oxygen is represented by a dot, corresponds to $BaOS^2O^5 + 2H^3C^4 + 5$ Aq. It should also be mentioned that throughout their article, Dumas and Boullay used the terms *atoms* and *volumes* synonymously. Hence, the term *atom* here means a portion of a substance.

8. Ibid., 305, emphasis mine. The term *dilute acid* in the last sentence refers to untransformed sulfuric acid diluted by water produced in the reaction.

9. The mixed verbal and symbolic form of Dumas and Boullay's balancing schema tacitly presupposed the following assumptions about the constitution of alcohol (left side): four volumes (vols.) of alcohol vapor corresponds to four vols. of bicarbonated hydrogen and four vols. of water vapor, if according to the formula $H^2C^2 + 1/2\ H^2O$ one vol. of alcohol is made up of one vol. of each of these immediate constituents (ibid., 309); transformed into a Berzelian formula, this is $4\ C^2H^2 + 2\ H^2O$. If the balancing is undertaken by means of Berzelian formulas, this results in $2\ SO^3 + (4\ C^2H^2 + 4$ vols. of water) $= S^2O^5 + 2\ H^3C^4 + 1\ O + 2\ H + 4$ vols. of water (whereby "4 vols. of water" on the right side corresponds to the "4 vols. of water released" in the schema, and $1\ O + 2\ H$ symbolizes the "two vols. of water formed" during the reaction).

10. Ibid., 294.

11. Ibid., 310.

12. Figure 4.3b is my visualization of that implicit building-block model.

13. See Hennell 1827.

14. See Dumas and Boullay 1827, 307.

15. See Dumas and Boullay 1828. If seen in the context of the later work of Dumas and Boullay, who mainly worked toward a new classification of organic substances, this change of opinion becomes immediately explicable (see next chapter). The new version, according to which sulfovinic acid was a binary compound of sulfuric acid and bicarbonated hydrogen, made it possible to classify sulfovinic acid in the newly formed substance class of compounds of bicarbonated hydrogen. To this end, only a small transformation of the preceding formula model was necessary. The formula for water, which stood for water of crystallization, had to be reintegrated into the formula for sulfovinic acid. For more on this, see the following chapter.

16. See Berzelius 1829, 286.

17. See Dumas and Boullay 1828; see also below.

18. Pelouze 1833, 589.

19. See Mitscherlich 1834.

20. Liebig 1834, 31.

21. See Williamson 1850.

22. For a more detailed analysis of this problem, see Chapter 9.

23. In the following chapter on classification, I discuss this problem in more detail.

24. In analyzing the text that accompanied Dumas and Boullay's formula model of the formation of sulfovinic acid, I need to point out a possible anachronistic misunderstanding. From today's perspective, we would understand the additions and regroupings of the symbols of the formulas in the model of the formation reaction of sulfovinic acid as molecular rearrangements. For example, $2\ SO^3 = S^2O^5 + O$ would be interpreted as representing the synthesis of two molecules of sulfur trioxide (SO^3) to a single molecule of S^2O^5 and the release of one atom of oxygen. Similarly, today's chemists would understand the formation of sweet wine oil as the synthesis of four molecules of bicarbonated hydrogen ($4\ H^2C^2$) into two molecules of sweet wine oil ($2\ H^3C^4$) with the removal of two atoms of hydrogen. Both Dumas and Boullay and other chemists of the period did not interpret the formula model in this way, however, at least not in the context of the experimental practice. They viewed these two reactions as analyses, that

is, as a release of oxygen and hydrogen, respectively. The reason for this is that chemists' scientific object was not the behavior of submicroscopic atoms but rather, in a traditional intellectual framework, the macroscopic level of substances or substance components and their recombinations.

25. For this notion, see F. L. Holmes 1993.

Chapter 5

1. See Kopp 1966a, 4:410 f.; Partington 1961–70, 4:240 f.

2. See Chevreul 1823; Kopp 1966a, 4:388 f.; Partington 1961–70, 4:248 f.

3. See also Chapter 2.

4. See Berzelius 1973 (1820). See also Chapter 2.

5. See Dumas and Boullay 1828. For more details on the concept of binary constitution, see Chapter 6.

6. See Chevreul 1823.

7. See also Kapoor (Kapoor 1969a), who pointed out the special relevance of analogy in the chemical classifications of organic substances and who studied Dumas's classification in the broader theoretical context of the concept of binary constitution and of the "atomic theory" (without considering the function of chemical formulas) as well as in the narrower context of experiments and "theories" of ethers. The general significance of analogies with inorganic compounds was also pointed out by Brooke (Brooke 1973 and 1987).

8. Liebig 1837, 26.

9. Liebig, Poggendorff and Wöhler 1842–1864, 8:673.

10. Kekulé 1861–1862, 1:63.

11. See Chapter 3.

12. See Dumas and Boullay 1828, 52.

13. Ibid., 50.

14. After Gay-Lussac had suggested that alcohol was made up of one volume of water and one volume of oil-building gas, Thenard made a revision of his earlier assumption about the constitution of muriatic acid ether (see Chapter 3), stating that water was expelled in the synthesis of muriatic ether; hence muriatic ether then was viewed as a combination of bicarbonated hydrogen and muriatic acid (Thenard 1817–1818, 3:283). This model was extended to all halogen acid ethers in 1824 (Thenard 1824).

15. See Boullay 1807c; Thenard 1807f. See also Chapter 3.

16. Dumas and Boullay 1828, 48.

17. See Thenard 1824, 4:137 ff.

18. See Dumas and Boullay 1828, 28.

19. Ibid., 28.

20. See Berzelius 1815, 176. Berzelius's formula for acetic acid was $H^6C^4O^3$. The difference between C^4 and C^8 arose because the French chemists correlated "atomic weights" with the theoretical volumes of the substances and their density, and thus obtained for the "atomic weight" of carbon a value (roughly 6) only half as large as Berzelius (see Dumas 1828–1846, 5:54).

21. See Dumas and Boullay 1828, 43. In principle, the new claim that acetic acid ether contains water of crystallization could be experimentally tested because water of crystallization usually could be separated by heating. However, Dumas and Boullay noted that it is difficult to remove water of crystallization from the salts of oxalic acids without decomposing them. In 1827, in connection with an analogous hypothesis about the binary structure of sulfovinic acid, they had in fact attempted to prove the existence of water of crystallization experimentally.

22. See Dumas and Boullay 1828, 18 and 33 ff.

23. Ibid., 36 f.

24. Ibid., 42.

25. Ibid., 41.

26. See Dumas and Boullay 1827.

27. In 1827, Dumas and Boullay reported an experiment for deciding between the two alternative formula models. The formula model $S^2O^5 + 2\,H^3C^4 + H^2O$ differed from the formula model preferred in 1828, $2\,SO^3 + 4\,H^2C^2$, in that the first implied the assumption that sulfovinic acid and its salts contained one "atom" of water of crystallization. Dumas and Boullay attempted to separate the hypothetical water of crystallization by heating the barium salt of sulfovinic acid to 150–160° C. However, the results of the experiment were unclear, in part because the salt changed chemically.

28. See Hennell 1827.

29. See Dumas and Boullay 1828, 42.

30. Ibid., 52.

31. See Chevreul 1823, 445.

32. See Dumas and Boullay 1828, 45.

33. Ibid., 47 f., emphasis mine.

34. It should be added here that the formula models of the constitution of sugar had consequences for existing hypotheses about fermentation. In 1815, Gay-Lussac had assumed that ordinary sugar was made up of three volumes of carbon vapor and three volumes of water vapor. By balancing this with the quantitative composition of alcohol and carbonic acid, he then concluded that in the fermentation of sugar, carbonic acid was removed to leave alcohol (Gay-Lussac 1815, 316 ff.). Dumas and Boullay's model of the constitution of cane sugar contradicted both hypotheses. Because it showed that cane sugar was a carbonate of bicarbonated hydrogen containing one "atom" of water of crystallization, the two chemists concluded that in fermentation water must be absorbed for the formation of alcohol (Dumas and Boullay 1828, 46). They linked this conclusion to the experimentally verifiable hypothesis that an increase in weight must take place (ibid., 46 f.).

35. In doing so, he also worked with chemical formulas; his own formula models differed from those of Dumas and Boullay only in that in his formulas, the partial formula for water of crystallization was integrated into the formula for ether (Berzelius 1829, 289 ff.).

36. Ibid., 292.

37. Ibid., 294.

38. Ibid., 294 f., emphasis mine.

39. Ibid., 296. Berzelius used the term *zu versinnlichen*.

40. Ibid., 297.

41. Berzelius 1833a, 623.

42. See Liebig 1834, 3.

43. Pelouze 1833, 142.

44. On the history of the classification of organic substances after 1828, see Fischer 1973a, 1973b, and 1974, as well as Kapoor 1969b.

Chapter 6

1. Dumas 1840, 170 f., emphasis mine.

2. Gerhardt 1844–1845, 1:12, emphasis mine.

3. In the following, I analyze only the meaning of the concept of binary constitution in the context of the experimental and taxonomic praxis of the European chemists. *Binarity* and the adjective *binary* also had other meanings, especially in atomic theories like that of Dalton where *binary* could also mean "made up of two atoms."

4. See Klein 1994a, 1994b, 1995, 1996.

5. See Geoffroy 1718.

6. Berzelius 1814b, 324, emphasis mine. See also Berzelius 1812, 324.

7. Berzelius 1814b, 324.

8. Berzelius 1973 (1820), 103.

9. Dumas 1840a, 168, emphasis mine.

10. Lavoisier also occasionally hypothesized that organic compounds were compounds analogous to the inorganic oxides. He did, however, not consistently support this opinion, sometimes hypothesizing that organic compounds are tertiary compounds, composed directly of carbon, hydrogen, and oxygen. On this, see Melhado 1981, 124 f., 139 f.; as well as Holmes 1985, 402.

11. This view was shared by the physiologists of the time. See Lenoir 1989, 122.

12. Gmelin 1817–1819, 3:936 f., emphasis mine.

13. An early experiment, carried out in 1811, played a role for this in which Berzelius attempted to divide organic compounds electrochemically, analogously to inorganic compounds, but without success (Berzelius 1811a, 471 f.).

14. Berzelius 1813b, 446.

15. Berzelius 1814b, 325.

16. Ibid., 328.

17. Thus, in 1814, after early doubts, Berzelius extended the applicability of his theory of proportions (or portions) to organic compounds, although this required major modifications in an essential aspect of the theory. His belief in the unity of nature played here a central role. See on this Chapter 1.

18. Berzelius 1973 (1820), 103.

19. Ibid., 104.

20. Ibid., 44.

21. See Wöhler and Liebig 1832. Berzelius, however, still did not assume in 1832 any general binarity because the benzoyl radical itself was not constituted binarily. Only in 1833 did he also begin to apply the binarity concept to compound radicals. See also Brooke 1973, 1992.

22. Berzelius 1832, 284.

23. Here we should point out that Liebig and Wöhler's interpretations were, in retrospect, false. From today's perspective, the phenyl group is indeed a relatively stable group of atoms within aromatic compounds, but neither do aromatic compounds have a binary constitution, nor is the phenyl group isolatable as a substance.

24. Liebig 1838, 3.

25. Ibid.

26. Ibid.

27. On this concept of organic radicals from the 1830s, see also Brooke 1992, 200. It should be mentioned that many historians of chemistry define the meaning of this concept as "a group of atoms" (see, for example, Holmes 1981, 332). This definition is incomplete (because it does not mention the additional meaning of a substance that can be isolated in experiments) or even misleading, if the particular historical meaning of *atom* in the context of experimental practice and the formulation of models of constitution and classification goes unmentioned. In this context, chemists, including Dumas and Liebig, understood an atom to be a scale-independent chemical portion.

28. The new substances were benzoyl chlorine, benzoyl bromine, benzoyl iodine, benzoyl sulfur, benzoyl cyanogen, and benzamide. The already known substances were benzoic acid, benzoic ether, and benzoin.

29. In retrospect, it should be pointed out that in this case several reaction products are also produced, but in small amounts. In the conditions of the time, Wöhler and Liebig found only two reaction products.

30. Wöhler and Liebig 1832, 266.

31. Ibid., 257.

32. Ibid., 263.

33. Hence, I disagree with F. L. Holmes's assertion that Wöhler and Liebig could "deduce" the formula of the benzoyl radical from the results of quantitative analysis: "From their elementary analyses they deduced that a 'benzoyl radical,' $C^{14}H^{10}O^{2}$, persisted unchanged through all these reactions" (Holmes 1981, 333).

34. The backdrop for this transformation was the generally accepted opinion that the gas bicarbonated hydrogen was an inorganic decay product of organic compounds. For Berzelius, who highlighted the difference between organic and inorganic compounds, it was not acceptable to have an inorganic substance as a characteristic building block for a whole class of organic compounds.

35. See Berzelius 1832, 286.

36. Berzelius 1833a; see more on this below.

37. See Liebig 1834, 26 f.

38. Ibid., 27.

39. Dumas 1828–1846, 5:73, emphasis mine. In this context, Dumas mentioned attacks against the concept of binarity. These were undertaken by Alexandre Baudrimont, a Parisian pharmacist, in his *Introduction à l'étude de la chimie par la théorie atomique* (1833). See Gerhardt 1844–1845, 1:10 f.

40. See Dumas and Liebig 1837.

41. Ibid., 568.

42. Ibid., 569.

43. See Berzelius 1833a.
44. See Chapters 4 and 5.
45. See Hennell 1827 and 1828.
46. See Serullas 1829. Hennell and Serullas did not use Berzelian formulas, but they transformed analytic data into theoretical "proportions" (Hennell) and "atoms" (Serullas) by dividing them by the "atomic weight" of each element. Poggendorff added the following chemical formula for sulfovinic acid (with the "atomic weight" C = 12 rather than 6) to Serullas's article: $2\ SO^3 + (2\ H^4C^2 + H^2O)$, which was supported by Serullas's verbal description of its composition of "atoms" (Serullas 1829, 49). Hennell's numbers of theoretical proportions can be translated into the formula $2\ SO^3 + C^4H^4$ or $2\ SO^3 + 2\ C^2H^2$ (with the "atomic weight" C = 6) (Hennell 1827, 18, 22).
47. Liebig and Wöhler determined the weight percentage of barium sulfate first by calcination and then by precipitation. In a second step, the weight percentage of carbon, hydrogen, and oxygen was determined by combustion analysis with copper oxide.
48. Liebig and Wöhler 1831, 489.
49. Liebig and Wöhler 1832, 40, emphasis mine.
50. Magnus 1833, 154. Magnus here followed the Berzelian way of writing in which oxygen is symbolized with a dot and a double atom of hydrogen with a line through the symbol H; he also adopted the name "etherin" for the substance denoted by the formula C^4H^8. He suppressed Wöhler and Liebig's model of constitution with ether as a component.
51. Ibid., 157f.
52. Magnus was convinced that in his experimental variations of the production of sulfovinic acid he had found two further reaction products, also acidic, which differed from sulfovinic acid only in that they contained one less "atom" of water. This resulted from comparison of the formula models, in which he represented sulfovinic acid with the formula model $2\ SO^3 + Ae + 2\ H^2O$ and the two new acids, which Magnus considered isomers and termed ethersulfuric acids, with the formula $2\ SO^3 + Ae + H^2O$ (Ae being the symbol introduced by Berzelius for "etherin," C^4H^8). The formula models agreed on the one hand with those of Dumas and Boullay in that they showed the substance indicated by the formula C^4H^8 as a component; on the other hand, they differed in that they hypothesized a higher water content (ibid., 172). In his experimental tests of water of crystallization, Magnus compared the two isomers with sulfovinic acid.
53. See Liebig and Wöhler 1831, 487.
54. Magnus 1833, 173.
55. Berzelius 1833a, 624.
56. See Boullay 1807c, 100.
57. See Pelouze 1833.
58. Ibid., 137f. Pelouze took the theoretical combining weight of carbon from Berzelius, C = 12.
59. See Berzelius 1833a, 625.
60. See Liebig 1833a, 150.
61. Ibid., 149.
62. Ibid., 151.
63. See Liebig 1833b.

64. Ibid., 608.

65. Liebig wanted this name to express that the substance could have also been produced "from alcohol in the formation of acetic acid" (ibid., 610).

66. Ibid., 608 f.

67. Ibid., 609.

68. See Berzelius 1832, 286.

69. See Liebig 1833b, 613 ff. Wood vinegar was produced in the distillation of wood. In 1835, Dumas and Peligot published the results of an extensive series of experiments on wood spirit, in which they showed that Liebig's wood spirit was not a pure substance (Dumas and Peligot 1835). Wood spirit was later given the systematic names *methyl alcohol* or *methanol*. According to Dumas and Peligot, pure wood spirit was discovered in 1812 by Philip Taylor, but this discovery was first published in 1822. Shortly afterward, it became commercially available (Dumas and Peligot 1835a, 54 f.).

70. Liebig 1833b., 614.

71. Ibid., 618.

72. See Berzelius 1833a, 627.

73. Wöhler and Liebig 1832, 252.

74. Dumas and Peligot 1835a, 57.

75. Berzelius 1832, 286.

76. Berzelius 1833b, 175.

77. See Berzelius 1833a, 627. In a similar way, the new formula model of wood spirit could "be reduced to the rational formula $C^2H^5 + O$" (ibid., 627), using Liebig's formula model (4 C + 10 H + O)+ O and dividing this by two. The new formula showed that wood spirit contained the same radical as ordinary ether, but with a different number of "atoms."

78. Ibid., 628. $\bar{A}$ is the abbreviated formula for acetic acid.

79. Ibid., 618, emphasis mine.

80. See Liebig 1834.

81. Liebig had previously always avoided using the Berzelian exponents, but he began at this time to use subscripts. In contrast to Berzelius's formula C^2H^5, Liebig preferred the formula C_4H_{10} for the postulated radical. He does not explain this difference. We might suppose that in this way he wanted to avoid Berzelius's use of a line to symbolize double radicals. Nevertheless, in this way a compound radical with a relatively large number of element "atoms" was hypothesized.

82. Ibid., 1.

83. Ibid., 7.

84. Ibid., 15.

85. Ibid.

86. Ibid. In an appendix, Liebig added a theoretical explanation of his failure. The chemical affinity of organic compounds, he wrote there, outweighs all other powers that might be effective in the decomposition of inorganic compounds, and this is the reason that the binary inorganic compounds are not just as easily decomposed with reagents as the analogous oxides and salts (ibid., 26 f.).

87. Ibid., 4.

88. Ibid., 14. Liebig repeated a second experiment, which Dumas and Boullay had

used to support their models of constitution, namely, that of Faraday with olefiant gas and sulfuric acid. Faraday had claimed that olefiant gas (or bicarbonated hydrogen) can as a whole combine with sulfuric acid, and Dumas and Boullay had interpreted this as evidence for the basic character of this substance. After repeating the experiments, Liebig was convinced that no compound formation takes place (Liebig 1834, 8 f.).

89. See Zeise 1831.

90. Liebig 1834, 9.

91. Zeise 1831, 533.

92. Liebig 1834, 11.

93. Ibid.

94. Ibid., 27.

95. Like Dumas and Boullay, Liebig argued for both models of constitution with an interpretation of alcoholic fermentation, which claimed that in this process the sugar is first decomposed into carbonic acid and ether and that the ether then is turned into alcohol by the absorption of water (ibid., 22).

96. Ibid.

97. Ihde, for example, writes, "Despite the confusion in deciding on suitable radicals, the radical theory continued to gain support during the 1830s" (Ihde 1964, 189).

98. See Brock 1993; Brooke 1992; Holmes 1981, 1993; Rocke 1984.

99. See Holmes 1993, 125 f.; see also Holmes 1981, 336 f.

100. Berzelius 1973 (1820), 82.

101. See Berzelius 1833a, 621.

102. See Chapter 5.

103. See Chapter 3.

104. See Dumas 1828–1846, 5:3, 78.

105. See Liebig 1841, 203 f. and Chapter 8. The role played here by religious beliefs may have been different for the two men. Berzelius became more religious when he grew older. Lindroth writes that after 1820 "he gives us a world view imbued with religious feeling" (Lindroth 1992, 319). He also connects this with Berzelius's insistence that between "organic and inorganic matter there is an uncrossable line" (ibid.).

106. Berzelius 1973 (1820), 49, emphasis mine.

Chapter 7

1. "L'épreuve de l'experience"; Dumas 1834, 114.

2. See Liebig 1832.

3. See Chapter 3.

4. See Berthollet 1785.

5. Liebig 1832, 184.

6. Ibid., 207 and 209; see also below.

7. See Berzelius 1833a.

8. See Liebig 1832, 184.

9. Dumas 1828–1846, 5:611.

10. See Dumas 1834, 140; emphasis mine.

11. See Chapter 4.
12. Dumas 1828–46, 5:70.
13. Dumas 1834, 140 f.
14. Dumas 1828–1846, 1:XXXI.
15. Dumas 1834, 143.
16. Dumas 1834, 115.
17. Ibid., 141.
18. Dumas 1828–1846, 5:613.
19. See Liebig 1835, 143.
20. For the concept of accommodation of models, see Pickering 1995.
21. See Dumas 1828–1846, 5:99 and 608.
22. Ibid., 99.
23. Dumas 1828–1846, 5:101; emphasis mine.
24. Dumas exemplified his model of stepwise substitution by different examples of reactions (see Dumas 1840, 154 and 160 ff.). I have reformulated his examples for the case of the production of chloral.
25. See Klein 1994a, 1995.
26. I thank William Brock for pointing out to me the importance of the interaction between Dumas and Laurent. In 1840, Laurent made priority claims concerning the "discovery" of the theory of substitution (Laurent 1840). Although I do not wish to investigate further the question of priority, it should be noted that Laurent did not have a concept of equivalent substitution prior to 1834. In his 1833 experiments with naphthalene and chlorine, Laurent obtained two reaction products, a solid and an oily one (see Laurent 1833, 276). Based on the quantitative analysis of the two products and the transformation of the analytical data into Berzelian formulas, he interpreted the formation of the oily product, which he denoted by the formula $C^{10}H^4 + Ch$, not as a replacement reaction but as a composition or synthesis leaving naphthalene $C^{10}H^4$ intact (ibid., 283 f.). In contrast, he interpreted the formation of the solid product, denoted by the formula $Ch^2 + C^{10}H^3$, as a "decomposition," in which *two* volumes of chlorine (Ch) replace *one* volume of hydrogen contained in naphthalene $C^{10}H^4$ yielding $Ch^2 + C^{10}H^3$ and 2 volumes of hydrochloride acid (corresponding to 1 HCh) (ibid., 201). On Laurent's work, see also Blondel-Mégrelis 1996; Kapoor 1969a, 1969b; Novitski 1992. It should also be noted that Liebig and Wöhler's 1832 paper did not contain a concept of substitution that implied all aspects of meaning analyzed and defined above (see Wöhler and Liebig 1832, and Brock 1993, 215 f.). In their 1832 paper, Liebig and Wöhler allowed an equivalent replacement of 2 H for 2 Cl in oil of bitter almond (which was based on work on paper with Berzelian formulas, too), but they did not consider the possibility of a stepwise replacement.
27. See Laurent 1835a, 196, and Laurent 1835b, 388ff. Laurent used the term *substitution* only in 1835b (see Laurent 1835b, 389).
28. Dumas 1834, 115.
29. Latour 1990, 23.
30. See Pickering 1995.
31. See Dumas and Peligot 1835b, 1835c.
32. See Laurent 1835a and 1835 b.

33. See Malaguti 1837a, 1837b.
34. See Poggendorff 1836, 97.
35. Liebig and Pelouze 1836, 157.
36. See Liebig 1835.
37. Ibid., 161.
38. Ibid., 160.
39. Ibid., 161.
40. Ibid., 160.
41. Ibid., 161.
42. Berzelius 1838, 633.
43. See Dumas 1838b.
44. See Dumas 1839a, 1839b, 1840a–c.
45. At the end of the eighteenth century, René-Just Haüy introduced the term *type* to classify crystals. Dumas explicitly referred to this tradition of crystallography. His concept of type also bears common features with Cuvier's and de Candolle's concepts of type in the sense of an invisible plan of construction. For the latter influence, see also Nye 1993, 66. The term *type* had been introduced into chemistry earlier by Laurent, based on work with Berzelian formulas; its meaning differed slightly from Dumas's concept of type (see Laurent 1835b, 383).
46. Dumas 1840a, 164.
47. See Hacking 1992a; 1999, 71 ff., and Pickering 1989, 1990, 1995.
48. Hacking 1992a, 30.
49. Pickering 1995, 22.
50. Ibid.
51. See Chapter 6.
52. See Chapter 5.
53. Liebig 1832, 207 f., emphasis mine.
54. Ibid., 209; emphasis mine.
55. Ibid., 205 f.; emphasis mine.
56. Dumas 1834, 133.
57. Dumas 1828–46, 5:70, emphasis mine.
58. Ibid., 70.
59. Ibid., emphasis mine.
60. Hacking 1992a, 54.

Chapter 8

1. F. L. Holmes has pointed out before that the formation of ordinary ether was "one of the most important reactions in organic chemistry" of the 1830s (Holmes 1981, 333).

2. See Kohler 1994, 2 f. Other well-known biological model objects or organisms are maize, *Escherichia coli*, white mice, or rats (see Clause's history of the standard Wistar rat; Clause 1993). In eighteenth-century chemistry, the salts and later oxides (or calces) played the role of model objects.

3. Ibid., 3.
4. See Chapter 3.
5. Volhard 1909, 1:252.
6. Kohler 1994, 5 f.
7. Ibid., 19 ff.
8. See Wallach 1901, 1:639.
9. Berthelot 1872, 13.
10. I distinguish between *model objects* and *models*. The former are the material things of research, the latter the inscriptions referring to them and fitting them into the existing intellectual framework.
11. See discussions in Chapters 2 and 3.
12. See Chapter 6.
13. See Wöhler 1828.
14. For details on the history of the urea synthesis, see Benfey 1975, 14 ff.
15. It should be noted that Wöhler used Berzelian formulas, which became only common from that time on, for reconstructing the synthesis of urea.
16. Ramberg, 2000, 173.
17. On "disciplinary identity" of mid-nineteenth-century organic chemistry, see also Nye 1993, 19.
18. Wöhler 1828, 253; emphasis mine.
19. Wallach 1901, 1:208, emphasis mine.
20. See Pickering 1995, 19.
21. Ibid.
22. Schlossberger 1857, 252.
23. Liebig 1861, 673.
24. Berthelot 1860, CIV.
25. Rheinberger 1997, 109.
26. Liebig 1841, 203 f.; emphasis mine.
27. Brock 1997, 92.
28. "The field of investigation which has very recently been opened to experiment in such organic metamorphoses is apparently boundless; and it has been taken possession of by labourers of the highest talent and activity; the fruits of whose industry are so abundant as to cover the ground with somewhat of that confusion which might be expected from a number of independent workmen actively engaged in the same track. The bearing of these investigations upon the theory of organic radicals is extremely important, as tending to show what are the most permanent groups of the organic elements, and what the most capable of substitution for each other, either in inorganic or organic combinations, without changing the type of the compound" (Daniell 1843, 642).
29. Ibid., 677.
30. See Rocke 1993.
31. On these later processes, see ibid., 1ff.
32. Poggendorff 1836, 1.
33. See Wallach 1901, 1:604.
34. Wöhler and Liebig 1832, 249.
35. Berzelius 1839, 289; emphasis mine.

36. For the formation of scientific pharmacy in France, see Simon 1998; for that of pharmaceutical botany, see Löw 1977, 9.

37. For this notion, see Mittelstraß 1992.

38. I address only an aspect of Kuhn's incommensurability thesis, which is one of the most discussed parts of his theory. See Hoyningen-Huene 1993.

39. Kuhn 1996, 198.

40. See Kuhn 1987 and 1989.

41. Buchwald 1992, 42 f.

42. See the the discussion of this concept in Chapter 2.

43. I have avoided Kuhn's term *scientific revolution* mainly because I do not take over his concept of an overarching general paradigm (or theory).

44. Pickering makes a similar observation with respect to the formation of high-energy physics and the coming into being of quarks in the 1970s (see Pickering 1995, 191).

45. This alternative view has been suggested by Holmes (see Holmes 1971).

46. Helge Kragh has coined the nice term *European commonwealth of chemistry*. He has rightly emphasized, "One may debate whether modern chemistry was the invention of the French, the Germans, or the English, but it is beyond discussion that it was a European science" (Kragh 1998, 329).

47. See Homburg 1998, 1999.

48. See Baird 1993.

49. See Chapter 5.

50. See Chapter 6.

51. See Chapter 1.

52. Holmes 1993, 129 f., emphasis mine.

53. Rheinberger 1997, 74. It is a nice incident that Auguste Laurent also compared organic chemistry with a labyrinth (Laurent 1854, IX).

54. For the historical transformations in nineteenth-century natural history and the emergence of life sciences, see Appel 1987; Caron 1988; Farber 1982; Larson 1994; Lenoir 1989; Lepenies 1976; Nyhart 1995 and 1996. For the historical development of the interplay of chemistry and biology since 1800, see Fruton 1990.

55. See Nye 1993, 66 f.

56. See Dumas 1828–1845, 5:78.

57. Ibid., 2.

58. Dumas and Liebig 1837, 569.

59. See Dumas 1841.

60. On this issue, see, for example, see Geison 1978; Lenoir 1989 and 1997.

61. With respect to the rise of experimental physiology in mid-nineteenth-century Germany, Timothy Lenoir has convincingly argued that it was conditioned by "external constraints" of the larger social and cultural systems, such as science policy and materialist ideology (Lenoir 1997, 76 f.).

62. See, for example, Travis 1993, Rheinhardt and Travis 2000.

63. Lenoir 1997, 77.

64. See Volhard 1909, 1:255.

65. I thank William Brock for making me aware of the latter.

66. See Holmes 1971.

67. Holmes 1993, 122.

68. Ibid., emphasis mine.

69. Holmes describes the mode of research of the individual "investigative pathway" as follows: "This activity [of the individual experimenter] must be followed in fine detail if we are to understand how significant increments of scientific knowledge are gained. . . . One must penetrate to the more intimate level of daily activity if one wishes to identify fully the complex interplay of intellectual factors, craft practices, and social complexes that direct the investigator along his or her pathway to new knowledge. Nor is it enough to reconstruct vignettes from such a pathway at critical or high points along it. To understand its dynamic one must have the patience to sustain the detailed trace over the years or decades that form the investigator's outlook and shape the discoveries that occur along the way" (Holmes 1993, 122).

70. See Bloor 1991; Barnes, Bloor, and Henry 1996.

71. See, for example, Holmes 1985.

72. See Liebig 1841 and Daniell 1843.

73. This question has been studied by Fruton with respect to the development of biochemical sciences in the nineteenth century (Fruton 1990).

74. See Lenoir 1997, 76 f.

Chapter 9

1. See Nye 1993, 76.

2. Dagognet 1969, 186.

3. The term *long-term concept* is borrowed from Galison. Galison has distinguished three different intellectual levels entering into experimentation, termed *long-term constraints*, *middle-term constraints*, and *short-term constraints* (Galison 1987, 243 ff.). The conceptual network of chemical compound and a reaction, which comprised the concept of binary constitution and reaction types, built at the beginning of the eighteenth century, was a long-term constraint of eighteenth- and early nineteenth-century chemical experimentation and interpretive modeling. European chemists' shared goal of extending this concept to organic compounds in the 1830s was a middle-term constraint, and the more particular analogies with inorganic classes, which entered into interpretive models of experiments, were short-term constraints.

4. See Latour 1987, 1990, 1999.

5. The meaning of the notion of *reference* (or *signification*) depends on a broader theory of reference. Here and in the following I add *meaning* to avoid a common misunderstanding: that *reference* is an abstract, formal relation between a sign and a signifier, devoid of *meaning* stemming from the entire system of signs and their practical functions in forms of life (in the Wittgensteinian sense).

6. I described elsewhere the emergence, reference, and meaning of this conceptual system. See Klein 1994a, 1994b, 1995, 1996.

7. See Chapter 3

8. With respect to twentieth-century experimental physics, Peter Galison pointed to

an analogous fact: "This asymmetry between experiment and theory is often hidden because experimentalists express their public claims in a language that suggests experimental results are independent of the researcher's judgements, experience, or skills. Reading an article, one could conclude that an effect would follow from an experimental setup with the inexorability of logical implication. But lurking behind the confidence of the experimental paper lies a body of work that relies on a subtle judgement that is notoriously ill-suited for the prose of hypothesis and deduction" (Galison 1987, 244).

9. Often, the subsequent quantitative analysis was also a test for purity; however, quantitative analysis yielded certain results for impurity only so that impure substances could be selected and further prepared, whereas certain evidence for purity was lacking.

10. For the notion of "inscription device," see Latour 1987, 64 ff.

11. See Galison 1987, 256.

12. For an analogous case in the history of physics, see Galison 1987, 245.

13. I use the term *mechanism* not in the contemporary chemical sense, but rather to denote the varieties of quasi-mechanical combinations and recombinations of the building blocks of the initial substances into the reaction products.

14. See Dumas and Boullay 1827.

15. See Dumas 1834.

16. See Dagognet 1973, 114 ff.

17. See Latour 1987, 1990, 1999.

18. Latour 1990, 22.

19. See Chapter 6.

20. Latour 1990, 45.

21. Ibid., 23.

22. For this approach, I have benefited from the work of a group of Berlin philosophers on the concept of "scientific labor" that was oriented at Hegel's and Marx's concept of "labor" (see, e.g., Arndt 1983 and 1985; Damerow et al. 1983; Ruben 1978).

23. The concept of valence, which constrained the reshuffling, was introduced gradually from the 1850s onward.

24. See Chapter 1.

LITERATURE CITED

Alborn, Timothey L. 1989. Negotiating notation: chemical symbols and British Society, 1831–1835. *Annals of Science* 46:437–460.

Allén, Sture, ed. 1989. *Possible Worlds in Humanities, Arts, and Sciences: Proceedings of Nobel Symposium 65*. Berlin: de Gruyter.

Ampère, André Marie. 1814. Lettre de M. Ampère à M. le Comte Berthollet, sur la détermination des proportions dans lesquelles les corps se combinent d'après le nombre et la disposition respective des molécules dont leurs particules intégrantes sont composées. *Annales de Chimie* 90:43–86.

Analyses of Books. 1815. An attempt to establish a pure scientific system of mineralogy, by the application of the electro-chemical theory and the chemical proportions. By J. Jacob Berzelius. Translated from the Swedish original by John Black. *Annals of Philosophy* 5:302–306.

Appel, Toby A. 1987. *The Cuvier-Geoffroy Debate: French Biology in the Decades before Darwin*. Oxford: Oxford University Press.

Arndt, Andreas. 1983. Bibliographische Notiz zum Stichwort Arbeitskonzept. In *Arbeit und Philosophie: Symposium über philosophische Probleme des Arbeitsbegriffs*. Edited by Peter Damerow, Peter Furth, and Wolfgang Lefèvre. Bochum: Germinal, 241–244.

———. 1985. Zur Herkunft und Funktion des Arbeitsbegriffs in Hegels Geistesphilosophie. *Archiv für Begriffsgeschichte* 39:99–115.

Arnheim, Rudolf. 1969. *Visual Thinking*. Berkeley: University of California Press.

Avogadro, Amedeo. 1811. Essai d'une manière de déterminer les masses relatives des molécules élémentaires des corps, et les proportions selon lesquelles elles entrent dans ces combinaisons. *Journal de Physique* 73:58–76.

Bachelard, Gaston. 1973. *Le Pluralisme Cohérent de la Chimie Moderne*, 2nd ed. Paris: Librairie Philosophique J. Vrin.

Baird, Davis. 1993. Analytical chemistry and the "big" scientific Instrumentation Revolution. *Annals of Science* 50:267–290.

Barnes, Barry, David Bloor, and John Henry. 1996. *Scientific Knowledge: A Sociological Analysis*. Chicago: University of Chicago Press.

Benfey, Otto Theodor. 1975. *From Vital Force to Structural Formulas*. Washington, D.C.: American Chemical Society.

Berthelot, Pierre Eugène Marcellin. 1860. *Chimie Organique Fondée sur la Synthèse*. 2 vols. Paris: Mallet-Bachelier.

———. 1864. *Leçons sur les Méthodes Générales de Synthèse en Chimie Organique: Professées en 1864 au Collège de France*. Paris: Gauthier-Villars.

———. 1872. *Traité Elémentaire de Chimie Organique*. Paris: Dunod.

———. 1876. *La Synthèse Chimique*. Paris: Germer Baillière.

Berthollet, Claude Louis. 1785. Mémoire sur la décomposition de l'esprit de vin et de l'éther, par le moyen de l'air vital. *Histoire de l'Académie Royale des Sciences. Avec des Mémoires de Mathématique & de Physique, pour la même Année*, 308–315.

———. 1816. Note relative au Mémoire de MM. Colin et Robiquet. *Annales de Chimie et de Physique* 1:426–429.

Berzelius, Jöns Jacob. 1811a. Versuch, die bestimmten und einfachen Verhältnisse aufzufinden, nach welchen die Bestandtheile der unorganischen Natur miteinander verbunden sind. *Annalen der Physik* 37:249–334, 415–472.

———. 1811b. Versuch, die bestimmten und einfachen Verhältnisse aufzufinden, nach welchen die Bestandtheile der unorganischen Natur miteinander verbunden sind. *Annalen der Physik* 38:161–226.

———. 1812. Versuch einer lateinischen Nomenclatur für die Chemie, nach electrisch-chemischen Ansichten. *Annalen der Physik* 42:37–89.

———. 1813a. Experiments on the nature of azote, of hydrogen, and of ammonia, and upon the degrees of oxidation of which azote is susceptible. *Annals of Philosophy* 2:276–284, 357–368.

———. 1813b. Essay on the cause of chemical proportions, and on some circumstances relating to them; together with a short and easy method of expressing them. *Annals of Philosophy* 2:443–454.

———. 1814a. Essay on the cause of chemical proportions, and on some circumstances relating to them; together with a short and easy method of expressing them. *Annals of Philosophy* 3:51–62, 93–106, 244–257, 353–364.

———. 1814b. Experiments to determine the definite proportions in which the elements of organic nature are combined. *Annals of Philosophy* 4:323–331, 401–409.

———. 1814c. Essai sur les proportions déterminées dans lesquelles se trouvent réunis les élémens de la nature organique. *Annales de Chimie* 92:141–159.

———. 1814d. *An Attempt to Establish a Pure Scientific System of Mineralogy, by the Application of the Electro-chemical Theory and the Chemical Proportions*. London: Robert Baldwin.

———. 1815. Experiments to determine the definite proportions in which the elements of organic nature are combined. *Annals of Philosophy* 5:93–101, 174–184, 260–275.

———. 1819. *Essai sur la théorie des proportions chimiques et sur l'influence chimique de l'électricité*. Paris: Méquignon-Marvis.

———. 1821. *Von der Anwendung des Löthrohrs in der Chemie und Mineralogie*. Aus der Handschrift übersetzt von Heinrich Rose. Nürnberg: Joh. Leonhard Schrag.

———. 1829. *Jahresbericht über die Fortschritte der physischen Wissenschaften*. Vol. 8.

———. 1832. Schreiben von Berzelius an Wöhler und Liebig über Benzoyl und Benzoesäure. *Annalen der Pharmacie* 3:282–287.

———. 1833a. Betrachtungen über die Zusammensetzung der organischen Atome. *Annalen der Physik und Chemie* 28:617–630.
———. 1833b. Ueber die Constitution organischer Zusammensetzungen. *Annalen der Pharmacie* 6:173–176.
———. 1833–41. *Lehrbuch der Chemie*, 3rd ed. Translated by F. Wöhler. 10 vols. Dresden: Arnold.
———. 1838. Lettre de M. Berzelius à M. Pelouze. *Compte Rendu* 6:629–648.
———. 1839. Ueber einige Fragen des Tages in der organischen Chemie. *Annalen der Physik und Chemie* 47:289–322. *Annalen der Pharmacie* 31:1–35.
———. 1843–48. *Lehrbuch der Chemie*, 5th ed. 5 vols. Dresden: Arnold.
———. 1973. *Versuch über die Theorie der chemischen Proportionen und über die chemischen Wirkungen der Electricität; nebst Tabellen über die Atomgewichte der meisten unorganischen Stoffe und deren Zusammensetzungen.* Dresden: Arnold, 1820. Reprint. Hildesheim: Gerstenberg.
Blondel-Mégrelis, Marika. 1996. *Dire les Choses: Auguste Laurent et la Méthode Chimique.* Paris: Librairie Philosophique J. Vrin.
Bloor, David. 1991. *Knowledge and Social Imagery*, 2nd ed. Chicago: University of Chicago Press.
Boullay, Pierre François Guillaume. 1807a. Mémoire sur la Formation de l'Ether phosphorique à l'aide d'un apparail particulier. *Annales de Chimie* 62:192–197.
———. 1807b. Observations sur l'éther sulfurique et sa préparation. *Annales de Chimie* 62:242–247.
———. 1807c. Mémoire sur le mode de composition des éthers muriatique et acétique. *Annales de Chimie* 63:90–101.
———. 1811. Nouvel Éther résultant de l'action de l'acide arsenique sur l'alcool. *Annales de Chimie* 78:284–297.
Brock, William H. 1986. The British Association Committee on Chemical Symbols 1834: Edward Turner's Letter to British Chemists and a Reply by William Prout. *Ambix* 33(1):33–42.
———. 1993. *The Norton History of Chemistry.* New York: W. W. Norton & Company.
———. 1997. *Justus von Liebig: The Chemical Gatekeeper.* Cambridge: Cambridge University Press.
Brooke, John Hedley. 1971. Organic synthesis and the unification of chemistry—a reappraisal. *The British Journal for the History of Science* 5(20):363–392.
———. 1973. Chlorine substitution and the future of organic chemistry: methodological issues in the Laurent-Berzelius correspondence (1843–1844). *Studies in History and Philosophy of Science* 4(1):47–94.
———. 1987. Methods and Methodology in the Development of Organic Chemistry. *Ambix* 34(3):147–155.
———. 1992. Berzelius, the dualistic hypothesis, and the rise of organic chemistry. In *Enlightenment Science in the Romantic Era.* Edited by Evan M. Melhado and Tore Frängsmyr. Cambridge: Cambridge University Press, 180–221.
Buchwald, Jed Z. 1992. Kinds and the wave theory of light. *Studies in History and Philosophy of Science* 23(1):39–74.

———. 2000. How the Ether Spawned the Microworld. In *Biographies of Scientific Objects*. Edited by Lorraine Daston. Chicago: University of Chicago Press, 203–225.

———, ed. 1995. *Scientific Practice: Theories and Stories of Doing Physics*. Chicago: University of Chicago Press.

Butlerov, Aleksandr M. 1868. *Lehrbuch der Organischen Chemie; zur Einführung in das specielle Studium derselben*. German ed., rev. and enl., transl. from the Russian. Leipzig: Quandt & Händel.

Cardwell, D. S. L., ed. 1968. *John Dalton & the Progress of Science*. Manchester: Manchester University Press.

Caron, Joseph A. 1988. "Biology" in the life sciences: a historiographical contribution. *History of Science* 26:223–268.

Carrier, Martin. 1993. What is right with the miracle argument: establishing a taxonomy of natural kinds. *Studies in History and Philosophy of Science* 24(3): 391–409.

Cartwright, Nancy. 1983. *How the Laws of Physics Lie*. Oxford: Oxford University Press.

Cartwright, Nancy, Towfic Shomar, and Mauricio Suárez. 1995. The tool box of science: tools for the building of models with a superconductivity example. In *Theories and Models in Scientific Processes*. Edited by William E. Herfel, Wladyslaw Krajewski, Ilkka Niiniluoto, and Ryszard Wójcicki. Amsterdam: Rodopi, 137–149.

Cavanna, D., and S. Rocchietta. 1959. Il linguaggio simbolica della chimica. *Minerva Farmaceutica* 8:204–208.

Caven, Robert Martin, and John Arnold Cranston. 1928. *Symbols and Formulae in Chemistry: An Historical Study*. London: Blackie & Son.

Chevreul, Michel Eugène. 1823. *Recherches Chimiques sur les Corps Gras d'Origine Animale*. Paris: Levrault.

Christie, Maureen. 1994. Philosophers versus chemists concerning "laws of nature." *Studies in History and Philosophy of Science* 25(4):613–629.

Clause, Bonnie T. 1993. The Wistar rat as a right choice: establishing mammalian standards and the ideal of a standardized mammal. *Journal of the History of Biology* 26:329–49.

Cooke, Josiah P. 1874. *The New Chemistry*. New York: D. Appleton & Company.

Couper, Archibald S. 1858a. On a new chemical theory. *Philosophical Magazine* 16:104–116.

———. 1858b. Sur une nouvelle théorie chimique. *Annales de Chimie et de Physique* 53(3):469–489.

Crosland, Maurice. 1962. *Historical Studies in the Language of Chemistry*. London: Heinemann Educational Books.

Dabit. 1799. Extrait du mémoire du cit: Dabit sur l'éther. *Annales de Chimie* 34:289–305.

Dagognet, François. 1969. *Tableaux et Languages de la Chimie*. Paris: Éditions du Seuil.

———. 1973. *Écriture et Iconographie*. Paris: Librairie Philosophique J. Vrin.

Dalton, John. 1808–27. *A New System of Chemical Philosophy*. 2 vols. Vol. 1, part 1.

Manchester: S. Russell, 1808, part 2. Manchester: Russell & Allen, 1810, vol. 2. Manchester: Executers of S. Russell, 1827.

———. 1814. Remarks on the essay of Dr. Berzelius on the cause of chemical proportions. *Annals of Philosophy* 3:174–180.

Damerow, Peter, Peter Furth, and Wolfgang Lefèvre, eds. 1983. *Arbeit und Philosophie: Symposium über philosophische Probleme des Arbeitsbegriffs.* Bochum: Germinal.

Daniell, J. Frederic. 1843. *An Introduction to the Study of Chemical Philosophy: Being a Preparatory View of the Forces Which Concur to the Production of Chemical Phenomena*, 2nd ed. London: John W. Parker.

Daston, Lorraine, ed. 2000. *Biographies of Scientific Objects.* Chicago: University of Chicago Press.

Deiman, Johann Rudolf, Adrien Paets-Van-Troostwyck, Nicolas Bondt, and Anthoni Lauwerenburgh. 1794. Recherches sur les diverses espèces des gaz qu'on obtient en mêlant l'acide sulfurique concentré avec l'alcool. *Journal de Physique, de Chimie, d'Histoire Naturelle et des Arts* 2:178–191.

Diderot, Denis J., and Jean LeRond d'Alembert. 1966. Encyclopédie ou Dictionnaire Raisonné des Sciences, des Arts, et des Métiers. 35 vols. Reprint of the first edition. Stuttgart: Frommann, 1751–1780.

Döbereiner, Johann Wolfgang. 1821–25. *Zur pneumatischen Chemie.* 5 vols. Jena: Crökersche Buchhandlung.

Douglas, Mary, and David Hull, eds. 1992. *How Classification Works: Nelson Goodman among the Social Sciences.* Edinburgh: Edinburgh University Press.

Dumas, Jean-Baptiste André. 1826. Mémoire sur quelques Points de la Théorie atomistique. *Annales de Chimie et de Physique* 33:337–391.

———. 1828–46. *Traité de Chimie appliqué aux Arts.* 8 vols. Paris: Béchet jeune.

———. 1830. Ueber die Chloroxalsäure. *Annalen der Physik und Chemie* 20: 166–169.

———. 1833. Sur les Camphres artificiels des essences de Térébenthine et de Citron. *Annales de Chimie et de Physique* 52:400–410.

———. 1834. Recherches de Chimie organique. *Annales de Chimie et de Physique* 56:113–150.

———. 1835. Ueber die Wirkung des Chlors auf den Alkohol. *Annalen der Pharmacie* 16:164–171.

———. 1837. *Leçons sur la Philosophie Chimique.* Paris: Béchet jeune.

———. 1838a. Recherches de chimie organique relative à l'action du chlore sur l'alcool. *Mémoires de l'Académie Royale des Sciences* 15:519–556.

———. 1838b. Réponse de M. Dumas à la lettre de M. Berzelius. *Comptes Rendus* 6:689–702.

———. 1839a. Mémoire sur la constitution de quelques corps organiques et sur la théorie des substitutions. *Comptes Rendus* 8:609–622.

———. 1839b. Note sur la constitution de l'acide acétique et de l'acide chloracétique. *Comptes Rendus* 9:813–815.

———. 1840a. Mémoire sur la loi des substitutions et la théorie des types. *Comptes Rendus* 10:149–178.

———. 1840b. Note relative aux réclamations de M. Laurent. *Comptes Rendus* 10:511–524.

———. 1840c. Les types chimiques. *Annales de Chimie et de Physique* 73:73–100.

———. 1842. *Essai de Statique Chimique des Etres Organisés*. Paris: Fortin, Masson & C.

Dumas, Jean-Baptiste André, and Polydore Boullay. 1827. Mémoire sur la formation de l'Ether sulfurique. *Annales de Chimie et de Physique* 36:294–310.

———. 1828. Mémoire sur les ethers composés. *Annales de Chimie et de Physique* 37:15–53.

Dumas, Jean-Baptiste André, and Justus Liebig. 1837. Note sur l'état actuel de la Chimie organique. *Compte Rendu* 5:567–572.

Dumas, Jean-Baptiste André, and Eugène Melchior Peligot. 1835a. Ueber den Holzgeist und die verschiedenen ätherartigen Verbindungen, welche er bildet. *Annalen der Pharmacie* 15:1–58.

———. 1835b. Ueber das Zimmtoel. *Annalen der Pharmacie* 13:76–78.

———. 1835c. Organisch-chemische Untersuchungen über das Zimmtöl, die Hippursäure und die Fettsäure. *Annalen der Pharmacie* 14:50–74.

Duncan, Alistair. 1996. *Laws and Order in Eighteenth-Century Chemistry*. Oxford: Clarendon Press.

Ecco, Umberto. 1976. *A Theory of Semiotics*. Bloomington: Indiana University Press.

Eriksson, Gunnar. 1992. Berzelius and the atomic theory: the intellectual background. In *Enlightenment Science in the Romantic Era: The Chemistry of Berzelius and Its Cultural Setting*, Uppsala Studies in History of Science. Edited by Evan M. Melhado and Tore Frängsmyr, 10: 56–84. Cambridge: Cambridge University Press.

Farber, Paul L. 1982. The transformation of natural history in the nineteenth century. *Journal of the History of Biology* 15:145–152.

Fischer, N. W. 1973a. Organic classification before Kekulé. *Ambix* 20(1):106–131.

———. 1973b. Organic classification before Kekulé. *Ambix* 20(3):209–233.

———. 1974. Kekulé and organic classification. *Ambix* 21(1):29–52.

Foucault, Michel. 1997. *The Order of Things: An Archaeology of the Human Sciences*. Translated from the French. Reprint of the first English edition of 1970. London: Routledge.

Fourcroy, Antoine-François de. 1797. Extrait d'un mémoire sur trois espèces differentes de Gaz hydrogène carboné, retirées de l'éther et de l'alcool par différens procédés, envoyés à l'Institut par la Société des Chimistes hollandois; tiré d'un rapport lu à la 1ere classe de l'Institut de France, par le citoyen Fourcroy, séance du 26 frimaire, an 5e (16 déc. 1796). *Annales de Chimie* 21:48–71.

———. 1801–02. *Système des Connaissances chimiques: et de leurs Applications aux Phénomènes de la Nature et de l'art*. 11 vols. Paris: Baudouin.

———. 1807. *Chemical Philosophy; or, The Established Bases of Modern Chemistry*, 3rd ed. Translated by W. Desmond. London: H. D. Symonds.

Fourcroy, Antoine-François de, and Nicolas L. Vauquelin. 1797. De l'action de l'acide sulfurique sur l'alcool, et de la formation de l'ether. *Annales de Chimie* 23:203–215.

Francoeur, Eric. 1997. The forgotten tool: the design and use of molecular models. *Social Studies of Science* 27:7–40.

Francoeur, Eric, and Jérôme Segal. n.d. From physical to virtual models: macromolecular structures and the origins of interactive molecular graphics. In *Displaying the Third Dimension: Models in Science, Technology and Medicine.* Edited by Soraya de Chadarevian and Nick Hopwood. Stanford: Stanford University Press. (In press).

Frängsmyr, Tore, John Lewis Heilbron, and Robin E. Rider, eds. 1990. *The Quantifying Spirit in the Eighteenth Century.* Berkeley: University of California Press.

Fruton, Joseph S. 1990. *Contrasts in Scientific Style: Research Groups in the Chemical and Biological Sciences.* Philadelphia: American Philosophical Society.

Galison, Peter. 1987. *How Experiments End.* Chicago: University of Chicago Press.

———. 1997. *Image and Logic: A Material Culture of Microphysics.* Chicago: University of Chicago Press.

———. 1998. Feynman's war: modelling weapons, modelling nature. *Studies in History and Philosophy of Modern Physics* (Special Issue: *Cultures of Theory.* Edited by Peter Galison and Andrew Warwick) 29B(3):391–434.

Galison, Peter, and Andrew Warwick. 1998. Introduction: cultures of theory. *Studies in History and Philosophy of Modern Physics* (Special Issue: *Cultures of Theory.* Edited by Peter Galison and Andrew Warwick) 29B(3):287–294.

———, eds. 1998. *Cultures of Theory. Studies in History and Philosophy of Modern Physics* (Special Issue) 29B(3).

Gay-Lussac, Joseph Louis. 1809. Mémoire sur la combinaison des substances gazeuses, les unes avec les autres. *Mémoires de physique et de chimie de la Société d'Arceuil* 2:207–234, 252–253.

———. 1815. Lettre de M. Gay-Lussac à M. Clément, sur l'analyse de l'alcool et de l'éther sulfurique, et sur les produits de la fermentation. *Annales de Chimie* 95:311–318.

———. 1820. Sur l'alteration qu'éprouve l'acide sulfurique en agissant sur l'alcool. *Annales de Chimie et de Physique* 13:62–79.

———. 1828. *Cours de Chimie.* 2 vols. Paris: Pichon & Didier.

Gay-Lussac, Joseph Louis, and Louis Jacques Thenard. 1811. *Recherches Physico-Chemiques.* 2 vols. Paris: Deterville.

Geison, Gerald L. 1978. *Michael Foster and the Cambridge School of Physiology: The Scientific Enterprise in Late Victorian Society.* Princeton: Princeton University Press.

Geoffroy, Etienne François. 1718. Table des differentes rapports observés en chimie entre differentes substances. In *Histoire de l'Académie Royale des Sciences. Avec des Mémoires de Mathématique & de Physique pour la même Année, Mémoires,* 202–212.

Gerhardt, Charles Fréderic. 1844–1845. *Précis de Chimie Organique.* 2 vols. Paris: Fortin, Masson & Cie.

———. 1853–1856. *Traité de Chimie Organique.* 4 vols. Paris: Firmin Didot frères.

Gillispie, Charles Coulston, ed. 1981. *Dictionary of Scientific Biography.* 16 vols. in 8. New York: Charles Scribner's Sons.

Gmelin, Leopold. 1817–1819. *Handbuch der theoretischen Chemie.* 3 vols. Frankfurt am Main: Varrentrapp.

———. 1843–1870. *Handbuch der Chemie*, 4th ed. 13 vols. Heidelberg: Karl Winter.

Gombrich, Ernst. 1969. *Art and Illusion: A Study in the Psychology of Pictorial Representation.* Princeton: Princeton University Press.

Gooding, David, Trevor Pinch, and Simon Schaffer, eds. 1989. *The Uses of Experiment: Studies of Experimentation in the Natural Sciences.* Cambridge: Cambridge University Press.

Goodman, Nelson. 1976. *Languages of Art: An Approach to a Theory of Symbols.* 2nd ed. Indianapolis: Hackett Publishing Company.

———. 1978. *Ways of Worldmaking.* Indianapolis: Hackett Publishing Company.

Graebe, Carl. 1991. *Geschichte der organischen Chemie. 1920.* Reprint. Berlin: Springer.

Guyton de Morveau, Louis Bernard, Nicolas L. Vauquelin, and Claude Louis Berthollet. 1807. Sur un Mémoire de M. Thenard, sur l'Ether nitreux. *Annales de Chimie* 61:282–302.

Hacking, Ian. 1983. *Representing and Intervening: Introductory Topics in the Philosophy of Natural Science.* Cambridge: Cambridge University Press.

———. 1992a. The self-vindication of the laboratory sciences. In *Science as Practice and Culture.* Edited by Andrew Pickering. Chicago: University of Chicago Press, 29–64.

———. 1992b. "Style" for historians and philosophers. *Studies in History and Philosophy of Science* 23(1):1–20.

———. 1993. Working in a new world: the taxonomic solution. In *World Changes: Thomas Kuhn and the Nature of Science.* Edited by Paul Horwich. Cambridge, Mass.: MIT Press, 275–310.

———. 1999. *The Social Construction of What?* Cambridge, Mass.: Harvard University Press.

Haugeland, John. 1998. *Having Thought: Essays in the Metaphysics of Mind.* Cambridge, Mass.: Harvard University Press.

Heidelberger, Michael, and Friedrich Steinle, eds. 1998. *Experimental Essays—Versuche zum Experiment.* Baden-Baden: Nomos Verlagsgesellschaft.

Hennell, Henry. 1827. Ueber die Wirkung zwischen Schwefelsäure und Alkohol, nebst Bemerkungen über die Zusammensetzung und Eigenschaft der dabei entstehenden Verbindungen. *Annalen der Physik und Chemie* 9:12–22.

———. 1828. Ueber die Wirkung zwischen Schwefelsäure und Alkohol, und über die Natur des Prozesses der Aetherbildung. *Annalen der Physik und Chemie* 14:273–285.

Herfel, William E., Wladyslaw Krajewski, Ilkka Niiniluoto, and Ryszard Wójcicki, eds. 1995. *Theories and Models in Scientific Processes.* Amsterdam: Rodopi.

Hjelt, Edvard. 1916. *Geschichte der Organischen Chemie von ältester Zeit bis zur Gegenwart.* Braunschweig: Vieweg & Sohn.

Hoff, Jacobus Henricus van't. 1874. *Voorstel tot uitbreiding der tegenwoordig in de scheikunde gebruikte structuur-formules in de ruimte; benevens een daarmeê saamenhangende opmerking omtrent het verband tusschen optisch actief vermogen en chemische constitutie van organische verbindingen*. Utrecht: J. Greven.

Hoffmann, Roald, and Pierre Laszlo. 1989. Representation in chemistry. *Diogenes* 147:23–51.

———. 1991. Representation in chemistry. *Angewandte Chemie* 30 (1):1–16.

Holmes, Frederic L. 1971. Analysis by fire and solvent extractions: the metamorphosis of a tradition. *Isis* 62(2):129–148.

———. 1981. Justus von Liebig. In *Dictionary of Scientific Biography*. Edited by Charles Coulston Gillispie. Vol 7. New York: Charles Scribner's Sons. 329–350.

———. 1985. *Lavoisier and the Chemistry of Life: An Exploration of Scientific Creativity*. Madison: University of Wisconsin Press.

———. 1989. *Eighteenth-Century Chemistry as an Investigative Enterprise*. Berkeley: University of California Press.

———. 1993. Justus Liebig and the construction of organic chemistry. In *Chemical Sciences in the Modern World*. Edited by Seymour H. Mauskopf. Philadelphia: University of Pennsylvania Press, 119–134.

———. 1998. *Antoine Lavoisier—The Next Crucial Year: Or the Sources of His Quantitative Method in Chemistry*. Princeton: Princeton University Press.

Holmes, Frederic L., and Trevor H. Levere, eds. 2000. *Instruments and Experimentation in the History of Chemistry*. Cambridge, Mass.: MIT Press.

Homburg, Ernst. 1998. Two factions, one profession: the chemical profession in German society 1780–1870. In *The Making of the Chemist: The Social History of Chemistry in Europe, 1789–1914*. Edited by David M. Knight and Helge Kragh. Cambridge: Cambridge University Press, 39–76.

———. 1999. The rise of analytical chemistry and its consequences for the development of the German chemical profession (1780–1860). *Ambix* 46(1):1–31.

Horwich, Paul, ed. 1993. *World Changes: Thomas Kuhn and the Nature of Science*. Cambridge, Mass.: MIT Press.

Hoyningen-Huene, Paul. 1993. *Reconstructing Scientific Revolutions: Thomas S. Kuhn's Philosophy of Science*. Translated by Alexander T. Levine. With a foreword by Thomas S. Kuhn. Chicago: University of Chicago Press.

Hufbauer, Karl. 1982. *The Formation of the German Chemical Community (1720–1795)*. Berkeley: University of California Press.

Hughes, Jeff. 1998. "Modernists with a vengeance": changing cultures of theory in nuclear science, 1920–1930. *Studies in History and Philosophy of Modern Physics* (Special Issue: *Cultures of Theory*. Edited by Peter Galison and Andrew Warwick) 29B(3):339–368.

Ihde, Aaron J. 1964. *The Development of Modern Chemistry*. New York: Harper & Row.

Jardine, Nicholas, J. Anne Secord, and Emma C. Spary, eds. 1996. *Cultures of Natural History*. Cambridge: Cambridge University Press.

Jardine, Nicholas, and Emma C. Spary. 1996. The natures of cultural history. In *Cultures of Natural History*, edited by Nicholas Jardine, J. Anne Secord, and Emma C. Spary. Cambridge: Cambridge University Press, 3–13.

Kaiser, David. 2000. Stick-figure realism: conventions, reification, and the persistance of Feynman diagrams, 1948–1964. *Representations* 70:49–86.

Kapoor, Satish C. 1969a. Dumas and organic classification. *Ambix* 16(1,2):1–65.

———. 1969b. The origins of Laurent's organic classification. *Isis* 60(4):477–527.

Kekulé, Friedrich August. 1861–1882. *Lehrbuch der organischen Chemie, oder der Chemie der Kohlenstoffverbindungen.* 3 vols. Vols. 1 and 2. Erlangen: Ferdinand Enke, 1861 and 1866. Vol. 3. Stuttgart: Ferdinand Enke, 1882.

———. 1869. Ueber die Constitution des Benzols. In *Berichte der Deutschen Chemischen Gesellschaft.* Vol 2. Berlin: Ferd. Dümmler, 362–365.

Kim, Mi Gyung. 1992a. The layers of chemical language, I: constitution of bodies v. structure of matter. *History of Science* 30:69–96.

———. 1992b. The layers of chemical language, II: stabilizing atoms and molecules in the practice of organic chemistry. *History of Science* 30:397–437.

———. 1995. Essay review of labor and mirage: writing the history of chemistry. *Studies in History and Philosophy of Science* 26(1):155–165.

———. 1996. Constructing symbolic spaces: chemical molecules in the Académie des Sciences. *Ambix* 43(1):1–31.

Klein, Ursula. 1994a. *Verbindung und Affinität. Die Grundlegung der neuzeitlichen Chemie an der Wende vom 17. zum 18. Jahrhundert.* Basel: Birkhäuser.

———. 1994b. Origin of the concept of chemical compound. *Science in Context* 7(2):163–204.

———. 1995. E. F. Geoffroy's table of different "rapports" observed between different chemical substances—a reinterpretation. *Ambix* 42(2):79–100.

———. 1996. The chemical workshop tradition and the experimental practice—discontinuities within continuities. *Science in Context* 9(3):251–287.

———. 1998. Paving a way through the jungle of organic chemistry—experimenting within changing systems of order. In *Experimental Essays—Versuche zum Experiment.* Edited by Michael Heidelberger and Friedrich Steinle. Baden-Baden: Nomos Verlagsgesellschaft, 251–271.

———. 1999. Techniques of modelling and paper tools in classical chemistry. In *Models as Mediators: Perspectives on Natural and Social Sciences.* Edited by Mary Morgan and Margaret Morrison. Cambridge: Cambridge University Press, 146–167.

Knight, David M. 1978. *The Transcendental Part of Chemistry.* Folkestone, Kent: Dawson.

———. 1993. Pictures, diagrams and symbols: visual language in nineteenth-century chemistry. In *Non-Verbal Communication in Science Prior to 1900.* Edited by Renato G. Mazzolini. Florence: Biblioteca di Nuncius, Studi e Testi, 321–344.

———. 1995. Ideas in chemistry: a history of the science. 2nd printing. New Brunswick: Rutgers University Press.

Knight, David M., and Helge Kragh, eds. 1998. *The making of the chemist: The social history of chemistry in Europe, 1789–1914.* Cambridge: Cambridge University Press.

Kohler, Robert E. 1994. *Lords of the Fly:* Drosophila *Genetics and the Experimental Life.* Chicago: University of Chicago Press.

Kopp, Hermann Franz Moritz. 1966a. *Geschichte der Chemie*. 4 vols. Braunschweig: Vieweg, 1843–1847. Reprint. Hildesheim: Olms.

———. 1966b. *Die Entwicklung der Chemie in der neueren Zeit*. München: Oldenbourg, 1873. Reprint. Hildesheim: Olms.

Kragh, Helge. 1998. Afterword: The European Commonwealth of Chemistry. In *The Making of the Chemist: The Social History of Chemistry in Europe, 1789–1914*. Edited by David M. Knight and Helge Kragh. Cambridge: Cambridge University Press, 329–341.

Krätz, Otto. 1973. Zur Geschichte der organisch-chemischen Formelschreibweise: Ein Brief von C. W. Blomstrand an H. Kolbe. *Physis* 15:157–177.

Kuhn, Thomas S. 1987. The presence of past science. In *Sherman Memorial Lectures* at University College, London.

———. 1989. Possible worlds in history of science. In *Possible Worlds in Humanities, Arts, and Sciences: Proceedings of Nobel Symposium 65*. Edited by Sture Allén. Berlin: de Gruyter, 9–32, 49–51.

———. 1996. *The Structure of Scientific Revolutions*. 3rd ed. Chicago: University of Chicago Press.

Larson, James L. 1994. *Interpreting Nature: The Science of Living Form from Linnaeus to Kant*. Baltimore: Johns Hopkins University Press.

Laszlo, Pierre. 1993. *La parole des choses ou la language de la chimie*. Paris: Hermann.

Latour, Bruno. 1987. *Science in Action: How to Follow Scientists and Engineers through Society*. Cambridge, Mass.: Harvard University Press.

———. 1990. Drawing things together. In *Representation in Scientific Practice*. Edited by Michael Lynch and Steve Woolgar, 19–68. Cambridge, Mass.: MIT Press.

———. 1993. *We Have Never Been Modern*. Translated by Catherine Porter. Cambridge, Mass.: Harvard University Press.

———. 1999. *Pandora's Hope: Essays on the Reality of Science Studies*. Cambridge, Mass.: Harvard University Press.

Latour, Bruno, and Steve Woolgar. 1986. *Laboratory Life: The Construction of Scientific Facts*. Princeton: Princeton University Press.

Laudet. 1799. Observations sur les éthers. *Annales de Chimie* 34:282–288.

Laurent, Auguste. 1833. Sur les chlorures de naphthaline. *Annales de Chimie et de Physique* 52:275–285.

———. 1835a. Sur de nouveaux chlorures et brômures d'Hydrogène carboné. *Annales de Chimie et de Physique* 59:196–220.

———. 1835b. Sur la nitronaphtalase, le nitronaphtalèse et la naphtalase. *Annales de Chimie et de Physique* 59:376–397.

———. 1840. Réclamation de priorité relativement à la théorie des substitutions, et à celles des types ou radicaux dérivés. *Comptes Rendus* 10:409–417.

———. 1854. *Méthode de Chimie*. Paris: Mallet-Bachelier.

Lavoisier, Antoine-Laurent. 1789. *Traité Elémentaire de Chimie: Présenté dans un Ordre Nouveau et d'après les Découvertes Modernes*. 2 vols. Paris: Cuchet.

Lefèvre, Wolfgang, Jürgen Renn, and Urs Schöpflin, eds. n.d. *The Creation of the Scientific Image*. Basel: Birkhäuser. (In press).

Le Grand, Homer E., ed. 1990. *Experimental Inquiries: Historical, Philosophical and*

Social Studies of Experimentation in Science. Vol. 8. Australian Studies in History and Philosophy of Science. Dordrecht: Kluwer.

Lenoir, Timothy. 1989. *The Strategy of Life: Teleology and Mechanics in Nineteenth-Century German Biology*. Dordrecht: D. Reidel Pub. Co., 1982. Reprint. Chicago: University of Chicago Press.

———. 1994. Was the last turn the right turn? The semiotic turn and A. J. Greimas. *Configurations* 2(1):119–136.

———. 1997. *Instituting Science: The Cultural Production of Scientific Disciplines*. Stanford: Stanford University Press.

———. 1999. Shaping biomedicine as an information science. In *Proceedings of the 1998 Conference on the History and Heritage of Science Information Systems*, ASIS Monograph series. Edited by Mary Ellen Bowden, Trudi Bellardo Hahn, and Robert V. Williams. Medford, N.J.: Information Today, Inc. 27–45.

———, ed. 1998. *Inscribing Science: Scientific Texts and the Materiality of Communication*. Stanford: Stanford University Press.

Lenoir, Timothy, and Yehuda Elkana, eds. 1988. Practice, context, and the dialogue between theory and experiment. *Science in Context* 2(1).

Lepenies, Wolf. 1976. *Das Ende der Naturgeschichte: Wandel kultureller Selbstverständlichkeiten in den Wissenschaften des 18. und 19. Jahrhunderts*. München: Hanser.

Levere, Trevor H. 1993. *Affinity and Matter: Elements of Chemical Philosophy 1800–1865*. Vol. 12. Classics in the History and Philosophy of Science. Oxford University Press, U.K., 1971. Reprint. Yverdon: Gordon & Breach Science Publishers.

———. 1994. *Chemists and Chemistry in Nature and Society 1770–1878*. Brookfield: Variorum.

Liebig, Justus. 1831a. Ueber die Zersetzung des Alkohols durch Chlor. *Annalen der Physik und Chemie* 23:444.

———. 1831b. Ueber einen neuen Apparat zur Analyse organischer Körper, und über die Zusammensetzung einiger organischer Substanzen. *Annalen der Physik und Chemie* 21:1–43.

———. 1832. Ueber die Verbindungen, welche durch die Einwirkung des Chlors auf Alkohol, Aether, Ölbildendes Gas und Essiggeist entstehen. *Annalen der Pharmacie* 1:182–230.

———. 1833a. Ueber die Zusammensetzung der Weinphosphorsäure. *Annalen der Pharmacie* 6:149–151.

———. 1833b. Ueber Acetal (Sauerstoffäther), Holzgeist und Essigäther. *Annalen der Physik und Chemie* 27:605–626.

———. 1834. Ueber die Constitution des Aethers und seiner Verbindungen. *Annalen der Pharmacie* 9:1–39.

———. 1835. Ueber die Producte der Oxydation des Alkohols. *Annalen der Pharmacie* 14:133–167.

———. 1837. Ueber die Aethertheorie, in besonderer Rücksicht auf die vorhergehende Abhandlung Zeise's. *Annalen der Chemie und Pharmacie* 23:12–42.

———. 1838. Ueber Laurent's Theorie der organischen Verbindungen. *Annalen der Pharmacie* 25:1–31.

———. 1841. Anmerkung zu vorstehender Abhandlung [von J. Dumas und J. S. Staß]. *Annalen der Chemie und Pharmacie* 38:195–216.

———. 1853. *Anleitung zur Analyse organischer Körper*. 2nd ed. Braunschweig: Vieweg & Sohn.

———. 1861. Theorie. In *Handwörterbuch der reinen und angewandten Chemie*. Edited by Justus Liebig, Johann Christian Poggendorff, and Friedrich Wöhler. Vol. 8. Braunschweig: Vieweg & Sohn, 666–699.

Liebig, Justus, and Théophile Jules Pelouze. 1836. Notes diverses. *Annales de Chimie* 63:113–164.

Liebig, Justus, Johann Christian Poggendorff, and Friedrich Wöhler. 1842–64. *Handwörterbuch der reinen und angewandten Chemie*. 9 Vols. Braunschweig: Vieweg & Sohn.

Liebig, Justus, and Friedrich Wöhler. 1831. Ueber die Zusammensetzung der Schwefelweinsäure. *Annalen der Physik und Chemie* 22:486–491.

———. 1832. Ueber die Zusammensetzung der Schwefelweinsäure. *Annalen der Pharmacie* 1:37–43.

Lindroth, Sten. 1992. Berzelius and his time. In *Enlightenment Science in the Romantic Era: The Chemistry of Berzelius and Its Cultural Setting*. Edited by Evan M. Melhado and Tore Frängsmyr. Cambridge: Cambridge University Press, 9–34.

Löw, Reinhard. 1977. *Pflanzenchemie zwischen Lavoisier und Liebig*. Vol. 1. Münchner Hochschulschriften: Reihe Naturwissenschaften. Straubing: Donau-Verlag.

Lundgren, Anders. 1992. Berzelius, Dalton, and the chemical atom. In *Enlightenment Science in the Romantic Era: The Chemistry of Berzelius and Its Cultural Setting*. Edited by Evan M. Melhado and Tore Frängsmyr. Cambridge: Cambridge University Press, 85–106.

Lüthy, Christoph. n.d. The invention of atomist iconography. In *The Creation of the Scientific Image*. Edited by Wolfgang Lefèvre, Jürgen Renn, and Urs Schöpflin. Basel: Birkhäuser. (In press).

Lynch, Michael, and Steve Woolgar, eds. 1990. *Representation in Scientific Practice*. Cambridge, Mass.: MIT Press.

Magnus, Gustav. 1833. Ueber die Weinschwefelsäure, ihren Einfluss auf die Aetherbildung, und über zwei neue Säuren ähnlicher Zusammensetzung. *Annalen der Physik und Chemie* 27:367–388. *Annalen der Pharmacie* 6:152–173.

Malaguti, Faustino. 1837a. Note sur l'action du chlore sur les éthers composés à l'oxacide, et sur l'éther sulfurique. *Comptes Rendus* 5:334–335.

———. 1837b. Ueber die Wirkung des Chlors auf die Sauerstoffsäure-Aether und auf die Schwefeläther. *Annalen der Pharmacie* 24:40–43.

Marx, Karl. 1906. *Capital: A Critique of Political Economy*. Translated by Samuel Moore & Edward Aveling. Vol. 1. Chicago: Charles H. Kerr & Co.

Mason, H. S. 1943. History of the Use of Graphic Formulas in Organic Chemistry. *Isis* 34:346–354.

Mauskopf, Seymour H., ed. 1993. *Chemical Sciences in the Modern World*. Philadelphia: University of Pennsylvania Press.

Mazzolini, Renato G., ed. 1993. *Non-Verbal Communication in Science Prior to 1900*. Biblioteca di Nuncius, Studi e Testi 11. Firenze: Olschki.

Meinel, Christoph. 1999. Modelling a Visual Language for Chemistry, 1860–1875. Paper read at Conference: Types of Paper Tools and Traditions of Representations in the History of Chemistry, at Max Planck Institute for the History of Science, Berlin.

Melhado, Evan M. 1981. *Jacob Berzelius: The Emergence of His Chemical System*. Madison: University of Wisconsin Press.

———. 1992. Novelty and tradition in the chemistry of Berzelius (1803–1819). In *Enlightenment Science in the Romantic Era: The Chemistry of Berzelius and Its Cultural Setting*. Edited by Evan M. Melhado and Tore Frängsmyr. Cambridge: Cambridge University Press, 132–170.

Melhado, Evan M., and Tore Frängsmyr, eds. 1992. *Enlightenment Science in the Romantic Era: The Chemistry of Berzelius and Its Cultural Setting*. Cambridge: Cambridge University Press.

Meyer, Ernst von. 1895. *Geschichte der Chemie von den ältesten Zeiten bis zur Gegenwart: Zugleich Einführung in das Studium der Chemie*, 2nd ed. Leipzig: Veit & Co.

Mitchell, W. J. Thomas. 1987. *Iconology: Image, Text, Ideology*. Paperback edition. First published 1986. Chicago: University of Chicago Press.

———. 1995. *Picture Theory: Essays on Verbal and Visual Representation*. Paperback edition. First published 1994. Chicago: University of Chicago Press.

Mitscherlich, Eilhard. 1834. Sur la formation de l'Éther. *Annales de Chimie et de Physique* 56:433–439.

Mittelstraß, Jürgen. 1992. *Leonardo-Welt: Über Wissenschaft, Forschung und Verantwortung*. Frankfurt am Main: Suhrkamp.

Morfit, Campbell. 1850. *Chemical and Pharmaceutic Manipulations: A Manual of the Mechanical and Chemico-Mechanical Operations of the Laboratory*. London: Thomas Delf.

Morgan, Mary S., and Margaret Morrison. 1999. Introduction. In *Models as Mediators: Perspectives on Natural and Social Sciences*. Ideas in Context. Edited by Mary Morgan and Margaret Morrison. Cambridge: Cambridge University Press, 1–9.

———, eds. 1999. *Models as Mediators: Perspectives on Natural and Social Sciences*. Ideas in Context. Cambridge: Cambridge University Press.

Morin. 1830. Mémoire sur l'action du chlore sur l'hydrogène bicarboneé. *Annales de Chimie et de Physique* 43:225–244.

Morrell, J. B. 1972. The chemist breeders: the research schools of Liebig and Thomas Thomson. *Ambix* 19(1):1–46.

Mounin, George. 1981. A semiology of the sign system chemistry. *Diogenes* 113:218–228.

Novitski, Marya. 1992. *Auguste Laurent and the Prehistory of Valence*. Chur: Harwood Academic Publishers.

Nye, Mary Jo. 1993. *From Chemical Philosophy to Theoretical Chemistry: Dynamics of Matter and Dynamics of Disciplines, 1800–1950*. Berkeley: University of California Press.

———. 1996. *Before Big Science: The Pursuit of Modern Chemistry and Physics, 1800–1940*. New York: Twayne Publishers.

Nyhart, Lynn K. 1995. *Biology Takes Form: Animal Morphology and the German Universities, 1800–1900*. Chicago: University of Chicago Press.

———. 1996. Natural history and the "new" biology. In *Cultures of Natural History*. Edited by Nicholas Jardine, J. Anne Secord, and Emma C. Spary. Cambridge: Cambridge University Press, 426–443.

Odling, William. 1864. *Tables of Chemical Formulae*. London: Taylor & Francis.

Partington, James Riddick. 1936. The origin of modern chemical symbols and formulae. *The Journal of the Society of Chemical Industry* 55:759–762.

———. 1961–70. *A History of Chemistry*. 4 Vols. New York: St. Martin's Press.

Peirce, Charles Sanders. 1931–1958. *Collected Papers of Charles Sanders Peirce*. Edited by Charles Hartshorne and Paul Weiss. 8 Vols. Cambridge, Mass.: Harvard University Press.

Pelouze, Théophile Jules. 1833. Ueber die gegenseitige Einwirkung der Phosphorsäure und des Alkohols. *Annalen der Pharmacie* 6:129–146. *Annalen der Physik und Chemie* 27:575–590.

———. 1840. Observations sur la loi des substitutions de M. Dumas. *Comptes Rendus* 10:255–260.

Pickering, Andrew. 1989. Living in the material world: on realism and experimental practice. In *The Uses of Experiment: Studies of Experimentation in the Natural Sciences*. Edited by D. Gooding et al. Cambridge: Cambridge University Press, 275–297.

———. 1990. Knowledge, practice, and mere construction. *Social Studies of Science* 20:682–729.

———. 1995. *The Mangle of Practice: Time, Agency, and Science*. Chicago: University of Chicago Press.

———, ed. 1992. *Science as Practice and Culture*. Chicago: University of Chicago Press.

Poggendorff, Johann Christian. 1836. Elementar-Zusammensetzung der bisher zerlegten Substanzen organischen Ursprungs, nach den zuverlässigeren Angaben zusammengestellt vom Herausgeber. *Annalen der Physik und Chemie* 37:1–162.

Priesner, Claus. 1986. Spiritus aethereus—formation of ether and theories on etherification from Valerius Cordus to Alexander Williamson. *Ambix* 33(2/3):129–152.

———. 1989. How the language of chemistry developed. *Chemistry International* 11(6):216–224, 237–238.

Ramberg, Peter J. 1995. Arthur Michael's Critique of Stereochemistry, 1887–1899. *Historical Studies in the Physical and Biological Sciences* 26:89–138.

———. 2000. The death of vitalism and the birth of organic chemistry: Wöhler's urea synthesis and the disciplinary identity of organic chemistry. *Ambix* 47(3): 170–195.

———. n.d. *Chemistry in Space: The Aims and Achievements of Stereochemistry*. Aldershot: Ashgate. (In press.)

Ramsey, Bertrand O. 1974a. Molecules in three dimensions (I). *Chemistry* 47(1):6–9.

———. 1974b. Molecules in three dimensions (II). *Chemistry* 47(2):6–11.

———. 1981. *Stereochemistry*. London: Heyden.

Reinhardt, Carsten, and Anthony S. Travis. 2000. *Heinrich Caro and the Creation of Modern Chemical Industry*. Dordrecht: Kluwer.

Rheinberger, Hans-Jörg. 1995. From experimental systems to cultures of experimentation. In *Concepts, Theories and Rationality in the Biological Sciences: The Second Pittsburgh-Konstanz Colloquium in the Philosophy of Science; University of Pittsburgh, October 1–4, 1993*. Edited by Gereon Wolters and James G. Lennox. Pittsburgh: University of Pittsburgh Press, 107–121.

———. 1997. *Toward a History of Epistemic Things: Synthesizing Proteins in the Test Tube*. Stanford: Stanford University Press.

———. 1998. Experimental systems, graphematic spaces. In *Inscribing Science: Scientific Texts and the Materiality of Communication*. Edited by Timothy Lenoir. Stanford: Stanford University Press, 285–303.

Richter, Jeremias Benjamin. 1968. *Anfangsgründe der Stöchiometrie oder Meßkunst chemischer Elemente*. 3 Vols. Breslau & Hirschberg: Johann Friedrich Korn d. Ältere, 1792–1794. Reprint, 2 Vols. Hildesheim: Olms.

Robiquet, Pierre Jean, and Jean Jacques Colin. 1830. Note sur l'Huile du gaz oléfiant. *Annales de Chimie et de Physique* 2:206–213.

Robiquet, Pierre Jean, Théophile Jules Pelouze, and Jean-Baptiste Dumas. 1839. Rapport sur un Mémoire de M. Regnault, ingénieur des Mines, relativ à l'action du chlore sur les composés éthérés et à la théorie des éthers. *Comptes Rendus* 9:789–795.

Rocke, Alan J. 1984. *Chemical Atomism in the Nineteenth Century: From Dalton to Cannizzaro*. Columbus: Ohio State University Press.

———. 1992. Berzelius's Animal Chemistry: From Physiology to Organic Chemistry (1805–1814). In *Enlightenment Science in the Romantic Era: The Chemistry of Berzelius and Its Cultural Setting*. Edited by Evan M. Melhado and Tore Frängsmyr. Cambridge: Cambridge University Press, 107–131.

———. 1993. *The Quiet Revolution: Hermann Kolbe and the Science of Organic Chemistry*. Berkeley: University of California Press.

———. 2000. Organic analysis in comparative perspective: Liebig, Dumas, and Berzelius, 1811–1837. In *Instruments and Experimentation in the History of Chemistry*. Edited by Frederic L. Holmes and Trevor H. Levere. Cambridge, Mass.: MIT Press, 273–310.

———. 2001. *Nationalizing Science: Adolphe Wurtz and the Battle for French Chemistry*. Cambridge, Mass.: MIT Press.

Rose, Heinrich. 1800. Beweis, daß der durch Schwefelsäure bereitete Aether keine Schwefelsäure enthält. *Allgemeines Journal der Chemie* 9:253–260.

Ruben, Peter. 1978. *Dialektik und Arbeit der Philosophie*. Köln: Pahl-Rugenstein.

Russell, Colin A. 1968. Berzelius and the Development of the Atomic Theory. In *John Dalton & the Progress of Science*. Edited by D. S. L. Cardwell, 259–273. Manchester: Manchester University Press.

Saussure, Nicolas Theodore de. 1807. Mémoire: sur la composition de l'alcohol et de l'Éther sulfurique. *Journal de Physique, de Chimie, d'Histoire Naturelle et des Arts* 64:316–354.

———. 1811. Analyse du gaz oléfiant. *Annales de Chimie* 78:57–68.

———. 1814. Nouvelles observations sur la composition de l'alcool et de l'éther sulfurique. *Annales de Chimie* 89:273–305.

Schlossberger, Julius Eugen. 1857. *Lehrbuch der Organischen Chemie mit besonderer Rücksicht auf Physiologie und Pathologie, auf Pharmacie, Technik und Landwirthschaft.* 4th ed. Leipzig: Winter.

———. 1860. *Lehrbuch der Organischen Chemie mit besonderer Rücksicht auf Physiologie und Pathologie, auf Pharmacie, Technik und Landwirthschaft*, 5th ed. Leipzig: Winter.

Schödler, Friedrich. 1875. Das chemische Laboratorium unserer Zeit. *Jahrbuch der Illustrierten Deutschen Monatshefte* 38:21–47.

Schorlemmer, Carl. 1874. *A Manual of the Chemistry of the Carbon Compounds; or, Organic Chemistry.* London: Macmillan.

Sertürner, Friedrich Wilhelm. 1819. Bemerkungen über die Verbindungen der Säuren mit basischen und indifferenten Substanzen. *Annalen der Physik* 60:33–59.

Serullas, Georges Simon. 1829. Von der Wirkung der Schwefelsäure auf den Alkohol, und den aus ihr hervorgehenden Producten. *Annalen der Physik und Chemie* 15:20–52.

Simon, Jonathan. 1998. The chemical revolution and pharmacy: a disciplinary perspective. *Ambix* 45(1):1–13.

Sismondo, Sergio, and Snait Gissis, eds. 1999. Modeling and simulation. *Science in Context* 12(2).

Stemerding, Dirk. 1991. *Plants, Animals and Formulae: Natural History in the Light of Latour's "Science in Action" and Foucault's "The Order of Things."* Enschede: School of Philosophy and Social Sciences, University of Twente.

Stevens, Peter F. 1994. *The Development of Biological Systematics: Antoine-Laurent de Jussieu, Nature, and the Natural System.* New York: Columbia University Press.

Suckling, Colin J., Keith E. Suckling, and Charles W. Suckling. 1978. *Chemistry through Models: Concepts and Applications of Modelling in Chemical Science, Technology, and Industry.* Cambridge: Cambridge University Press.

Thackray, Arnold. 1970. *Atoms and Powers: An Essay on Newtonian Matter-Theory and the Development of Chemistry.* Harvard Monographs in the History of Science. Cambridge, Mass.: Harvard University Press.

———. 1972. *John Dalton, Critical Assessments of His Life and Science.* Cambridge, Mass.: Harvard University Press.

Thenard, Louis Jacques. 1807a. Mémoire sur les Éthers. *Mémoires de Physique et de Chimie de la Société d'Arcueil* 1:73–114.

———. 1807b. Deuxième mémoire sur les éthers: De l'éther muriatique. *Mémoires de Physique et de Chimie de la Société d'Arcueil* 1:115–135.

———. 1807c. Sur la découverte de l'éther muriatique. *Mémoires de Physique et de Chimie de la Société d'Arcueil* 1:135–139.

———. 1807d. Troisième mémoire sur les éthers: Des produits qu'on obtient en traitant l'alcool par les muriates métalliques, l'acide muriatique oxigéné et l'acide acétique. *Mémoires de Physique et de Chimie de la Société d'Arcueil* 1:140–160.

———. 1807e. Deuxième mémoire sur l'éther muriatique. *Mémoires de Physique et de Chimie de la Société d'Arcueil* 1:337–358.

———. 1807f. Nouvelles observations sur l'éther nitrique. *Mémoires de Physique et de Chimie de la Société d'Arcueil* 1:359–369.

———. 1809a. De l'action des acides végétaux sur l'alcool, sans l'intermède et avec l'intermède des acides minéraux. *Mémoires de Physique et de Chimie de la Société d'Arcueil* 2:5–22.

———. 1809b. Sur la combinaison des acides avec les substances végétales et animales. *Mémoires de Physique et de Chimie de la Société d'Arcueil* 2:23–41.

———. 1809c. Note sur la combinaison des matières végétales et animales avec les acides. *Mémoires de Physique et de Chimie de la Société d'Arcueil* 2:492–494.

———. 1817–18. *Traité de Chimie Elémentaire, Théorique et Pratique*, 2nd ed. 4 Vols. Paris: Crochard.

———. 1824. *Traité de Chimie Elémentaire, Théorique et Pratique*, 4th ed. 5 Vols. Paris: Crochard.

Thomson, Thomas. 1804. *A System of Chemistry*, 2nd ed. 4 Vols. Edinburgh: Bell & Bradfute, & E. Balfour.

———. 1808. On oxalic acid. *Philosophical Transactions of the Royal Society* 98:63–95.

———. 1813. On the Daltonian theory of definite proportions in chemical combinations. *Annals of Philosophy* 2:32–53.

Travis, Anthony S. 1993. *The Rainbow Makers: The Origins of the Synthetic Dyestuffs Industry in Western Europe*. Bethlehem, Pa.: Lehigh University Press.

Volhard, Jakob. 1909. *Justus von Liebig*. 2 Vols. Leipzig: Johann Ambrosius Barth.

Walden, Peter. 1927. Zur Entwicklungsgeschichte der chemischen Zeichen. In *Studien zur Geschichte der Chemie*, edited by Edmund O. von Lippmann. Festausgabe. Berlin: Springer, 80–105.

Walker, James. 1923. Chemical symbols and formulae. *Nature* 111:883–886.

Wallach, Otto, ed. 1901. *Briefwechsel zwischen J. Berzelius und F. Wöhler*. 2 Vols. Leipzig: Engelmann.

Warwick, Andrew. 1992. Cambridge Mathematics and Cavendish Physics: Cunningham, Campbell and Einstein's Relativity 1905–1911 (Part 1: The Uses of Theory). *Studies in History and Philosophy of Science* 23(4):625–656.

Webster, John W. 1826. *A Manual of Chemistry on the Basis of Professor Brande's*, 2nd ed. Boston: Richardson & Lord.

———. 1839. *A Manual of Chemistry on the Basis of Professor Brande's*. 3rd ed. Boston: Marsh, Capen, Lyon & Webb.

Weininger, Stephen J. 1998. Contemplating the finger: visuality and the semiotics of Chemistry. *Hyle* 4(1):3–27.

Whewell, William. 1831. On the employment of notation in chemistry. *Journal of the Royal Institution of Great Britain* 1:437–453.

Wightman, William P. D. 1961. Essay review of the language of chemistry. *Annals of Science* 17:259–267.

Williamson, Alexander. 1850. Theory of etherification. *The London, Edinburgh, and Dublin Philosophical Magazine and Journal of Science* 37:350–356.

Wise, Norton M., ed. 1995. *The Values of Precision.* Princeton: Princeton University Press.

Wislicenus, Johannes. 1887. Über die räumliche Anordnung der Atome in organischen Molekülen und ihre Bestimmung in geometrisch-isomeren ungesättigten Verbindungen. *Abhandlungen der mathematisch-physischen Classe der Königl. Sächsischen Gesellschaft der Wissenschaften* 14:1–77.

Wöhler, Friedrich. 1828. Ueber künstliche Bildung des Harnstoffs. *Annalen der Physik und Chemie* 12:253–256.

———. 1829. Ueber die Zersetzung des Harnstoffs und der Harnsäure durch höhere Temperatur. *Annalen der Physik und Chemie* 15:619–630.

Wöhler, Friedrich, and Justus Liebig. 1832. Untersuchungen über das Radikal der Benzoesäure. *Annalen der Pharmacie* 3:249–282.

Wollaston, William Hyde. 1808. On Super-acid and Sub-acid Salts. *Philosophical Transactions of the Royal Society* 98:96–102.

Wolters, Gereon, and James G. Lennox, eds. 1995. *Concepts, Theories and Rationality in the Biological Sciences: The Second Pittsburgh-Konstanz Colloquium in the Philosophy of Science; University of Pittsburgh, October 1–4, 1993.* Pittsburgh: University of Pittsburgh Press.

Zeise, William Christopher. 1831. Von der Wirkung zwischen Platinchlorid und Alkohol, und von den dabei entstehenden neuen Substanzen. *Annalen der Physik und Chemie* 21:497–541.

INDEX

Printed in the USA
CPSIA information can be obtained
at www.ICGtesting.com
JSHW021320221024
72173JS00001B/17

9 780804 743594